U.S. Marine Corps Close Combat

Plus

Technical Manual for 5.56MM, M16A2, M4 and M4A1 Carbine

Department of Defense

Pentagon
Publishing

DEPARTMENT OF THE NAVY
Headquarters United States Marine Corps
Washington, D.C. 20380-1775

18 February 1999

FOREWORD

1. PURPOSE

Today's Marines operate within a continuum of force where conflict may change from low intensity to high intensity over a matter of hours. Marines are also engaged in many military operations other than war, such as peacekeeping missions or noncombatant evacuation operations, where deadly force may not be authorized. During noncombative engagements, Marines must determine if a situation warrants applying deadly force. Sometimes Marines must decide in a matter of seconds because their lives or the lives of others depend on their actions. To make the right decision, Marines must understand both the lethal and nonlethal close combat techniques needed to handle the situation responsibly without escalating the violence unnecessarily. Marine Corps Reference Publication (MCRP) 3-02B, *Close Combat,* provides the tactics, techniques, and procedures of Marine Corps close combat. It also provides the doctrinal basis for the Marine Corps Close Combat Training Program (MCCCTP).

2. SCOPE

This publication guides individual Marines, unit leaders, and close combat instructors in the proper tactics, techniques, and procedures for close combat. MCRP 3-02B is not intended to replace supervision by appropriate unit leaders and close combat instruction by qualified instructors. Its role is to ensure standardization and execution of tactics, techniques, and procedures throughout the Marine Corps.

3. SUPERSESSION

MCRP 3-02B supersedes Fleet Marine Force Manual (FMFM) 0-7, *Close Combat,* dated 9 July 1993. There are significant differences between the two publications. MCRP 3-02B should be reviewed in its entirety.

4. WARNING

Techniques described in this publication can cause serious injury or death. Practical application in the training of these techniques will be conducted in strict accordance with approved Entry Level Close Combat, Close Combat Instructor (CCI), and Close Combat Instructor Trainer (CCIT) lesson plans. Where serious danger exists, the reac er is alerted by the following:

WARNING

---

5. CERTIFICATION

Reviewed and approved this date.

BY DIRECTION OF THE COMMANDANT OF THE MARINE CORPS

J. E. RHODES
Lieutenant General, U.S. Marine Corps
Commanding General
Marine Corps Combat Development Command

DISTRIBUTION: 14400006600

OVERVIEW OF CLOSE COMBAT

1. Purpose of Close Combat

Close combat is the physical confrontation between two or more opponents. It involves armed and unarmed and lethal and nonlethal fighting techniques that range from enforced compliance to deadly force. The purpose of close combat is to execute armed and unarmed techniques to produce both lethal and nonlethal results. Unarmed techniques include hand-to-hand combat and defense against hand-held weapons. Armed techniques include techniques applied with a rifle, bayonet, knife, baton, or any weapon of opportunity.

2. Continuum of Force

Marines will find themselves in both combative and noncombative situations. The threat level in these situations can rise and fall several times based on the actions of both Marines and the people involved. The escalation of force stops when one person complies with the demands imposed by another person. This range of actions is known as a continuum of force. Continuum of force is the concept that there is a wide range of possible actions, ranging from voice commands to application of deadly force, that may be used to gain and maintain control of a potentially dangerous situation (MCO 5500.6_, *Arming of Security and Law Enforcement [EE] Personnel and the Use of Force).* Continuum of force consists of five levels that correspond to the behavior of the people involved and the actions Marines use to handle the situation (see the table below). Close combat techniques are executed in levels three, four, and five.

Level One: Compliant(Cooperative)

The subject complies with verbal commands. Close combat techniques do not apply.

Level Two: Resistant (Passive)

The subject resists verbal commands but complies immediately to any contact controls. Close combat techniques do not apply.

Continuum of Force		
Level	**Description**	**Actions**
1	Compliant (Cooperative)	Verbal commands
2	Resistant (Passive)	Contact controls
3	Resistant (Active)	Compliance techniques
4	Assaultive (Bodily Harm)	Defensive tactics
5	Assaultive (Serious Bodily Harm/ Death)	Deadly force
Note: Shading indicates levels in which Marines use close combat techniques.		

Level Three: Resistant (Active)

The subject initially demonstrates physical resistance. Marines use compliance techniques to control the situation. Level three incorporates close combat techniques to physically force a subject to comply. Techniques include—

- Come-along holds.
- Soft-handed stunning blows.
- Pain compliance through joint manipulation and the use of pressure points.

Level Four: Assaultive (Bodily Harm)

The subject may physically attack Marines, but he does not use a weapon. Marines use defensive tactics to neutralize the threat. Defensive tactics include the following close combat techniques:

- Blocks.
- Strikes.
- Kicks.
- Enhanced pain compliance procedures.
- Nightstick blocks and blows.

Level Five: Assaultive (Serious Bodily Harm/Death)

The subject usually has a weapon and will either kill or seriously injure someone if he is not stopped immediately and brought under control. Typically, to control the subject, Marines apply deadly force through the use of a firearm, but they may also use armed and unarmed close combat techniques.

3. Marine Corps Tactical Concepts

Close combat techniques support the following key Marine Corps tactical concepts. The concepts are not standalone ideas but are to be combined to achieve an effect that is greater than their separate sum.

Achieving a Decision

Achieving a decision is important in close combat. An indecisive fight wastes energy and possibly Marines' lives. Whether the intent is to control an opponent through restraint or defend themselves in war, Marines must have a clear purpose before engaging in close combat and act decisively once engaged.

Gaining an Advantage

A basic principle of martial arts is to use the opponent' s strength and momentum against him to gain more leverage than one's own muscles alone can generate, thereby gaining an advantage. In close combat, Marines must exploit every advantage over an opponent to ensure a successful outcome. This can include employing various weapons and close combat techniques that will present a dilemma to an opponent. Achieving surprise can also greatly increase leverage. Marines try to achieve surprise through deception, stealth, and ambiguity.

Speed

Marines use speed to gain the initiative and advantage over the enemy. In close combat, the speed and violence of the attack against an opponent provides Marines with a distinct advantage. Marines must know and understand the basics of close combat so they can act instinctively with speed to execute close combat techniques.

Adapting

Close combat can be characterized by friction, uncertainty, disorder, and rapid change. Each situation is a unique combination of shifting factors that cannot be controlled with precision or certainty. For example, a crowd control mission may call for Marines to employ various techniques ranging from nonlethal restraint to more forceful applications. Marines who adapt quickly will have a significant advantage.

Exploiting Success

Typically, an enemy will not normally surrender simply because he was placed at a disadvantage. Marines cannot be satisfied with gaining an advantage in a close combat situation. They must exploit any advantage aggressively and ruthlessly until an opportunity arises to completely dominate the opponent. Marines must exploit success by using every advantage that can be gained.

(reverse blank)

Close Combat

Table of Contents

Page

Overview of Close Combat

Chapter 1. Fundamentals of Close Combat

1. Ranges of Close Combat 1-1
2. Weapons of the Body 1-2
3. Target Areas of the Body 1-3
4. Pressure Points of the Body 1-4
5. Basic Warrior Stance 1-6
6. Angles of Approach and Movement 1-7
7. Balance and Off-Balancing 1-8
8. Falls 1-9

Chapter 2. Lethal and Nonlethal Weapons Techniques

1. Bayonet Techniques 2-1
2. Nonlethal Rifle and Shotgun Retention Techniques 2-6
3. Nonlethal Handgun Retention Techniques 2-8
4. Firearm Disarmament Techniques 2-12

Chapter 3. Hand-Held Weapons

1. Fundamentals of Knife Fighting 3-1
2. Knife Fighting Techniques 3-3
3. Weapons of Opportunity 3-6
4. Fundamentals of Combative Stick 3-6
5. Combative Stick Techniques 3-7
6. Blocking Techniques 3-9
7. Unarmed Against Hand-Held Weapons 3-11
8. Counters to Hand-Held Weapon Attacks 3-13

Chapter 4. Strikes

1. Principles of Punches 4-1
2. Punches 4-2
3. Strikes with the Upper Body 4-4
4. Strikes with the Lower Body 4-10
5. Counters to Strikes 4-12

Page

Chapter 5. Throws

1. Turning Throw 5-1
2. Hip Throw 5-2
3. Leg Sweep 5-3

Chapter 6. Chokes and Holds

1. Types of Chokes 6-1
2. Chokes 6-1
3. Counters to Chokes and Holds 6-5

Chapter 7. Ground Fighting

1. Offensive Ground Fighting 7-1
2. Defensive Ground Fighting 7-3
3. Ground Fighting Chokes 7-5

Chapter 8. Nonlethal Techniques

1. Unarmed Restraints and Manipulation 8-1
2. Nonlethal Baton 8-9

Appendices

A. Pugil Stick Training A-1
B. Safety Precautions During Training B-1

CHAPTER 1

FUNDAMENTALS OF CLOSE COMBAT

This chapter describes all techniques for a right-handed person. However, all techniques can be executed from eittwr side.

The Marine is depicted in camouflage utilities. The opponent is depicted without camouflage.

The fundamentals of close combat include ranges, weapons of the body, target areas of the body, and pressure points of the body. These fundamentals form the basis for all close combat techniques. They provide Marines with a common framework regardless of the type of confrontation or the techniques used. If Marines apply these fundamentals properly in a close combat situation, they may save their lives or the lives of fellow Marines.

1. Ranges of Close Combat

Close combat engagements occur within three ranges: long range, midrange, and close range. During any engagement, these ranges may blur together or may rapidly transition from one to another until either the opponent is defeated or the conflict is resolved.

Long Range

During long range engagements, combatants engage each other with rifles, bayonets, sticks, or entrenching tools. See figure below.

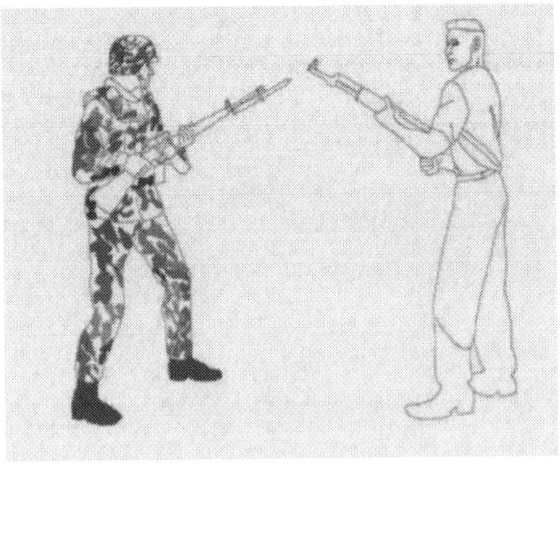

Midrange

During midrange engagements, combatants engage each other with knives, punches, or kicks.

Close Range

During close range engagements, combatants grab each other. Close range engagements also involve elbow strikes, knee strikes, and grappling.

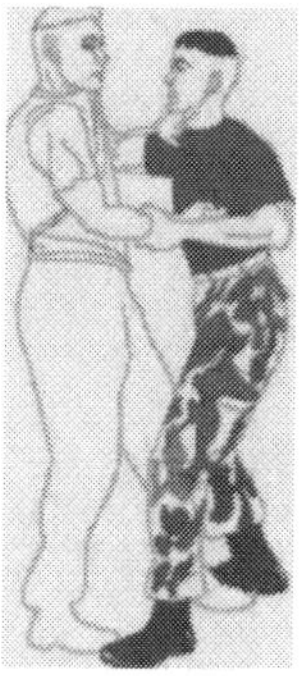

2. Weapons of the Body

Hands and Arms

The hands, forearms, and elbows are the arm's individual weapons. The hands consist of several areas that can be used as weapons: fists, edges of hands, palms, and fingers.

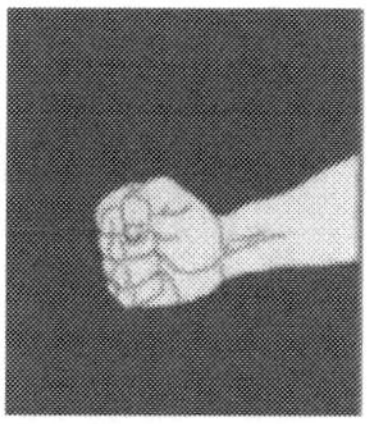

Fists. To minimize injury to the fists, Marines use their fists as weapons to target soft tissue areas such as the throat. The fists' striking surfaces are the first two knuckles of the hands or the meaty portions of the hands below the little fingers.

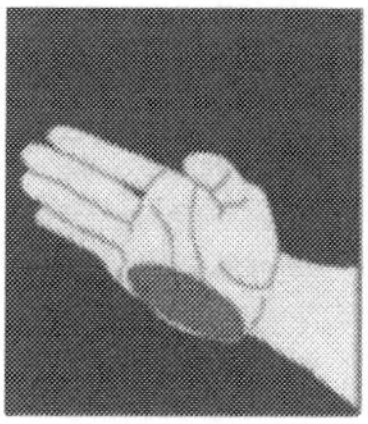

Edge of Hand. Marines use the edge of the hand (knife edge) as a weapon. Marines use the edge of the hand to strike soft tissue areas.

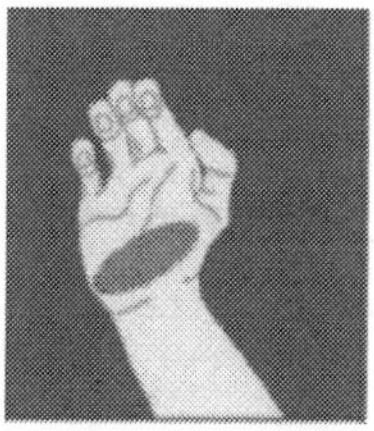

Palms. Because of the palm's padding, Marines use the heels of the palms to strike, parry, and/or block.

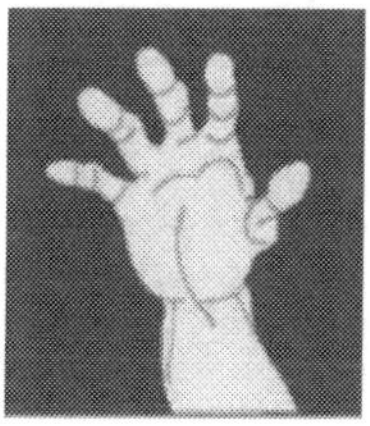

Fingers. Marines use the fingers to gouge, rip, and tear soft tissue areas (e.g., eyes, throat, groin).

Forearms. Marines use the forearms as a defensive tool to deflect or block attacks. Forearms can also be used as striking weapons to damage or break an opponent's joints and limbs. Marines sustain less self-injury when strikes are conducted with the forearms than when strikes are conducted with fists and fingers.

Elbows. Marines use the elbows as striking weapons. Because of the short distance needed to generate power, elbows are excellent weapons for striking during the close range of close combat.

Legs

The legs are more powerful than any other weapon of the body, and they are less prone to injury when striking. The feet are protected by boots and are the preferred choice for striking.

Feet. Marines use the balls of the feet, the insteps, and the toes to kick an opponent. Marines use the cutting edge of the heels and the heels to stomp on an opponent. Marines must be wearing boots when striking with the toes.

Knees. Like elbows, knees are excellent weapons in the close range of close combat. Knee strikes are most effective while fighting close to an opponent where kicks arc impractical. The opponent's groin area is an ideal target for the knee strike if he is standing upright. Knee strikes can deliver a devastating secondary attack to an opponent's face following an initial attack that caused him to bend at the waist.

3. Target Areas of the Body

During close combat, Marines strive to attack the accessible target areas of an opponent's body. The readily accessible areas will vary with each situation and throughout the engagement. The target areas are divided into five major groups: head, neck, torso, groin, and extremities. The figure below illustrates target areas of the body.

Head

The vulnerable regions of the head are the eyes, temple, nose, ears, and jaw. Massive damage to the head kills an opponent.

Eyes. The eyes are excellent targets because they are soft tissue areas that arc not protected by bone or muscle. Attacks to this area may cause the opponent to protect the area with his hands, allowing Marines to execute a secondary attack to other

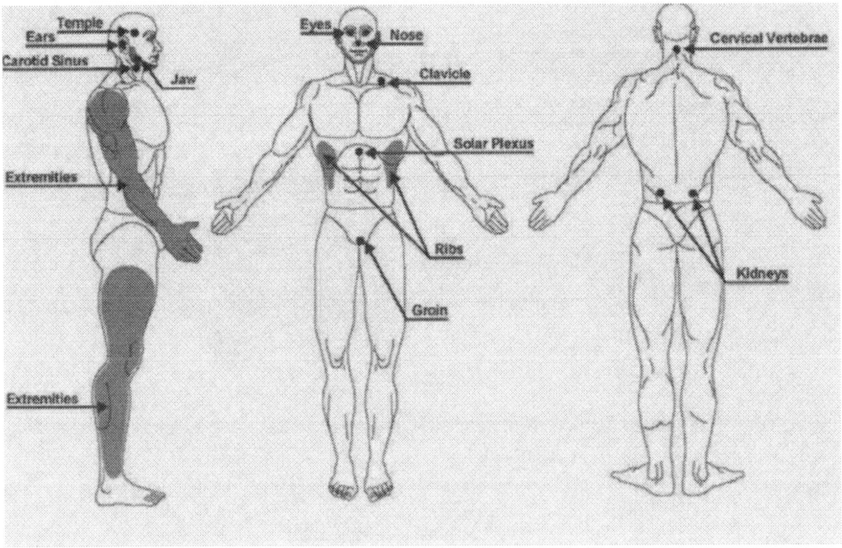

target areas while the opponent uses his hands to protect his eyes.

Temple. The temple is one of the most fragile areas of the head. Powerful strikes to the opponent's temple cause permanent damage and death.

Nose. The nose is very sensitive and easily broken. An attack to this area causes involuntary watering and closing of the opponent's eyes, rendering him vulnerable to secondary attacks. However, through training, individuals can condition themselves to withstand attacks to the nose. Therefore, any attack to the nose must be powerfully delivered.

Ears. Attacks to the ears may cause the eardrum to rupture. But this may not stop or even distract an opponent unless Marines powerfully deliver the strike.

Jaw. The jaw region, when struck forcefully, renders the opponent unconscious. Strikes to the jaw cause painful injuries to the teeth and surrounding tissues (e.g., lips, tongue), but the risk of self-injury is great unless Marines deliver strikes with a hard object such as a helmet, rifle butt, or boot heel.

Neck

The front of the neck, or throat area, is a soft tissue area that is not covered by natural protection. Damage to this region causes the opponent's trachea to swell, closing his airway, which can lead to death.

Carotid Sinus. The carotid sinus is located on both sides of the neck just below the jaw. Strikes to the carotid sinus restrict blood flow to the brain, causing loss of consciousness or death.

Cervical Vertebrae. The cervical vertebrae on the back of the neck, from the base of the skull to the top of the shoulders, contains the spinal cord, which is the nervous system's link to the brain. The weight of the head and the lack of large muscle mass allow damage to the cervical vertebrae and spinal cord. Excessive damage to this area causes pain, paralysis, or death.

Torso

Clavicle. The opponent's clavicle (or collar bone) can be easily fractured, causing immobilization of the arm.

Solar Plexus. Attacks to the opponent's solar plexus or center of the chest can knock the breath out of him and immobilize him.

Ribs. Damage to the opponent's ribs immobilizes him. It may also cause internal trauma.

Kidneys. Powerful attacks to the opponent's kidneys cause immobilization, permanent damage, or death.

Groin

The groin area is another soft tissue area not covered by natural protection. Any damage to this area causes the opponent to involuntarily protect his injured area, usually with his hands or legs. In male opponents, the scrotum is the main target since even a near miss causes severe pain, contraction of the lower abdominal muscles, deterioration of his stance, and possible internal trauma.

Extremities

Rarely will an attack to the opponent's extremities (arms and legs) cause death, but they are important target areas in close combat. Damage to an opponent's joints causes immobilization.

4. Pressure Points of the Body

There are nerves in the human body that, when pressure is applied or when they are struck, allow Marines to control a subject through pain compliance. Marines use pressure points to control an opponent when deadly force is not authorized. They also use pressure points to soften or distract an opponent so a lethal or nonlethal technique can be employed. The figure on page 1-5 illustrates the body's pressure points. Marines execute attacks to pressure points by—

: Rapidly kicking or striking pressure points.
: Slowly applying steady pressure to pressure points.

Infraorbital Nerve

The infraorbital nerve is just below the nose. Marines apply pressure to this nerve with an index finger to control the opponent.

Mastoid Process

The mastoid process is behind the base of the ear and beneath the edge of the jaw. Marines apply inward and upward pressure to this pressure point with the fingers to distract and control the opponent.

Jugular Notch

The jugular notch is at the base of the neck in the notch formed at the center of the clavicle. Marines apply pressure in a quick, stabbing motion with the index finger. Strikes to the jugular notch cause serious damage.

Brachial Plexus (Tie In)

The brachial plexus (tie in) is on the front of the shoulder at the joint. Strikes and pressure applied with the hand are effective on this nerve.

Radial Nerves

Radial nerves are on the inside of the forearms along the radius bones. Strikes and pressure applied with the hand to the radial nerve serve as a softening technique.

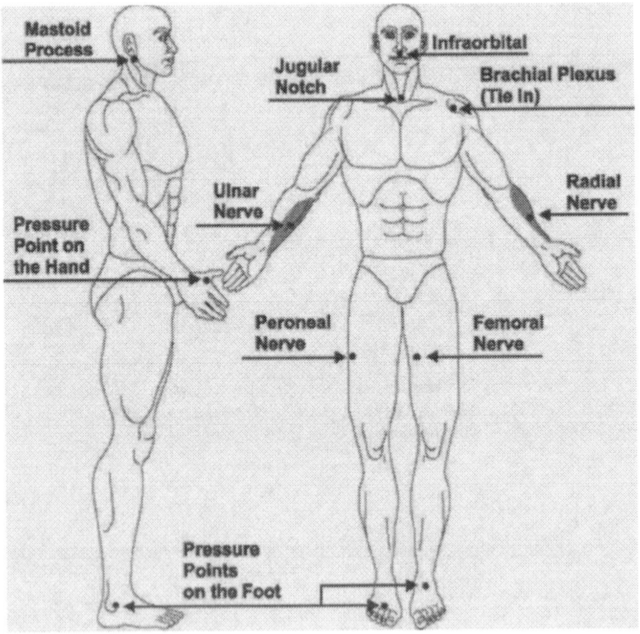

Ulnar Nerve

Ulnar nerves are on the outside of the forearms along the ulnar bones. Strikes and pressure applied with the hand to the ulnar nerve serve as a softening technique.

Pressure Point on the Hand

The hands contain a pressure point on the webbing between the thumbs and index fingers where the two bones of the fingers meet.

To force an opponent to soften or release his grip, Marines apply pressure with their index fingers to this pressure point or strike this pressure point with their fists.

Femoral Nerves

Femoral nerves are on the inside of the thighs along the femur bones. Strikes to the femoral nerve serve as a softening technique.

Peroneal Nerves

Peroneal nerves are on the outside of the thighs along the femur bones. Strikes to the peroneal nerve serve as a softening technique.

Pressure Points on the Feet

There are pressure points on the feet that, when pressure is applied or when they are struck, serve to soften or distract the opponent. Marines apply pressure with the toe, edge, or heel of their boots to the following points:

: The notch below the ball of the ankle.
: The top center of the foot, above the toes.
: The top of the foot where the leg and foot meet.

5. Basic Warrior Stance

Achieving a solid stance is critical to stability and movement throughout any close combat confrontation. The basic warrior stance provides the foundation for stability and movement that is needed to execute close combat techniques. To execute the basic warrior stance, Marines put their feet apart, hands up, elbows in, and chin down.

Feet Apart

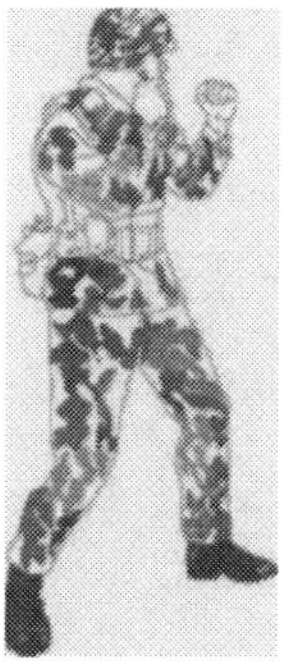

Place feet shoulder-width apart.

Keep the head forward and eyes on the opponent, take a half step forward with the left foot, and pivot on the heels so the hips and shoulders are at approximately a 45-degree angle to the right.

Distribute body weight evenly on both legs. Bend the knees slightly.

Hands Up

Curl the fingers naturally into the palm of the hand. Position the thumb across the index and middle fingers. Do not clench the fists. Clenching the fists increases muscular tension in the forearms and decreases speed and reaction time.

Bring the hands up to the face at chin level, with the palms facing each other. Hold the fists up high enough to protect the head, but not so high that they block the field of vision. Ensure continuous eye contact with the opponent.

Elbows In

Tuck the elbows in close to protect the body.

Chin Down

Tuck the chin down to take advantage of the natural protection provided by the shoulders.

6. Angles of Approach and Movement

Marines use movement to control a confrontation and to retain a tactical advantage. Movement increases power and maximizes momentum. By moving around the opponent, Marines gain access to different target areas of the opponent's body. Movement allows Marines to use different weapons of the body and different close combat techniques to attack specific target areas.

Angles of Approach

Marines move anywhere within a 360-degree circle around the opponent to gain a tactical advantage. This circle provides access to different target areas of the opponent's body.

When facing an opponent, Marines move in a 45-degree angle to either side of the opponent. Moving at a 45-degree angle avoids an opponent's strike and puts Marines in the best position to attack the opponent. Marines should avoid being directly in front of an opponent during a confrontation. If a Marine is directly in front of an opponent, the opponent can rely on his forward momentum and linear power to seize the tactical advantage.

Movement

Marines must know how to move in all directions while maintaining the basic warrior stance. During any movement, the legs or feet should not be crossed. Once a movement is completed, the basic warrior stance should be resumed. Maintaining the basic warrior stance protects Marines and puts them in the proper position to launch an attack against an opponent.

Note: Before body movement begins. Marines turn their heads quickly to the new direction. The faster the head turns, the faster the body moves, and the quicker Marines attain visual contact with the opponent.

Forward to the Left. To move forward to the left from the basic warrior stance, Marines-

: Move the left foot forward at a 45-degree angle from the body (approximately 12 to 15 inches), keeping the toe pointed toward the opponent.

Bring the right foot behind the left foot as soon as the left foot is in place. This returns Marines to the basic warrior stance.

Forward to the Right. To move forward to the right from the basic warrior stance, Marines-

Move the right foot forward at a 45-degree angle from the body (approximately 12 to 15 inches).

Bring the left foot, toe pointing toward the opponent, in front of the right foot as soon as the right foot is in place. This returns Marines to the basic warrior stance.

Backward to the Left. To move backward to the left from the basic warrior stance, Marines execute the forward movement in reverse. Marines-

Move the left foot backward at a 45-degree angle from the body (approximately 12 to 15 inches), keeping the toe pointed toward the opponent.

Bring the right foot behind the left foot as soon as the left foot is in place. This returns Marines to the basic warrior stance.

Backward to the Right. To move backward to the right from the basic warrior stance, Marines execute the forward movement in reverse. Marines-

Move the right foot backward at a 45-degree angle from the body (approximately 12 to 15 inches).

Bring the left foot, toe pointing toward the opponent, in front of the right foot as soon as the right foot is in place. This returns Marines to the basic warrior stance.

7. Balance and Off-Balancing

Balance

In any close combat situation, Marines must strive to maintain balance. The last place to be in a close combat situation is on the ground. Marines must maintain a strong base and a low center of balance, their feet must be a shoulder-width apart, and they must stay on their toes to enable quick movement.

Off-balancing

Marines use off-balancing techniques to control an opponent. These techniques are used to throw an opponent to the ground while Marines remain standing, or they are used to put Marines in a position for an offensive attack.

Off-balancing techniques use the opponent's momentum to move or throw him. For example, if the opponent is charging a Marine, the Marine pulls the opponent to drive him to the ground. Likewise, if the opponent is pulling a Marine, the Marine pushes the opponent to drive him to the ground.

Off-balancing techniques also rely on the power generated by the opponent. For example, during combat a Marine may be tired or outnumbered. Depending on the generated energy and momentum of the opponent, the Marine employs off-balancing techniques with very little effort and still provides effective results.

Because off-balancing techniques rely on the momentum and power generated by the opponent, these techniques are particularly effective for Marines who may be outsized by their opponent or lack their opponent's strength.

Angles of Off-balancing

There are eight angles or directions in which an opponent can be off-balanced: forward, rear, right, left, forward right, forward left, rear right, and rear left.

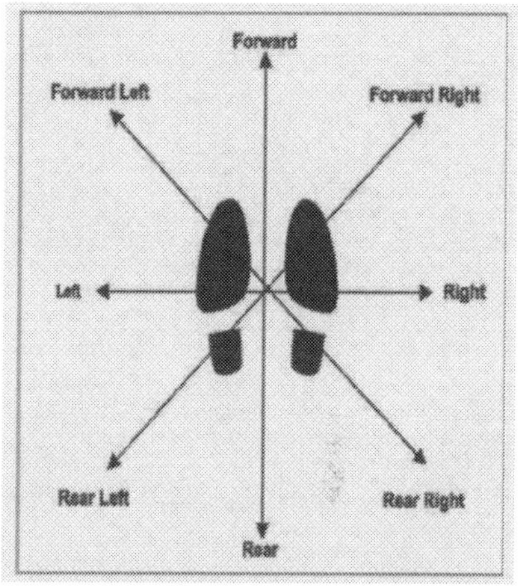

Note: The angles correspond to the Marine's perspective, not the opponent's. Forward, rear, right, and left are straight angles. Forward right, forward left, rear right, and rear left are considered quadrants that are at a 45-degree angle in either direction to the front or rear.

Off-balancing Techniques

Marines off-balance an opponent by pushing, pulling, or bumping the opponent with their hands, arms, or bodies.

To pull, Marines grab an opponent with their hands and drive him forcefully to one of the rear quadrants or to the right or left.

To push, Marines grab the opponent with their hands and drive him forcefully into one of the front quadrants or to the right or left.

Marines execute bumping in the same manner as pushing, but use their shoulders, hips, and legs instead of their hands to off-balance the opponent.

8. Falls

Marines may lose their balance or be thrown to the ground during encounters with an opponent. Marines use falling techniques to absorb the impact of a fall and to quickly return to their feet following an opponent's attack.

Whether falling or being thrown by an opponent, Marines strive to reduce the force of the impact, to prevent serious personal injury, and to increase the chances of survival. Falling techniques use the body's large muscles (back, thighs, buttocks) to protect vital organs and bones from injury and immobilization.

Front Fall

Marines execute a front fall to break a fall on the front. To execute the front fall, Marines-

Bend the elbows and place the palms facing out in a position to spread and absorb the impact of the fall.

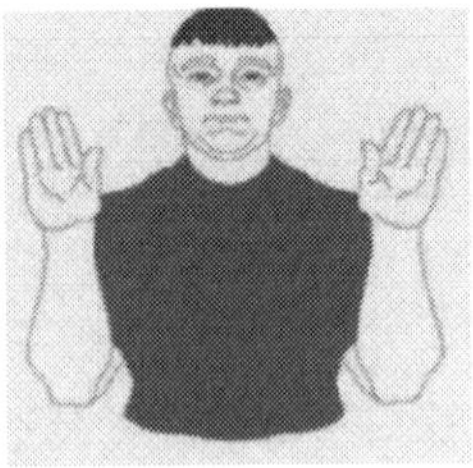

Fall forward, breaking the fall with the forearms and palms. The forearms and hands, down to the fingertips, should strike the ground simultaneously.

Offer resistance with the forearms and hands to keep the head raised off the ground.

Side Fall

Marines execute a side fall to break a fall on the side. To execute the side fall, Marines-

Bring the right arm across the body so the hand is next to the left shoulder with the palm facing inboard.

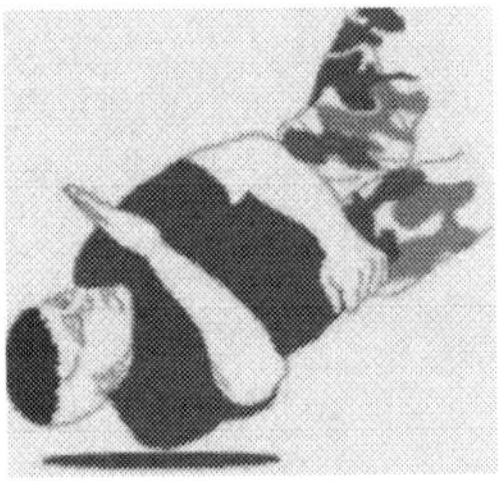

Fall to the side, breaking the fall with the right arm by slapping the ground and making contact from the shoulder or forearm down to the hand. At the same time, tuck the chin and keep the head raised off the ground. The chin should be tucked to the chest at all times to prevent whiplash.

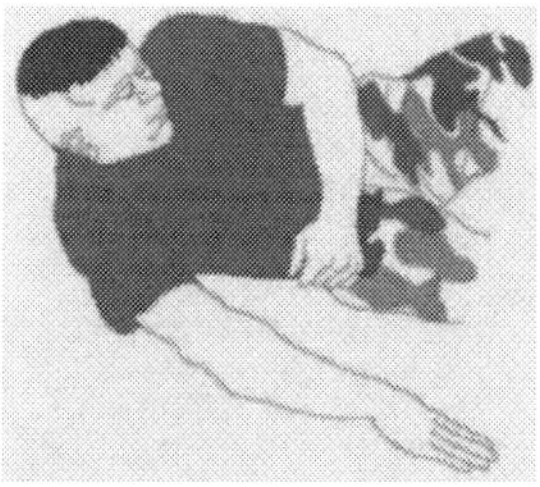

Stretch the right leg out to make contact with the ground and to distribute and absorb the impact. Bend the left leg, allowing the foot to make contact with the ground.

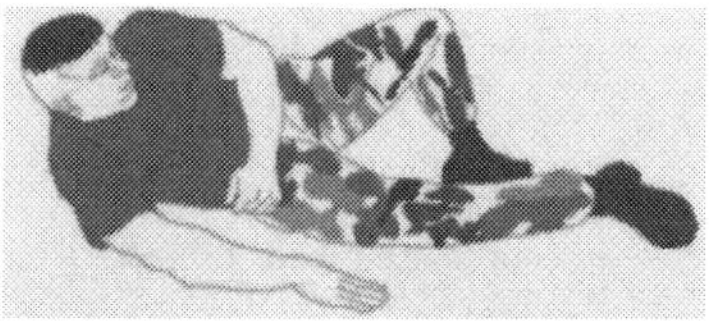

Back Fall

Marines execute a back fall to break the fall when being thrown or falling backward. To execute the backfall, Marines-

Cross the hands in front of the chest and tuck the chin.

Fall backward and slap the ground with the forearms and hands to absorb the impact of the fall and keep the head off the ground.

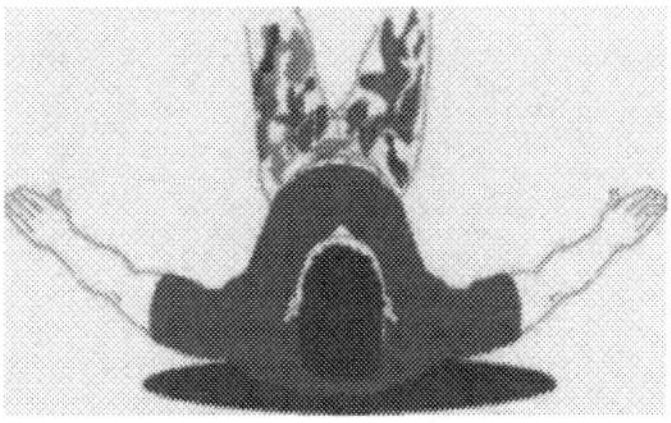

Forward Shoulder Roll

Marines use the forward shoulder roll to break a fall from an opponent's attack and to use the momentum of the fall to get back on their feet quickly. Ideally, Marines execute the forward shoulder roll to a standing position so they can continue fighting. To execute the forward shoulder roll, Marines-

Contact the ground with the back of the right forearm and upper arm. Tuck the chin into the chest.

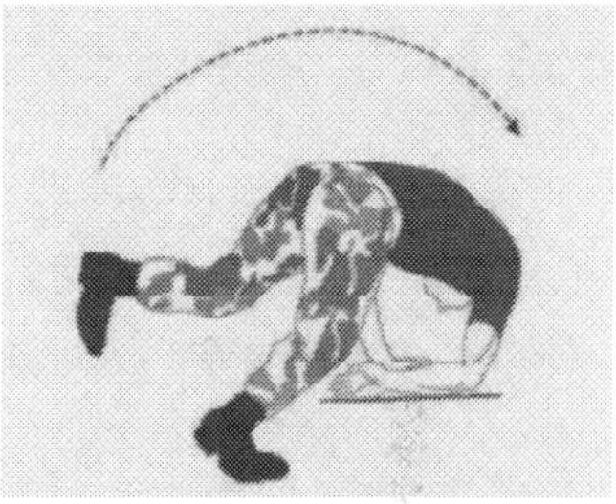

Roll onto the right shoulder, rolling diagonally across the back to land on the left hip.

Slap the ground with the left arm, absorbing the impact from the shoulder to the hand, palm down.

Keep the left leg straight to absorb as much of the impact as possible. The right leg is bent and the foot hits flat on the ground.

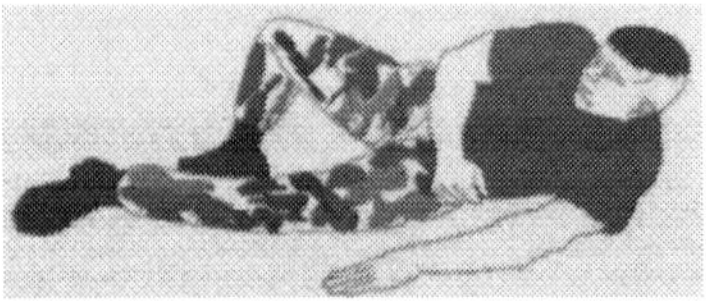

Bend the left leg upon impact to push off with the left knee and leg to a squatting and then a standing position. Forward momentum should carry the Marine to a standing position.

(reverse blank)

CHAPTER 2

LETHAL AND NONLETHAL WEAPONS TECHNIQUES

This chapter describes all techniques for a right-handed person. However, all techniques can be executed from either side.

In drawings, the Marine is depicted in woodland camouflage utilities; the opponent is depicted without camouflage. In photographs, the Marine is depicted in woodland camouflage utilities; the opponent is depicted in desert camouflage utilities.

1. Bayonet Techniques

WARNING

During training, Marines must have bayonets sheathed. Marines use bayonet dummies to practice bayonet techniques. When practicing offensive and defensive bayonet techniques student-on-student, Marines use pugil sticks.

All Marines armed with a rifle carry a bayonet. The bayonet is an effective weapon if Marines are properly trained in offensive and defensive bayonet techniques. An offensive attack, such as a thrust, is a devastating attack that can quickly end a fight. Defensive techniques, such as the block and parry, can deter the opponent's attack and allow Marines to regain the initiative. Through proper training, Marines develop the courage and confidence required to effectively use a bayonet to protect themselves and destroy the enemy. In situations where friendly and enemy troops are closely mingled and rifle fire and grenades are impractical, the bayonet becomes the weapon of choice.

Holding the Rifle

To execute bayonet techniques, Marines hold the rifle in a modified basic warrior stance. All movement begins and ends with the basic warrior stance. To hold the rifle, Marines—

Use an overhanded grasp to grab the small of the rifle's stock. Use an underhanded grasp to grab the handguards of the rifle.

Lock the buttstock of the rifle against the hip with the right forearm.

Orient the blade end of the rifle toward the opponent.

Offensive Bayonet Techniques

Straight Thrust. Marines use the straight thrust to disable or kill an opponent. It is the most deadly offensive technique because it causes the most trauma to an opponent. Target areas are the opponent' s throat, groin, or face. The opponent's chest and stomach are also excellent target areas if not protected by body armor or combat equipment. To execute the straight thrust, Marines—

Lift the left leg and lunge forward off the ball of the right foot while thrusting the blade end of the weapon forward, directly toward the opponent.

Retract the weapon and return to the basic warrior stance.

Slash. Marines use the slash to kill an opponent or to create an opening in his defense. The target area is the opponent's neck. To execute the slash, Marines—

Extend the left hand back toward the left shoulder.

Thrust the left hand forward and swing it to the right, bring the right hand back toward the hip, and turn the cutting edge of the blade toward the opponent's neck. The movement is a slashing motion so the blade cuts across the opponent's neck.

Horizontal Buttstroke. Marines use the horizontal buttstroke to weaken an opponent's defenses, to cause serious injury, or to set him up for a killing blow. Target areas are the opponent's head, neck, and legs. To execute the horizontal buttstroke, Marines—

Step forward with the right foot and drive the right hand forward. Rotate the hips and shoulders into the strike. Move the left hand back toward the left shoulder.

Strike the opponent with the butt of the weapon.

Vertical Buttstroke. Marines use the vertical buttstroke to weaken an opponent's defenses, to cause serious injury, or to set him up for a killing blow. Target areas are the opponent's groin and face. To execute the vertical buttstroke, Marines—

Step forward with the right foot and drive the right hand straight up.

Pull the left hand back over the left shoulder.

Strike the opponent with the butt of the weapon.

Smash. Marines use the smash as a follow-up technique to the horizontal or vertical buttstroke, particularly if they missed the target. The target area is the opponent's head. To execute the smash following a buttstroke, Marines-

Step forward with the right foot and place the blade end of the weapon over the left shoulder and elevate the right elbow above the shoulders.

Drive the arms straight forward, striking the opponent with the butt of the weapon.

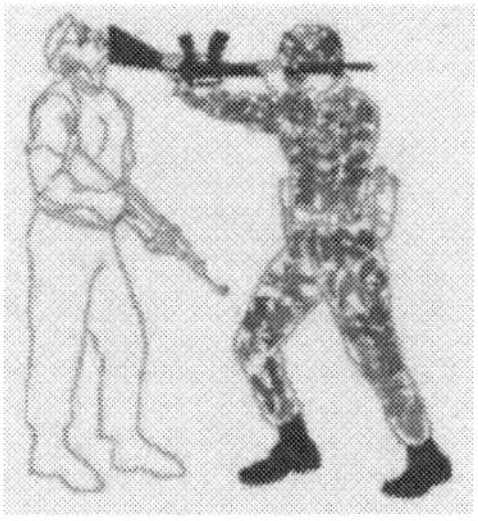

Defensive Bayonet Techniques

Parry. Marines use a parry as a defensive technique to redirect or deflect an attack. A parry is a slight redirection of a linear attack by an opponent; e.g., a straight thrust or a smash. To execute the parry, Marines-

Use the bayonet end of the rifle to redirect the barrel or bayonet of the opponent's weapon.

Lock the weapon against the hip with the right forearm.

Rotate to the right or left, moving the bayonet end of the rifle to parry the opponent's attack. Rotation is generated from the hips.

Redirect or guide the opponent's weapon away from the body by exerting pressure against the opponent's weapon.

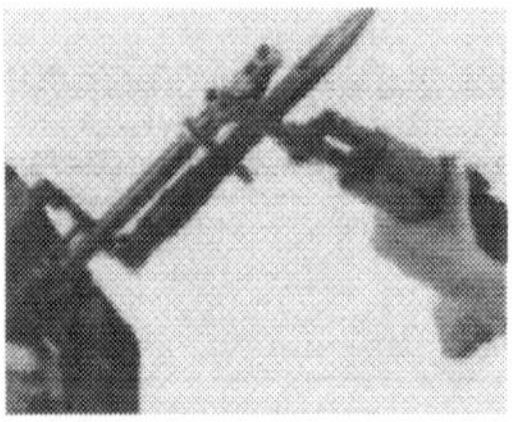

High Block. Marines execute a high block against a vertical attack coming from high to low. To execute the high block, Marines—

Thrust the arms up forcefully at approximately a 45-degree angle from the body. The weapon should be over the top of the head and parallel to the ground. The elbows are bent, but there is enough muscular tension in the arms to absorb the impact and deter the attack.

Low Block. Marines execute the low block

against a vertical attack coming from low to high. To execute the low block, Marines—

Thrust the arms down forcefully at approximately a 45-degree angle from the body. The weapon

should be at or below the waist and parallel to the ground. The elbows are bent, but there is enough muscular tension in the arms to absorb the impact and deter the attack.

Left and Right Block. Marines execute a left or right block against a horizontal buttstroke or a slash. To execute the left or right block, Marines-

Thrust the arms forcefully to the right or left, holding the rifle vertically in the direction of the attack. The elbows are bent, but there is enough muscular tension in the arms to absorb the impact and deter the attack.

Counter Action Following the Block. After deflecting an opponent's attack with a block, Marines counter with a slash or a horizontal buttstroke to regain the initiative. However, the objective in any bayonet fight is to thrust forward with the blade end of the weapon to immediately end the fight.

Group Strategy

On occasion, Marines may engage an opponent as a member of a group or numerous opponents by one's self or as a member of a group. By combining bayonet fighting movements and simple strategies, Marines can effectively overcome their opponent or opponents.

Offensive Strategy: Two Against One. If two bayonet fighters engage one opponent, the fighters advance together.

Fighter 1 engages the opponent while fighter 2 swiftly and aggressively attacks the opponent's exposed flank and destroys the opponent.

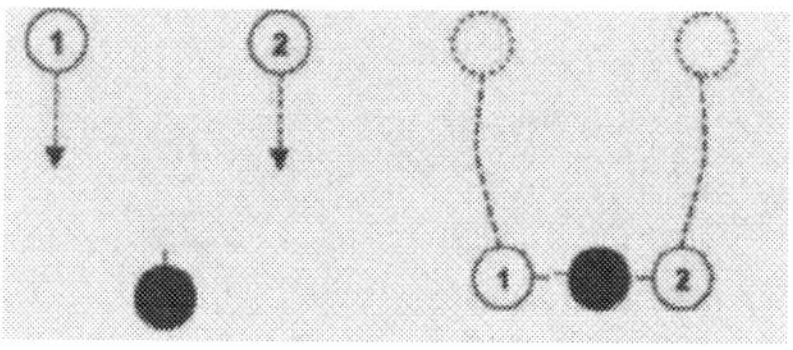

Offensive Strategy: Three Against Two. If three bayonet fighters engage two opponents, the fighters advance together keeping their opponents to the inside.

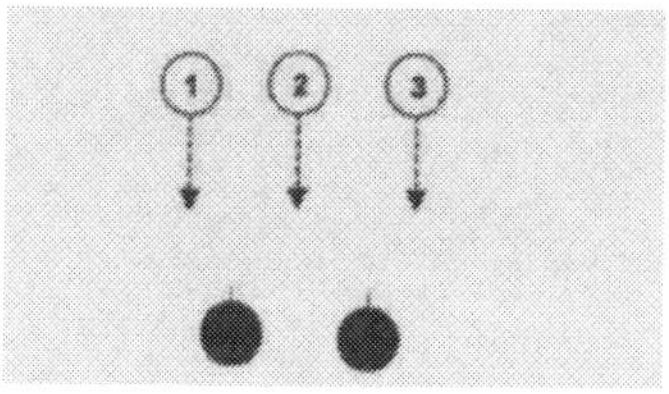

Fighters 1 and 3 engage opponents. Fighter 2 attacks the opponent's exposed flank engaged by fighter 1 and destroys the opponent.

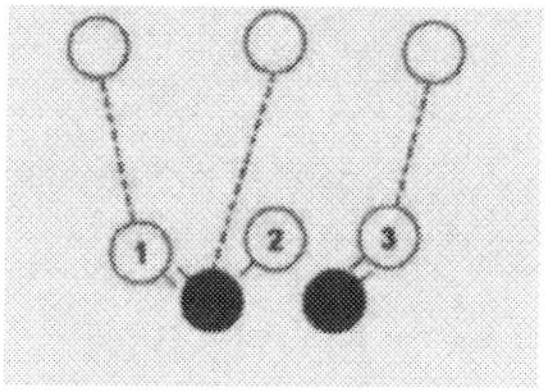

Fighters 1 and 2 turn and attack the exposed flank of the opponent engaged by fighter 3 and destroy the opponent.

Defensive Strategy: One Against Two. If a fighter is attacked by two opponents, the fighter immediately positions himself at the flank of the nearest opponent and keeps that opponent between himself and the other opponent.

Using the first opponent's body as a shield against the second opponent, the fighter destroys the first opponent quickly before the second opponent moves to assist.

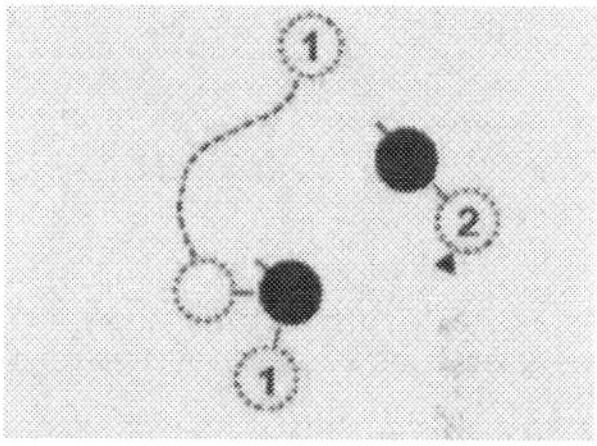

Then, the fighter engages and destroys the second opponent.

Defensive Strategy: Two Against Three. If two fighters are attacked by three opponents, the fighters immediately move to the opponent's flanks.

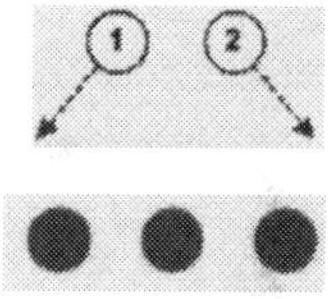

Fighters 1 and 2 quickly attack and destroy their opponents before the third opponent closes in.

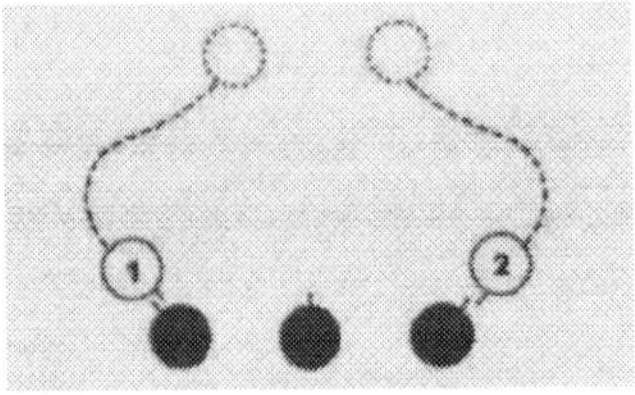

Fighter 1 engages the third opponent while fighter 2 attacks the opponent's exposed flank and destroys the opponent.

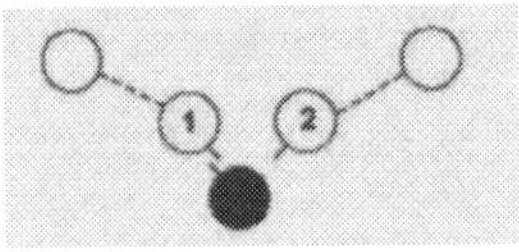

2. Nonlethal Rifle and Shotgun Retention Techniques

Most Marines are armed with the M16A2 service rifle. Marines are taught to keep their weapon with them at all times. Marines must be constantly alert to their surroundings and the people moving in and around their environment. Marines may be confronted by an individual who tries to take theii weapons. If this happens, Marines should not struggle with the individual. To retain positive control of their weapons, Marines must understand and apply weapons retention techniques, otherwise known as armed manipulation. The fol lowing techniques can be used with either the rifle or the shotgun.

Blocking Technique

To execute a blocking technique, Marines-

Stand in a defensive posture.

Use the weapon to block the opponent by thrusting it out firmly, with the elbows bent. Do not try to hit the opponent with the rifle. The rifle is used as a barrier.

Technique if Opponent Grabs Weapon Underhanded

Usually an opponent will try to grab the weapon or block it as an instinctive action. If the opponent uses an underhand grab to seize the handguards of the weapon, Marines—

Trap the opponent's closest finger(s) above the knuckle with the thumb so he cannot release his grip-

Apply bone pressure on the opponent's finger to initiate pain compliance.

Rotate the barrel of the weapon up or down quickly while maintaining pressure on the opponent's hand. At the same time, quickly pivot to off-balance the opponent.

Technique if Opponent Grabs Weapon Overhanded

If the opponent uses an overhand grab to seize the weapon overhanded, Marines—

Trap the opponent's finger to hold his hand in place.

Rotate the barrel to place it across the opponent's forearm and apply downward pressure. This action is similar to an armbar.

Technique if the Opponent Grabs the Muzzle

If the opponent grabs the muzzle of the weapon, Marines—

Rotate the muzzle quickly in a circle motion, Slash downward with the muzzle to release the opponent's grip.

Butt Strikes

Strikes with the butt of the weapon control or ward off an attacker. During any of the retention techniques, Marines use the heel or cutting edge of the weapon to deliver butt strikes to the inside or the outside of the opponent's thigh. The inside butt strike targets the femoral nerve as illustrated in the photo to the left.

The outside butt strike targets the peroneal nerve as illustrated in the photo to the left. Strikes can be made to the outside or inside of the thighs. If a strike to one side of the thigh misses, Marines follow back through with the butt of the weapon on the other side of the thigh.

Off-balancing Techniques

Marines apply off-balancing techniques to throw an opponent to the ground and retain possession of the weapon.

If the opponent grabs the weapon and pushes, Marines should not push on the weapon. They should—

Move with the momentum and movement of the

opponent by pivoting in the direction of the movement by stepping back.

Throw the opponent to the ground with a quick jerking movement by lowering the mu/./le and swinging the butt of the weapon.

If the opponent grabs the weapon and pulls, Marines—

Step on the opponent's foot and push forward to off-balance him and drive him to the ground.

Sweep the opponent's feet out from under him by hooking his leg with the leg and kicking backward.

3. Nonlethal Handgun Retention Techniques

Many Marines are armed with the M9 service pistol. Marines must keep their weapons in their possession at all times. Marines must be constantly alert to their surroundings and the people moving in and around the environment. Marines may be confronted by an individual who tries to take their weapons. To retain positive control of the weapon, Marines must understand and apply handgun retention techniques.

Blocking Technique

Marines perform the following blocking techniques if an opponent attempts to grab their pistol in the holster. Marines—

Place the body between the weapon and the opponent by immediately pivoting so the weapon is away from the opponent.

Step back and away from the opponent while placing the hand on the pistol grip.

Extend the left hand and block, deflect, or strike the opponent's arm.

Armbar Technique

Marines use the armbar technique when an opponent uses his right hand to grab their pistol in the holster. To execute the armbar technique, Marines—

Trap the opponent's right hand by grasping the opponent's wrist or hand with the right hand and applying pressure against the body.

Step back with the right foot and pivot sharply to the right to be next to the opponent. Always pivot in a direction that keeps the weapon away from the opponent.

Straighten the opponent's arm to apply an armbar. The arm should be straight across the torso.

Continue pivoting to the right while pulling back on the opponent's shoulder. This action may break the opponent's arm.

Place the left hand across the opponent's face and apply pressure back and down to take the oppo-

nent to the ground. Pressure applied beneath the nose or on the trachea is equally effective.

Wristlock Technique

Marines use the wristlock technique when an opponent grabs their pistol while it is in the holster with his right hand. To execute the w ristlock technique, Marines—

Grasp the opponent's wrist or hand with the right hand and apply pressure against the body.

Step back and away from the opponent and pivot to the right so the weapon is away from the opponent. Always pivot in a direction that keeps the weapon away from the opponent.

Reach over the opponent's arm with the left hand and grab his hand, applying pressure against his arm with the left forearm. Execute a wristlock.

Incorporate the second hand into the wristlock and, stepping back with the left foot, pivot to the left.

Execute a two-handed wristlock by exerting downward pressure with the thumb and rotating his hand to the left.

Continue to pivot to off-balance the opponent and drive him to the ground.

Note: If the opponent grabs the pistol with his left hand, Marines execute the wristlock with one hand and step in toward the opponent, rather than

away from the opponent.

Softening Techniques

Handgun retention techniques use softening techniques applied to pressure points. Bone pressure

and strikes with the hands (i.e., hammer fist), knees, and feet are also effective softening techniques.

Pressure Points. Marines use pressure point techniques to get the opponent to loosen his grip. Marines use their finger tips to apply pressure to the webbing between the index finger and thumb, the jugular notch, and the brachial plexus tie in. The following figure illustrates pressure applied to the brachial plexus tie in.

Bone Pressure. Bone pressure is the application of pressure on a bone against a hard object to initiate pain compliance. To apply bone pressure, Marines use their hand to trap the opponent's hand on the weapon. Marines apply a slow, steady pressure to the opponent's hand and fingers until his grip is softened or he releases his hold.

Strikes. If it is difficult to apply a retention technique, Marines employ strikes or kicks to force the opponent to loosen his grip. Strikes to the

eyes, the arms (radial nerve), or shoulder (brachial plexus tie in) soften the opponent's grip on the weapon.

Kicks and knee strikes to the peroneal nerve, the femoral nerve, or the groin are effective because the opponent is typically unprepared to counter the strike.

Stomping on the top of the opponent's foot may distract him or loosen his grip on the weapon.

4. Firearm Disarmament Techniques

Marines use firearm disarmament techniques during a close-range confrontation if they are unarmed and the opponent has a firearm (pistol).

These techniques are equally effective if Marines are armed but do not have time to withdraw and present the weapon. The goal of firearm disarmament techniques is to gain control of the situation so Marines gain the tactical advantage. The goal is not necessarily to get control of the opponent's weapon.

Pistol to the Front

This technique is used when Marines are unarmed and the opponent has a pistol pointing at their front (e.g., head, chest). The technique is the same if the opponent sticks the pistol under the Marine's chin. To execute the counter when an opponent is pointing a pistol toward the front of a Marine, Marines—

Place the hands close to the weapon, about chest high, palms out.

Use the left hand to grab the opponent's forearm and push the opponent's hand with the pistol away to clear the body from in front of the weapon. At the same time, rotate the right shoulder back to clear the body from the weapon.

Maintain control of the opponent's arm.

Grasp the weapon with the right hand by placing the thumb underneath the pistol and the fingers over top of the pistol.

Keep the right hand wrapped tightly around the muzzle and quickly rotate the pistol in the opponent's hand so the muzzle is facing the opponent.

Grasp and pull the opponent's wrist or forearm away from the body while rotating the weapon.

Rotate the weapon toward the opponent while pulling it up and back and out of the opponent's grasp.

Pistol to the Rear

This technique is performed when Marines are unarmed and the opponent has a pistol pointing to the back of the Marine's head. To execute the counter when the opponent is pointing a pistol to the rear, Marines—

Place the hands close together about chest high, palms out.

Step back with the left foot, pivoting on the right foot so the side is against the opponent's front. This action clears the body from the weapon's line of fire. Keep the left hand up.

Pivot on the left foot to face the opponent and, at the same time, raise the left elbow and reach over the top of the opponent's arm with the left arm.

Wrap the left arm tightly around the opponent's arm above his elbow to control it.

Push on the opponent's shoulder with the right hand while pulling up with the left arm to achieve an armbar. This action releases the opponent's grip on the weapon. If necessary, the opponent can be taken to the ground with a leg sweep.

Note: To execute this technique, the weapon must be close to or touching the rear of the Marine. If the weapon is too far away from the body, this technique would be difficult to execute or it would be ineffective.

CHAPTER 3

HAND-HELD WEAPONS

This chapter describes all techniques for a right-handed person. However, all techniques can be executed from either side.

In drawings, the Marine is depicted in woodland camouflage utilities; the opponent is depicted without camouflage. In photographs, the Marine is depicted in woodland camouflage utilities; the opponent is depicted in desert camouflage utilities.

Marines must know how to defend against attacks when an opponent is either unarmed or armed with a held-held weapon. This chapter addresses the combative use of knives, specific weapons of opportunity, and sticks. However, virtually anything can be used as a hand-held weapon.

1. Fundamentals of Knife Fighting

Marines must be trained in knife fighting techniques. Marines experienced in offensive knife techniques can cause enough damage and massive trauma to stop an opponent. When engaged against each other, experienced knife fighters employ various maneuvers and techniques that are specific to knife fighting. Seldom, if ever, will Marines engage an opponent in a classical knife fight.

Note: When armed with a rifle, Marines are issued a bayonet. When armed with a pistol, Marine are issued a combat knife.

Angles of Attack

There are six angles from which an attack with a knife can be launched:

- Vertical strike coming straight down on an opponent.
- Forward diagonal strike coming in at a 45-degree angle to the opponent.
- Reverse diagonal strike coming in at a 45-degree angle to the opponent.
- Forward horizontal strike coming in parallel to the ground.
- Reverse horizontal strike coming in parallel to the ground.
- Forward thrust coming in a straight line to the opponent.

Target Areas of the Body

During any confrontation, the parts of the opponent's body that are exposed or readily accessible will vary. The goal in a knife fight is to attack the body's soft, vital target areas that are readily accessible (e.g, the face, the sides and front of the neck, the lower abdomen [or groin]).

Neck. Carotid arteries, located on either side of the neck, are good target areas because they are not covered by body armor or natural protection.

Lower Abdomen (or Groin). The lower abdomen (or groin region) is a good target area because it is not covered by body armor.

Heart. The heart, if not covered by body armor, is an excellent target which, if struck, can prove fatal in a matter of seconds or minutes.

Secondary Targets. There are secondary target areas that will cause substantial bleeding if an artery is severed. These target areas are not immediately fatal, but can become fatal if left unattended. Attacks to—

The legs can cause a great deal of trauma and prove fatal. For example, the femoral artery located on the inside of the thigh is a large artery which, if cut, will cause extensive blood loss.
The brachial artery, located between the biceps and triceps on the inside of the arm, can cause extensive bleeding and damage.
The arm's radial and ulnar nerves can cause extensive bleeding and damage.

Movement

Marines can move anywhere within a 360-degree circle around the opponent. This allows accessibility to different target areas of the opponent's body. Marines should avoid being directly in front of an opponent because the opponent can rely on his forward momentum to seize the tactical advantage. If Marines face an opponent, movement is made in a 45-degree angle to either side of the opponent. This angle avoids an opponent's strike and places Marines in the best position to attack an opponent.

Wearing the Combat Knife

Marines must wear the combat knife where it is easily accessible and where it can best be retained. It is recommended the combat knife be worn on the weak side hip, blade down, sharp edge facing forward. Marines can place it behind the magazine pouch where it is easily accessible to them, but not easily grabbed by an opponent.

Grip

The grip on the knife should be natural. Marines grasp the knife's grip with the fingers wrapped around the grip naturally as it is pulled out of its sheath. This is commonly known as a hammer grip. The blade end of the knife is always facing the opponent.

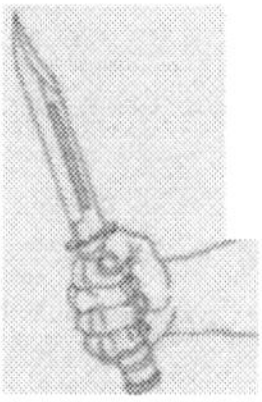

Stance

Marines use the basic warrior stance as the foundation for knife techniques. The left hand is a vertical shield that protects either the ribs or the head and neck. The right elbow is bent with the blade pointing forward toward the opponent' s head. This position serves as an index point, where all techniques are initiated.

Principles of Knife Fighting

The following are key principles of knife fighting:

Execute movements with the knife blade within a box, shoulder-width across from the neck down to the waistline. The opponent has a greater chance of blocking an attack if the blade is brought in a wide, sweeping movement to the opponent.

Close with the opponent, coming straight to the target.

Move with the knife in straight lines.

Point the knife's blade tip forward and toward the opponent.

Apply full body weight and power in each of the knife techniques. Full body weight should be put into the attack in the direction of the blade's movement (slash or thrust).

Apply constant forward pressure with the body and blade to keep the opponent off-balanced.

2. Knife Fighting Techniques

Slashing Techniques

Marines use slashing techniques to close with an enemy. Slashing techniques distract the opponent or damage the opponent so Marines can close in. Typically, Marines target the opponent's limbs, but any portion of the body that is presented can become a target.

Vertical Slash Technique. The vertical slash follows a vertical line straight down through the target. To execute the vertical slash, Marines—

Thrust the right hand out and bring the weapon straight down on the opponent, continuing to drag the knife down through the opponent's body.

Maintain contact on the opponent's body with the blade of the knife.

Forward Slash Technique. The forward slash follows a straight line in a forehand stroke, across the target areas of either the neck (high diagonal slash) or abdominal region (low horizontal slash). To execute the forward slash, Marines-

Extend the right hand while simultaneously rotating the palm up until the knife blade makes contact with the opponent.

Snap or rotate the wrist through the slashing motion to maximize blade's contact with the opponent.

Drag the knife across the opponent's body, from right to left, in a forehand stroke. The movement ends with the forearm against the body and the knife at the left hip with its blade oriented toward the opponent.

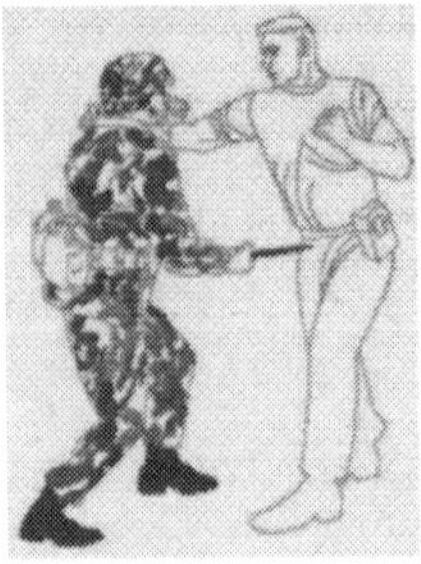

Reverse Slash Technique. The reverse slash is a follow-up technique to a forward attack. It allows Marines both a secondary attack and the ability to resume the basic warrior stance. The reverse slash follows a straight line in a backhand stroke, across the target areas of either the neck (high diagonal slash) or abdominal region (low horizontal slash). To execute the reverse slash, Marines—

Extend the right hand while simultaneously rotating the palm down until the knife blade makes contact with the opponent.

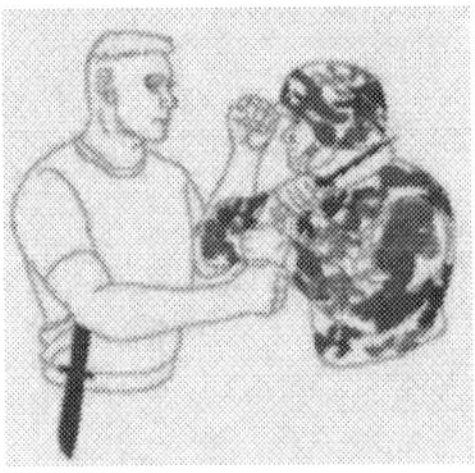

Snap or rotate the wrist through the slashing motion to maximize the blade's contact with the opponent.

Drag the knife across the opponent's body, from left to right, in a backhand stroke. Maintain contact on the opponent's body with the blade of the knife.

Thrusting Techniques

The primary objective of knife fighting is to insert the blade into an opponent to cause extensive damage and trauma. This is done with a thrusting technique. Thrusting techniques are more effective than slashing techniques because of the damage they can inflict. However, Marines use slashing techniques to close with the enemy so that they are closer to the opponent, which allows them to use the thrusting technique.

Vertical Thrust. The thrusting motion follows a vertical line straight up through the target (low into the abdomen region or high into the neck). To execute the vertical thrust, Marines—

Thrust the right hand toward the target, inserting the knife blade straight into the opponent.

Pull the knife out of the opponent.

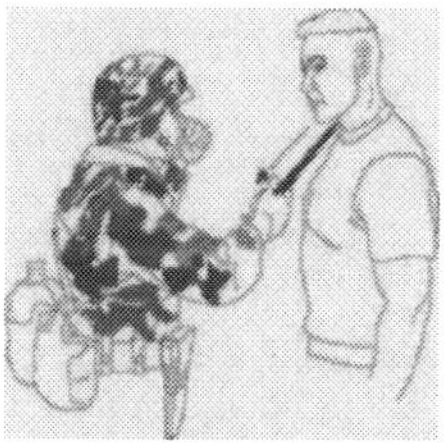

Forward Thrust. The forward thrust follows a straight line straight into the opponent's neck (high thrust) or abdominal region (low thrust). To execute the forward thrust, Marines—

Thrust the right hand, palm down, toward the target, inserting the knife blade straight into the opponent.

Rotate the palm up once the knife is inserted to twist the blade.

Drop the right elbow and bring the knife to the opposite side of the opponent's body from where it was inserted. At the same time, rotate the hips and shoulders downward to bring body weight to bear on the attack.

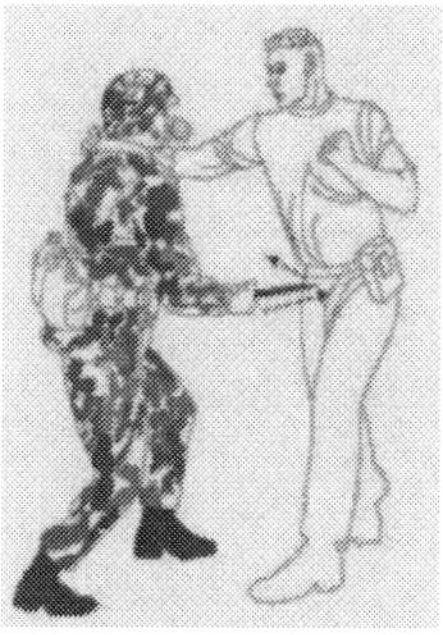

Reverse Thrust. The reverse thrust is a follow-up technique to a forward attack. It allows Marines both a secondary attack and the ability to resume the basic warrior stance. The reverse thrust follows a horizontal line straight into the opponent's neck (high thrust) or abdominal region (low thrust). To execute the reverse thrust, Marines-

Bend the right arm, crossing the arm to the left side of the body.

Thrust the right hand, palm up, toward the target, inserting the knife blade straight into the opponent.

Rotate the palm down to twist the blade once the knife is inserted.

Bring the knife to the opposite side of the opponent's body from where it was inserted.

3. Weapons of Opportunity

During an unarmed close combat situation, Marines use their bodies as weapons, but they should be ready and able to use anything around them as a weapon. For example, Marines could throw sand or liquid in an opponent's eyes to temporarily impair his vision or smash the opponent's head with a rock or helmet. Marines must use whatever means are available and do whatever it takes to take control of the situation and to win, or they face the possibility of losing their lives. Some weapons of opportunity are discussed in the following subparagraphs.

Entrenching Tool

An entrenching tool (E-tool) is commonly carried by Marines. It can be an excellent weapon, especially when sharpened. Marines can use the E-tool to block, slash, and thrust at an opponent.

Tent Pole and Pins

Marines can use tent poles and pins to block, strike, or thrust at an opponent.

Web Belt

Marines can stretch a web belt between their hands to block attacks by an opponent.

Battlefield Debris

Marines can use debris on the battlefield (e.g., sticks, glass, a sharp piece of metal) to cut, slash, or stab an opponent. They can also use other types of debris such as shovel or ax handles, boards, metal pipes, or broken rifles to strike an opponent or apply a choke.

Helmet

A helmet can be used to strike an opponent on an unprotected area like the head and face. Grasp the rim of the helmet and thrust the arms forward, striking the opponent with the top of the helmet.

4. Fundamentals of Combative Stick

On the battlefield, Marines must be ready and able to use anything as a weapon. They must learn and be able to use techniques that can be employed with most weapons of opportunity. Among these techniques are combative stick techniques. Combative stick techniques can be used with a stick, a club, a broken rifle, an E-tool, or even a web belt.

Angles of Attack

There are six angles from which an attack with a hand-held weapon can be launched:

Vertical strike coming straight down on an opponent.
Forward diagonal strike coming in at a 45-degree angle to the opponent.
Reverse diagonal strike coming in at a 45-degree angle to the opponent.
Forward horizontal strike coming in parallel to the ground.
Reverse horizontal strike coming in parallel to the ground.
Forward thrust coming in a straight line to the opponent.

Grip

Grasp the stick about 2 inches from its base.

Stance

The basic warrior stance serves as the foundation for combative stick techniques. The left hand becomes a vertical shield that protects the ribs or the head and neck. Depending on how heavy the weapon is, it should be held at a level approximately shoulder height.

Movement

Movement during combative stick techniques is the same as it is for other close combat techniques. Marines can move anywhere within a 360-degree circle around the opponent. This allows accessibility to different target areas of the opponent's body and gains a tactical advantage. Marines should avoid being directly in front of an opponent because the opponent can rely on his forward momentum to seize the tactical advantage. If Marines face an opponent, movement is made in a 45-degree angle to either side of the opponent. This angle avoids an opponent's strike and places Marines in the best position to attack an opponent.

5. Combative Stick Techniques

Strikes

Strikes are intended to inflict as much damage on an opponent as possible. Striking techniques apply to a weapon of opportunity such as a stick, a tent pole, a club, a broken rifle, an E-tool, or a Pipe-

Vertical Strike. To execute the vertical strike, Marines—

Bend the right arm, extending the weapon over the back of the right shoulder.

Rotate the forearm straight down off the elbow to bring the weapon down on the opponent.

Rotate the hips and shoulders forcefully toward the opponent.

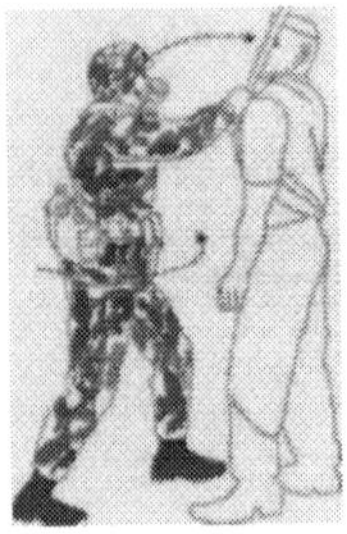

Follow through with the strike by allowing the weight of the weapon to go through the target area of the body.

Forward Strike. To execute a forward strike, Marines—

Step forward with the left foot in the direction of the strike.

Bend the right arm with the elbow extending out to the right and the weapon extended over the right shoulder.

Rotate the forearm to the right of the elbow to bring the weapon down onto the opponent. At the same time, forcefully rotate the hips and shoulders toward the opponent.

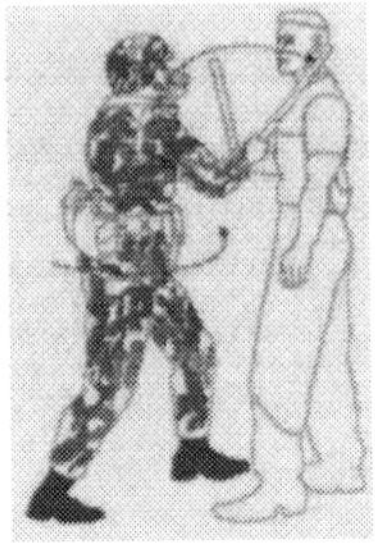

Follow through with the strike by allowing the weight of the weapon to go through the target area of the body.

Reverse Strike. The reverse strike is a follow-up technique to a forward strike. It allows Marines both a secondary attack and the ability to resume the basic warrior stance.To execute a reverse strike, Marines—

Step forward with the right foot in the direction of the strike.

Bend the right arm with the hand near the left shoulder. The weapon is extended over the left shoulder.

Rotate the forearm to the right of the elbow to bring the weapon down onto the opponent. At the same time, forcefully rotate the hips and shoulders toward the opponent.

Follow through with the strike by allowing the weight of the weapon to go through the target area of the body.

Forward Thrust. To execute the forward thrust, Marines—

Grasp the stick with the left hand, palm up, in a position where the stick can be controlled with two hands.

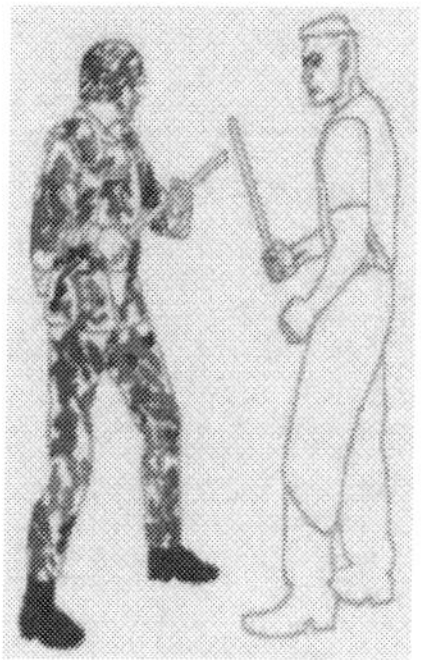

Lift the left leg and lunge forward off the ball of the right foot. At the same time, thrust the end of the weapon directly toward the opponent by thrusting both hands forward in a straight line.

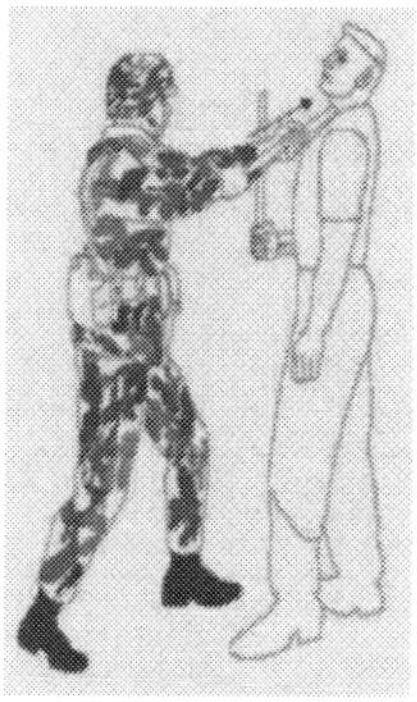

6. Blocking Techniques

A block is meant to deter or deflect an attack by an opponent. A block sets up Marines for a follow-on attack against the opponent. Blocks are executed by deflecting, rather than hitting or following through like a strike.

Blocks Against Unarmed Attacks

To block against an unarmed attack, Marines—

Step forward at a 45-degree angle with either the right or left foot. This moves the body out of the line of attack.

Raise the left arm and block the strike with the meaty portion of the forearm.

Employ the stick by one of two techniques:

Strike the opponent with the stick.
Use the stick to block on two points of contact. When the stick is used to block, it serves as an incidental strike. This technique is only used if Marines have closed with the opponent and are inside his strike.

Blocks Against Armed Attacks

Block for a Vertical Strike. To execute the block for a vertical strike, Marines—

Step forward with the left foot at a 45-degree angle to the left. This moves the body out of the line of attack.

Block on two points of contact to disperse the impact of the attack:

Block the opponent's stick by positioning the stick so it is perpendicular to the opponent's stick. If the stick is not perpendicular to the opponent's stick, the stick can slide through and make contact on the Marine.
Block the opponent's wrist or forearm with the back of the left forearm.

Note: Use the stick to block the opponent's arm if closer to the opponent. It is the same movement as blocking with the arm, except the opponent's arm is blocked with both the stick and the arm.

Block for a Forward Strike. To execute the block for a forward strike, Marines-

Step forward with the left foot at a 45-degree angle to the left. This moves the body out of the line of attack and inside the opponent's strike.

Block on two points of contact to disperse the impact of the attack:

- Block the opponent's wrist or forearm with the meaty portion of the back of the left forearm.
- Strike the opponent's attacking biceps with the stick.

Block for a Reverse Strike. To execute the block for a reverse strike, Marines—

Step forward at a 45-degree angle with the right foot to the right. This moves the body out of the line of attack and inside the opponent's strike.

Block on two points of contact to disperse the impact of the attack:

Block the opponent's stick by positioning the stick so it is perpendicular to the opponent's stick.

Block the opponent's forearm with the meaty portion of the left forearm.

Note: If closer to the opponent, block his triceps with the back of the left forearm and strike his forearm with the stick.

7. Unarmed Against Hand-Held Weapons

If Marines are engaged against an opponent with a knife, a stick, or some other weapon of opportunity, they must establish and maintain an offensive mindset, not a defensive mindset. Their survival depends on it. They cannot afford to think about getting cut or hurt.

Angles of Attack

Before Marines can learn to block or counter an attack with a hand-held weapon (e.g., knife, stick), they must know from what angle the opponent is attacking. There are six angles from which an opponent will typically attack with a hand-held weapon:

Vertical strike coming in straight down on the Marine.
Forward diagonal strike coming in at a 45-degree angle to the Marine.
Reverse diagonal strike coming in at a 45-degree angle to the Marine.
Forward horizontal strike coming in parallel to the ground.
Reverse horizontal strike coming in parallel to the ground.
Forward thrust coming in a straight line to the Marine.

Blocks

If the opponent has a hand-held weapon (e.g., knife or stick), Marines parry the opponent's hand or arm to block the attack.

Basic Block. To execute the basic block technique, Marines-

Step forward at a 45-degree angle to move out of the line of the attack. Always step in the direction of the strike.

Thrust the forearms forward, hands up, against the opponent's attacking arm. Contact is made on the opponent's arm with the backs of the fore-arms.

Block for a Vertical Strike. To execute the block against a vertical strike, Marines—

Step forward with the left foot at a 45-degree angle to the left to move out of the line of attack.

Thrust the forearms forward, hands up, against the outside of the opponent's attacking arm.

Block for a Forward Strike. To execute the block against a forward diagonal or forward horizontal strike, Marines-

Step forward with the left fool inside the opponent's attacking arm.

Block the attack with both arms bent so the fore-arms make contact with the opponent's biceps and forearm.

Block for a Reverse Strike. To execute the block against a reverse diagonal or reverse horizontal strike, Marines—

Step forward with the right foot to the outside of the opponent's attacking arm.

Block the attack with both arms bent so the fore-arms make contact with the opponent's triceps and forearm.

Block for a Forward Thrust. To execute the block against a forward thrust, Marines—

Bend at the waist, move the hips backward, and jump backward with both feet to move away from the attack. This action is known as "hollowing out."

Hollow out and block the attack with the arms bent and hands together on top of the attacking arm.

Overlap the hands slightly so one thumb is on top of the other hand's index finger. The other thumb should be under the other hand's index finger.

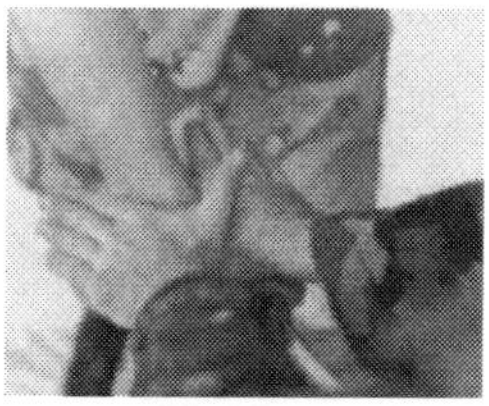

8. Counters to Hand-Held Weapon Attacks

Principles of Counters

A counter is used to control the situation to regain the tactical advantage and end the fight. Regardless of the type of weapon or angle of attack, the following principles apply to countering the attack with a hand-held weapon:

Move out of the line of attack. Movement is executed in a 45-degree angle forward to the left or right.

- Block the attack.

Note: The first two actions are taken simultaneously.

- Take control of the weapon by controlling the hand or arm that is holding the weapon. Never attempt to grab the opponent's weapon.
- Execute the appropriate follow-up to end the fight; e.g., strikes, joint manipulations, throws, or takedowns (see chap. 8). Marines should continue to attack the opponent until the fight ends.

Counter Techniques

There are two techniques that can be used to counter any armed attack: forward armbar counter and reverse armbar counter. These techniques can be used to counter a vertical attack, a forward diagonal strike, or a forward horizontal strike. With minor variations, the same techniques are used to counter reverse strikes. A third technique, the bent armbar counter, is used to counter a vertical attack.

Forward Armbar Counter. To execute the forward armbar counter to an attack coming from a forward strike, Marines-

Step forward with the left foot inside the opponent' s attacking arm.

Block the attack with both arms bent so the forearms make contact with the opponent's biceps and forearm.

Slide the left arm over the opponent's forearm and wrap the arm tightly around his arm, trapping the opponent's attacking arm between the biceps and torso.

Place the right hand on the opponent's shoulder or upper arm to further control his arm and to effect an armbar.

Execute an armbar and continue to exert steady pressure against the arm.

Reverse Armbar Counter. To execute the reverse armbar counter to an attack coming from a forward strike, Marines-

Step forward with the left foot inside the opponent' s attacking arm.

Block the attack with both arms bent so the forearms make contact with the opponent's biceps and forearm.

Control the opponent's arm with the left hand and pivot to the right so the back is against the opponent's side. Immediately slide the right arm over the opponent's biceps and wrap the arm tightly around his arm, trapping the opponent's attacking arm between the biceps and torso.

Grasp the opponent's wrist with the left hand and twist his thumb away from the body.

Take the opponent to the ground with an armbar takedown if he does not drop the weapon from the wristlock.

Bent Armbar Counter. This counter is particularly effective against a vertical attack. To execute the bent armbar counter, Marines—

Step forward with the left foot inside the opponent's attacking arm.

Block the attack with both arms bent so the forearms make contact with the opponent's biceps and forearm.

Grasp the opponent's forearm with the left hand. At the same time, slide the right arm underneath the opponent's triceps and grasp the opponent's forearm or wrist with the right hand.

Apply downward pressure with the hands against the opponent's forearm to off-balance the opponent.

CHAPTER 4

STRIKES

This chapter describes all techniques for a right-lmnded person. However, all techniques can be executed from either side.

In drawings, the Marine is depicted in woodland camouflage utilities; the opponent is depicted without camouflage. In photographs, the Marine is depicted in woodland camouflage utilities; the opponent is depicted in desert camouflage utilities.

Strikes are unarmed individual hitting techniques. Strikes use the hands, elbows, knees, feet, and, in some instances, other parts of the body as personal weapons. Marines must know how to execute strikes effectively. They must also know how to counter strikes from an opponent.

1. Principles of Punches

Muscle Relaxation

Muscle relaxation is crucial when executing punches. The natural tendency in a fight is to tense up, which results in rapid fatigue and decreased power generation. Marines who remain relaxed during a close combat situation generate greater speed, which results in greater generation of power. Relaxing the forearms generates speed and improves reaction time. At the point of impact, Marines clench the fist to cause damage to the opponent and avoid injury to the wrist and hand.

Weight Transfer

Weight transfer is necessary to generate power in a punch. Marines accomplish this by-

Rotating their hips and shoulders into the attack.

Moving their body mass straight forward or backward in a straight line.

Dropping their body weight into an opponent. Body mass can be transferred into an attack from high to low or from low to high.

Rapid Retraction

When Marines deliver a punch, rapid retraction of the fist is important. Once the hand has made contact with the target, Marines quickly return to the basic warrior stance. Rapid retraction-

Returns the hand and arm to the protection afforded by the basic warrior stance.

Prevents the opponent from being able to grab the hand or arm.

Permits the hand and arm to be "chambered" or "re-cocked" in preparation for delivering a subsequent punch.

Telegraphing

Telegraphing a strike occurs when body movements inform the opponent of the intention to launch an attack. Staying relaxed helps to reduce telegraphing.

Often, an untrained fighter telegraphs his intention to attack by drawing his hand back in view of his opponent, changing facial expression, tensing neck muscles, or twitching. These movements, however small, immediately indicate an attack is about to be delivered. If the opponent is a trained fighter, he may be able to evade or counter the attack. If the opponent is an untrained fighter, he may still be able to minimize the effect of an attack.

2. Punches

Punches may be thrown during any hand-to-hand confrontation. Most people resort to punching because it is a natural reaction to a threat. The purpose of a punch is to stun the opponent or to set him up for a follow-up finishing technique. However, punches should only be executed to the soft tissue areas of an opponent. A correctly delivered punch maximizes the damage to an opponent while minimizing the risk of injury to Marines.

Basic Fist

Punches are executed using the basic fist. To make the basic fist, the fingers are curled naturally into the palm of the hand and the thumb is placed across the index and middle fingers. Do not clench the fist until movement has begun. This reduces muscular tension in the forearm and increases speed and reaction time. Just before impact, Marines exert muscular tension on the hand and forearm to maximize damage to the opponent and reduce their chances of injury. Contact should be made with the knuckles of the index and middle fingers.

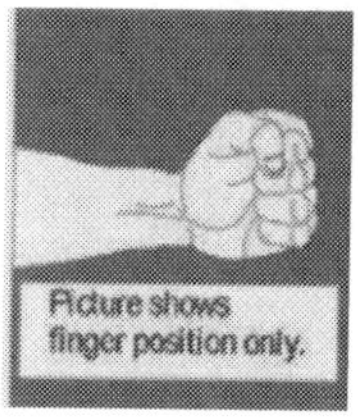

When striking with the basic fist, Marines must keep the hand straight, or in line, with the wrist to avoid injury to the wrist.

Lead Hand Punch

The lead hand punch is a snapping straight punch executed by the forward or lead hand. It is a fast punch designed to keep the opponent away and to set up an attack. A lead hand punch conceals movement and allows Marines to get close to the opponent. Lead hand punches should strike soft tissue areas, if possible. To execute the lead hand punch, Marines-

Snap the lead hand out to nearly full extension, while rotating the palm to the ground.

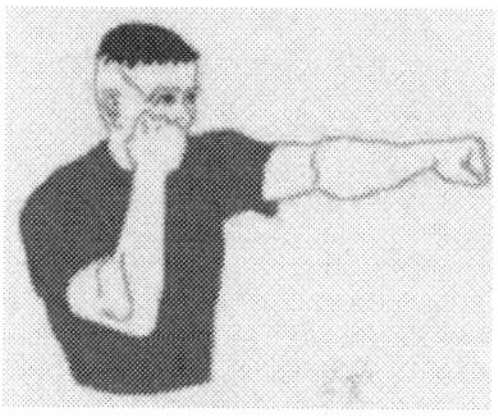

Contact the opponent with the first two knuckles of the fist.

Retract the hand immediately, resuming the basic warrior stance.

Rear Hand Punch

The rear hand punch is a snapping punch executed by the rear (right) hand. It is a power punch designed to inflict maximum damage on the opponent. Its power comes from pushing off the rear leg and rotating the hips and shoulders. To execute the rear hand punch, Marines—

Rotate the hips and shoulders forcefully toward the opponent and thrust the rear hand straight out, rotating the palm down, to nearly full extension.

Shift body weight to the lead foot while pushing off on the ball of the rear foot.

Contact the opponent with the first two knuckles of the fist.

Retract the hand immediately.

Uppercut

The uppercut is a powerful punch originating below the opponent's line of vision. It is executed in an upward motion traveling up the centerline of the opponent's body. It is delivered in close and usually follows a preparatory strike that leaves the target area unprotected. When delivered to the chin or jaw, the uppercut can render an opponent unconscious, cause extensive damage to the neck, or sever the tongue. To execute the uppercut, Marines—

Bend the arms, rotating the palm inboard. The distance the arms bend depends on how close the opponent is.

Rotate the hips and shoulders forcefully toward the opponent, thrusting the fist straight up toward the opponent's chin or jaw.

Contact the opponent with the first two knuckles of the fist.

Retract the hand immediately.

Hook

The hook is a powerful punch that is executed close in and is usually preceded by a preparatory strike. To execute the hook, Marines—

Thrust the right arm in a hooking motion toward the opponent, keeping the elbow bent while forcefully rotating the right shoulder and hip toward the opponent.

Contact the opponent with the first two knuckles of the fist. Continue rotating the shoulder and hip, following through with the fist to the target.

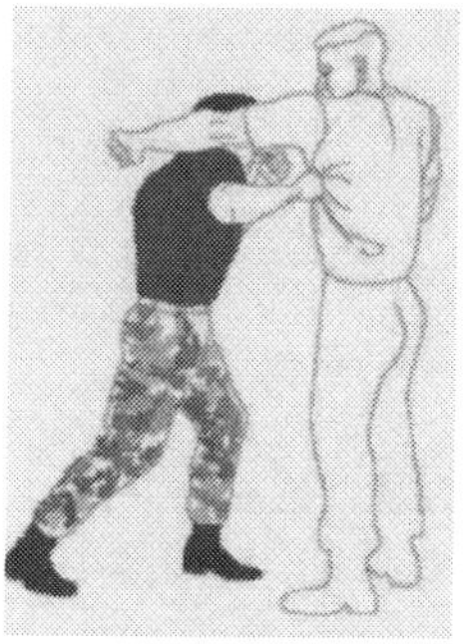

Retract the hand immediately.

3. Strikes with the Upper Body

Strikes stun the opponent or set him up for a follow-up finishing technique. The hands, forearms, and elbows are individual weapons of the arms that can be used to execute strikes including the hammer fist, chin jab, knife hand, eye gouge, and elbow strikes. These strikes provide a variety of techniques that can be used in any type of close combat situation.

Principles of Execution

There are several principles of execution that ensure a strike's effectiveness.

Generating Power. Weight transfer is necessary to generate maximum power in a punch. Marines accomplish this by-

Rotating the hips and shoulders into the attack.

Moving the body mass straight forward or backward in a straight line.

Dropping body weight into an opponent. Body mass can be transferred into an attack from high to low or from low to high.

Muscular Tension. The arms are relaxed until the moment of impact. At the point of impact, Marines apply muscular tension in the hand and forearm to maximize damage to the opponent and to avoid injury to the hand. The arms are relaxed until the moment of impact.

Hit and Stick and Follow-Through. A strike should be delivered so that the weapon (e.g., hand, elbow) hits and remains on the impact site (target) and follows through the target. This technique inflicts maximum damage on the opponent.

Strikes with the arms are executed with "heavy hands"; i.e., the strike is executed by driving through with the strike to allow the weight of the hand to go through the target area of the body. Contact with the opponent should be made with the arm slightly bent, and the arm extends as it moves through the target. This technique allows Marines to deliver strikes effectively without executing full force.

Movement. Movement puts Marines in the proper position for launching an attack against an opponent as well as providing protection.

Movement is initiated from the basic warrior stance and ends by resuming the basic warrior stance. Strikes can be performed with either the left or right arm depending upon—

The angle of attack.
The position of the opponent.
The opponent's available, vulnerable target areas.

Target Areas of the Body

For each strike, there are target areas of the body that, when struck, maximize damage to an opponent. Strikes use gross motor skills as opposed to fine motor skills. The target areas of the body are just that—areas. Pinpoint accuracy on a specific nerve is not needed for the strike to be effective.

Hammer Fist

Striking with the hammer fist concentrates power in a small part of the hand, which, when transferred to the target, can have a devastating effect. The striking surface of the hammer fist is the meaty portion of the hand below the little finger. To execute the hammer fist strike, Marines-

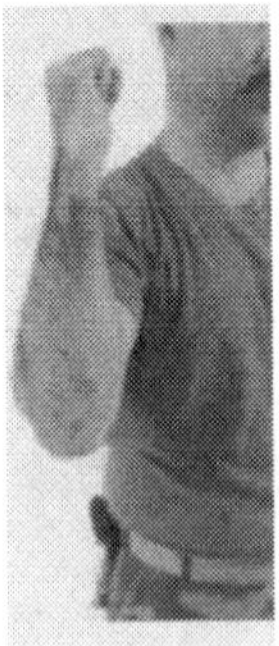

Make a fist and bend the arm at approximately a 45- to 90-degree angle. At the same time, rotate the right hip and right shoulder backward.

Thrust the fist forward onto the opponent while rotating the right hip and shoulder forward.

Rotate the wrist so the hammer fist makes contact on the opponent.

Chin Jab

The chin jab can immediately render an opponent unconscious and cause extensive damage to the neck and spine. The striking surface is the heel of the palm of the hand. To execute the chin jab, Marines—

Bend the right wrist back at a 90-degree angle with the palm facing the opponent and the fingers pointing up.

Keep the right arm bent and close to the body. Extend the hand into a concave position with the fingers slightly spread apart.

Step forward with the left foot toward the opponent, keeping the feet approximately shoulder-width apart and the knees bent. This is done to close with the opponent.

Keep the right arm bent and close to the side. Thrust the palm of the hand directly up under the opponent's chin. At the same time, rotate the right hip forward to drive the body weight into the attack to increase the power of the strike. The attack should travel up the centerline of the opponent's chest to his chin.

Knife Hand

The knife hand is one of the most versatile and devastating strikes. The striking surface is the cutting edge of the hand, which is the meaty portion of the hand below the little finger extending to the top of the wrist. The striking surface is narrow, allowing strikes on the neck between the opponent's body armor and helmet. The knife hand strike is executed from one of three angles: outside, inside, and vertical.

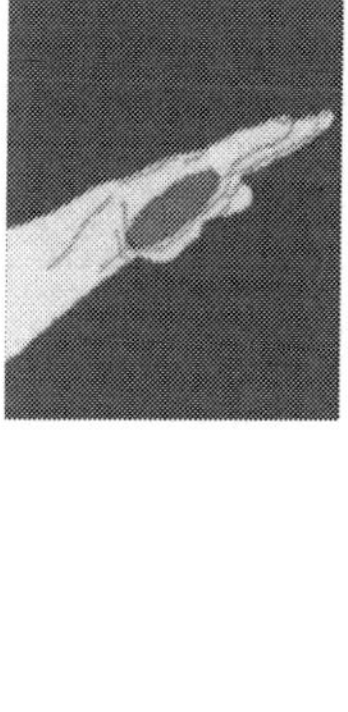

Outside Knife Hand. To execute the outside knife hand strike, Marines—

Execute a knife hand by extending and joining the fingers of the right hand and placing the thumb next to the forefinger (like saluting).

Retract the right hand. At the same time, rotate the right hip and right shoulder backward.

Thrust the knife hand forward (horizontally) onto the opponent while rotating the right hip and shoulder forward.

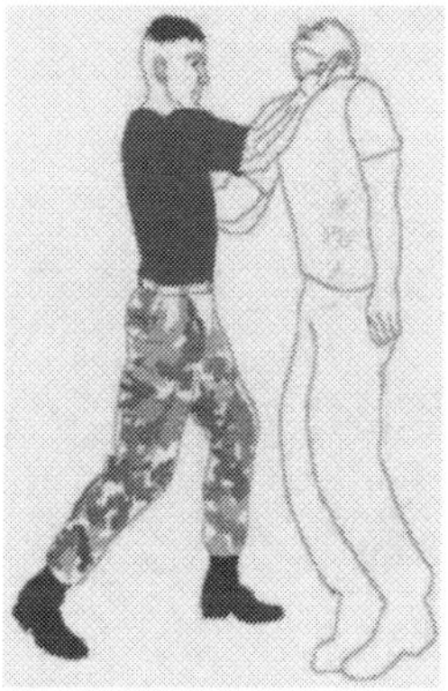

Inside Knife Hand. To execute the inside knife hand strike, Marines—

Execute a knife hand.

Bring the right hand over the left shoulder. At the same time, rotate the right shoulder forward and the left hip forward.

Thrust the knife hand forward (horizontally) onto the opponent while rotating the right hip and shoulder forward and the left shoulder backward.

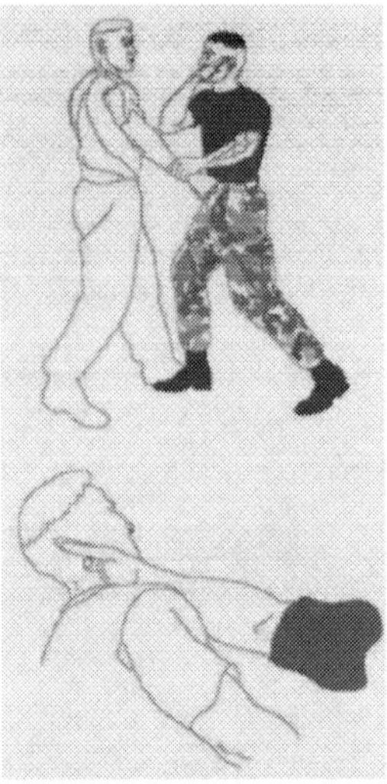

Vertical Knife Hand. When thrown vertically, the knife hand strike comes straight down in a straight line.

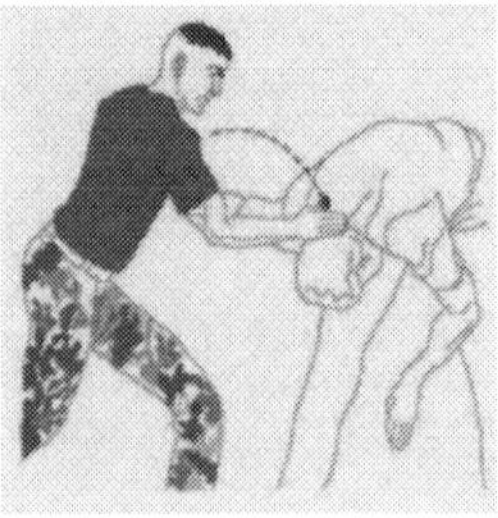

Eye Gouge

The eye gouge is used to attack an opponent's eyes, blinding him so follow-up strikes can be executed. The striking surface is the tips of the fingers or thumb. To execute the eye gouge, Marines—

Extend the right hand with the fingers slightly spread apart to allow entry into the eye sockets.

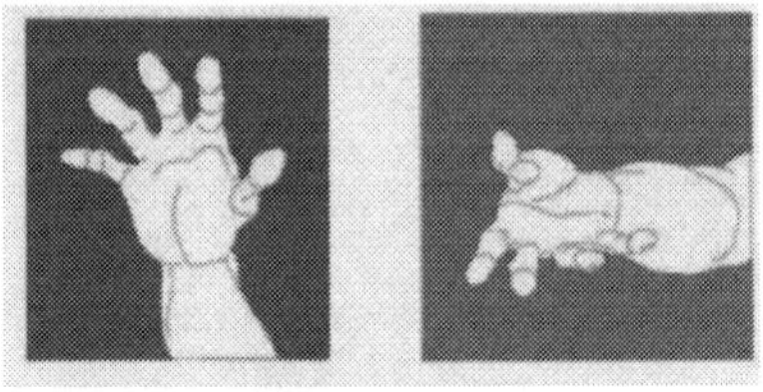

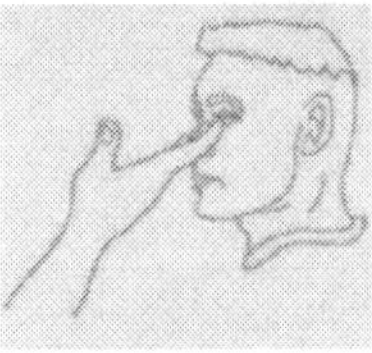

Place the palm of the hand either toward the ground or toward the sky and thrust the right hand forward into the opponent's eyes.

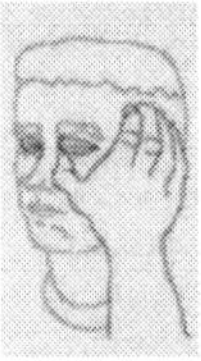

Thrust the hand forward at the opponent's nose level so the fingers or thumb slide naturally into the grooves of the opponent's eye sockets.

Elbow Strikes

The elbow is a powerful weapon that can be used in several different ways to attack virtually any part of an opponent's body. Elbow strikes can be performed either vertically (upward or downward) or horizontally (forward or reverse). The striking surface is 2 inches above or below the point of the elbow, depending upon the angle of attack, the opponent's attack angle, and the position of the opponent.

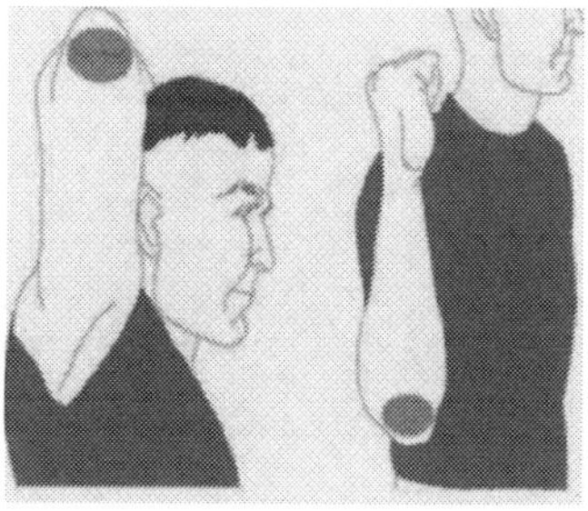

Above Elbow **Below Elbow**

Vertical Elbow Strike (Up). To execute an upward vertical elbow strike, Marines-

Bend the right elbow, keeping the fist close to the body. The fist is at shoulder level and the elbow is next to the torso.

Thrust the elbow vertically upward toward the opponent while rotating the right shoulder and hip forward to generate additional power.

Contact the opponent with the right forearm 2 inches above the point of the elbow.

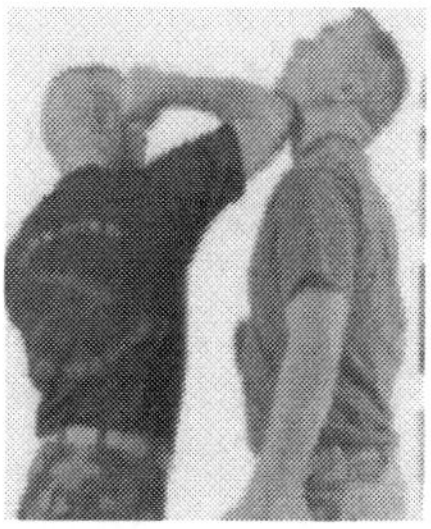

Vertical Elbow Strike (Down). To execute a downward vertical elbow strike, Marines—

Bend the right elbow, keeping the fist close to the body. The fist is on the shoulder and the elbow is raised well above the shoulder.

Thrust the elbow vertically downward toward the opponent while dropping body weight into the attack to generate additional power.

Contact the opponent with the right triceps 2 inches above the point of the elbow.

Horizontal Elbow Strike (Forward). To execute the forward horizontal elbow strike, Marines-

Tuck the right fist near the chest with the palm heel facing the ground.

Thrust the right elbow horizontally forward toward the opponent. The forearm is parallel to the ground.

Rotate the right shoulder and hip forward to generate additional power.

Contact the opponent with the right forearm 2 inches below the point of the elbow.

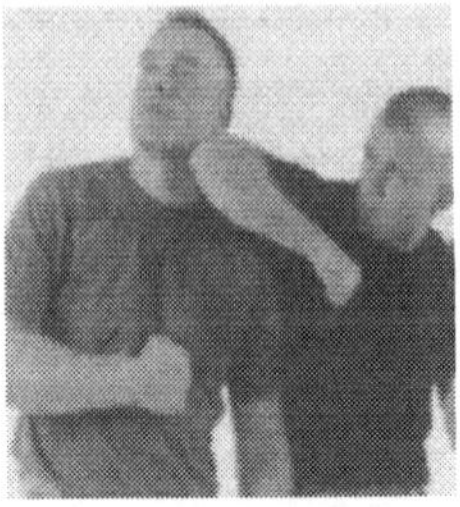

Horizontal Elbow Strike (Rear). To execute the rear horizontal elbow strike, Marines-

Tuck the right fist near the left shoulder with the palm heel facing the ground. At the same time, rotate the right shoulder forward and the left hip forward.

Thrust the right elbow horizontally rearward toward the opponent. The forearm is parallel to the ground and the hand moves toward the direction of the attack.

Rotate the right hip back and the right shoulder backward to generate additional power.

Contact the opponent with the right triceps 2 inches above the point of the elbow.

4. Strikes with the Lower Body

The legs are the body's most powerful weapons because they use the largest muscle groups to generate a strike. Legs are also less prone to injury. The feet are the preferred choice for striking because they are protected by boots. Marines use their feet, heels, and knees to execute kicks, knee strikes, and stomps.

Kicks

The purpose of a kick is to stop an opponent's attack or to create an opening in his defense in order to launch an attack. Kicks can be performed with the left (lead) leg or the right (rear) leg. Kicks with the rear leg have greater power because the hips are rotated into the attack. However, the rear leg is further away from the opponent so a strike with the rear leg will not contact the opponent as quickly as a strike with the lead leg.

Front Kick. The front kick is executed when the opponent is in front of the Marine. The front kick, delivered with either the rear or lead leg, is effective for striking below the waist. Attempting to kick higher results in diminished balance and provides the enemy with a greater opportunity to grab the leg or foot. The striking surfaces are the toe of the boot or the bootlaces. To execute the front kick, Marines-

Raise the left knee waist high, pivot the hips into the attack, and thrust the left foot forward toward the opponent.

Contact the opponent with the toe of the left boot or bootlaces.

Return to the basic warrior stance.

Side Kick. The side kick, delivered with the lead leg, is effective for striking the knees. The side kick is executed when the opponent is to the side of the Marine. The striking surface is the outside cutting edge of the boot near the heel. To execute the side kick, Marines—

Raise the right knee waist high and rotate the right hip forward.

Thrust the right foot to the right side toward the opponent, turning the foot at a 90-degree angle to maximize the striking surface on the opponent.

Contact the opponent with the cutting edge of the right boot.

Return to the basic warrior stance.

Knee Strikes

Knee strikes are excellent weapons during the close range of close combat fighting. The knee strike is generally delivered in close.

Vertical Knee Strike. The striking surface is the thigh, slightly above the knee. To execute the vertical knee strike, Marines-

Raise the right knee and drive it up forcefully into the opponent. Power is generated by thrusting the leg upward.

Contact the opponent 2 inches above the right knee.

Horizontal Knee Strike. The horizontal knee strike is executed with the leg generally parallel with the ground while rotating the hips to generate power. It is often delivered to the peroneal nerve. The striking surface is the front of the leg, slightly above or below the knee. To execute the horizontal knee strike, Marines-

Raise the right knee, rotate the right hip forward while pivoting on the left foot, and drive the knee horizontally into the opponent.

Contact the opponent 2 inches above the right knee.

Stomps

Stomps are delivered with the feet, usually when the opponent is down. Remember, when the opponent is down, Marines take whatever target is available.

Vertical Stomp. The vertical stomp allows Marines to remain upright and balanced, to rapidly deliver multiple blows with either foot, and to quickly and accurately attack the target. It is the preferred stomp. The striking surface is the flat bottom of the boot or the cutting edge of the heel. To execute the vertical stomp, Marines—

Raise the right knee above the waist with the right leg bent at approximately a 90-degree angle.

Drive the flat bottom of the right boot or the cutting edge of the right heel down onto the opponent forcefully. At the same time, bend the left knee slightly to drop the body weight into the strike.

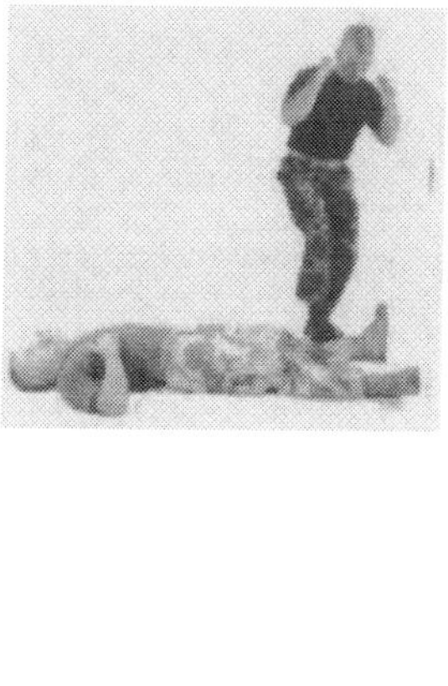

Ax Stomp. The striking surface of the ax stomp is the cutting edge of the heel. To execute the ax stomp, Marines-

Raise the right heel above the waist, keeping the right leg straight.

Drive the cutting edge of the right heel down onto the opponent forcefully. At the same time, bend the left knee slightly to drop the body weight into the strike.

5. Counters to Strikes

In a close combat situation, an opponent will attempt to strike Marines with punches and kicks. When an opponent uses a strike, Marines must first avoid the strike. This is accomplished by moving quickly and blocking. Next, Marines must get into an offensive position. This allows Marines to use offensive strikes to attack the opponent. Regardless of the strike, the counter to a strike requires Marines to move, block, and strike.

Move

The first step in countering a strike is to move out of the way of the strike's impact. Movement removes Marines from the opponent's intended strike point and positions Marines to attack. Movement is executed at approximately a 45-degree angle to the front or rear. Movement is always initiated from the basic warrior stance. Return to the basic warrior stance with the toe of the lead foot pointing toward the opponent once the movement is complete.

Block

Different blocks are executed based on the strike. These will be covered by individual counters.

Strike

Any of the upper body or lower body strikes can be executed as a follow-on attack or part of the counter to an opponent's strike. The follow-on strike is determined by the angle to the opponent, the position of the opponent, and the opponent's available vulnerable target areas.

Counters to Punches

Counter to a Lead Hand Punch. This counter is used when the opponent throws a lead hand punch. To execute the counter to the lead hand punch, Marines-

Step forward and to the left at approximately a 45-degree angle, moving in to the outside of the opponent's attacking arm.

Raise the left arm and block or deflect the opponent's lead hand with the palm of the hand or the meaty portion of the forearm.

Hit and stick by leaving the left arm against the opponent's right arm while stepping forward and to the left at approximately a 45-degree angle to close with the opponent.

Execute a finishing technique, such as a strike or a kick, to the opponent's exposed target areas.

Counter to a Rear Hand Punch. This counter is used when the opponent throws a rear hand punch. To execute the counter to the rear hand punch, Marines—

Step forward and to the left at approximately a 45-degree angle, moving in to the outside of the opponent's attacking arm.

Raise the left arm and block or deflect the opponent' s rear hand with the palm of the hand or the meaty portion of the forearm.

Hit and stick by leaving the left arm against the opponent's right arm while stepping forward and to the right at approximately a 45-degree angle to close with the opponent.

Execute a finishing technique, such as a strike or a kick, to the opponent's exposed target areas.

Counters to Kicks

Counter to a Front Kick (Left or Lead Leg). This counter is used when the opponent executes a front kick with his left leg. To execute the counter to a front kick, Marines-

Step forward and to the right at approximately a 45-degree angle, moving in to the outside of the opponent's striking leg.

Lower the left arm and block or deflect the opponent' s leg with the palm of the hand or the meaty portion of the forearm.

Hit and stick by leaving the left arm against the opponent's leg while stepping forward and to the left at approximately a 45-degree angle to close with the opponent.

Execute a finishing technique, such as a strike or a kick, to the opponent's exposed target areas.

Counter to a Front Kick (Right or Rear Leg). This counter is used when the opponent executes a front kick with his right leg. To execute the counter to a front kick, Marines—

Step forward and to the left at approximately a 45-degree angle, moving in to the outside of the opponent's striking leg.

Lower the left arm and block or deflect the opponent' s leg with the palm of the hand or the meaty portion of the forearm.

Hit and stick by leaving the left arm against the opponent's leg while stepping forward and to the right at approximately a 45-degree angle to close with the opponent.

CHAPTER 5

THROWS

This chapter describes all techniques for a right-handed person. However, all techniques can be executed from either side.

The Marine is depicted in camouflage utilities. The opponent is depicted without camouflage.

Marines use throwing techniques to maintain the tactical advantage and to throw the opponent to the ground during close combat. Throws apply the principles of balance, leverage, timing, and body position to upset an opponent's balance and to gain control by forcing the opponent to the ground. Throwing techniques are effective because they are size- and gender-neutral, and they rely on the momentum and power generated by the opponent rather than the strength or size of the Marine. Marines also execute a throw as a devastating attack against an opponent, possibly causing unconsciousness or broken limbs. When Marines execute throws, they must maintain balance and, simultaneously, prevent the opponent from countering a throw or escaping after being forced to the ground.

1. Turning Throw

Marines use a turning throw to take the opponent to the ground while they remain standing. A turning throw can also be executed from a stationary position. It is particularly effective if Marines and the opponent are wearing gear. To execute the turning throw, Marines-

Grasp the opponent's right wrist with the left hand.

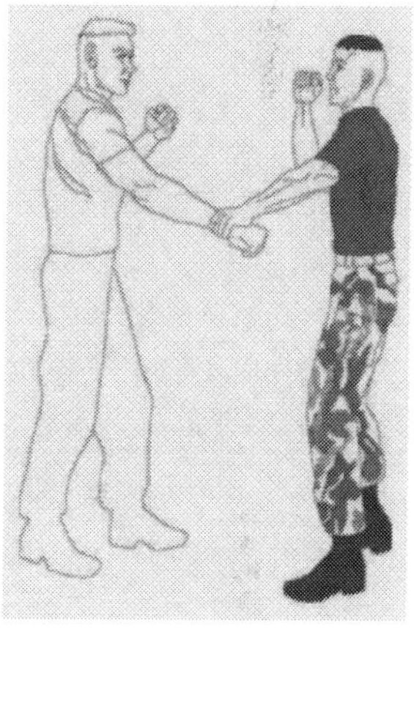

Step forward with the right foot, place it against the outside of the opponent's right foot, and pivot so the back of the heel is next to the middle of the opponent's foot.

Hook the opponent's right arm with the right arm and pinch his arm between the biceps and forearm, touching the opponent with the body.

Pull the opponent's wrist downward, keeping it close to the body.

Pivot to the left on the ball of the foot and continue pulling downward on the opponent's wrist while rotating the wrist outward to off-balance him.

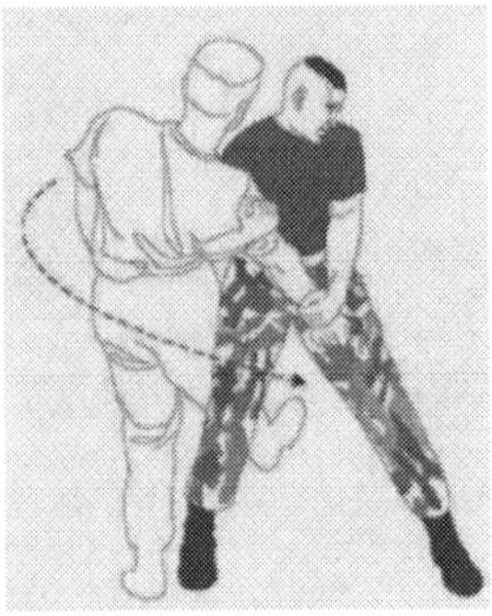

2. Hip Throw

Marines use a hip throw to take the opponent to the ground while they remain standing. A hip throw is particularly effective if the opponent is moving forward or pushing. Marines use the opponent's forward momentum to execute the hip throw. To execute the hip throw, Marines-

Grasp the opponent's right wrist with the left hand.

Step forward with the right foot and place it against the outside of the opponent's right foot.

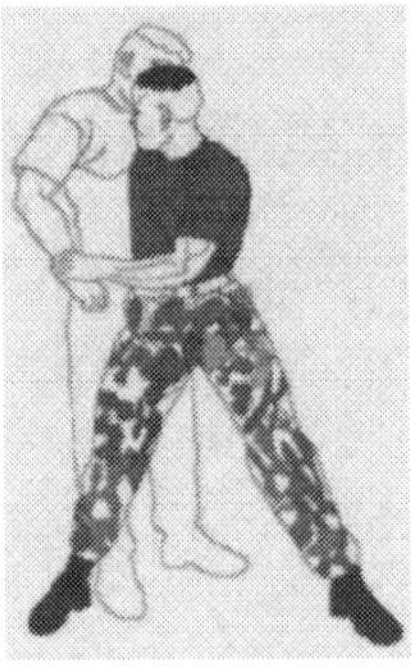

The back of the heel should be next to the opponent's foot.

Step back with the left foot and rotate on the ball of the foot. The back of the heel is next to the opponent's toe. The knees are bent.

Rotate at the waist and hook the right hand around the back of the opponent's body (anywhere from his waist to his head). The side and hip should be against the opponent.

Rotate the hip against the opponent. The hips must be lower than the opponent's.

Use the right hand to pull the opponent up on the hip to maximize contact.

Pull the opponent's arm across the body and, at the same time, lift the opponent off the ground slightly by bending at the waist, straightening the legs, and rotating the body to the left.

3. Leg Sweep

Marines use a leg sweep to take the opponent to the ground while they remain standing. A leg sweep is particularly effective if the opponent is already off-balanced and moving backward or pulling on the Marine. To execute the leg sweep, Marines—

Grab the opponent's right wrist with the left hand and grab the opponent's left shoulder with the right hand.

Note: Grab the opponent's clothing or gear if his wrist and shoulder cannot be grabbed.

Step forward with the left foot and place it on the outside of the opponent's right foot.

Pull the opponent's wrist downward, close to the body, and push his shoulder backward to off-balance him.

Raise the right knee no higher than the waist.

Kick the foot past the opponent's right leg.

Use the heel of the boot to make contact with the opponent's calf (anywhere from the top of the calf down to the Achilles tendon or on the inside of the calf).

Sweep the opponent to the ground.

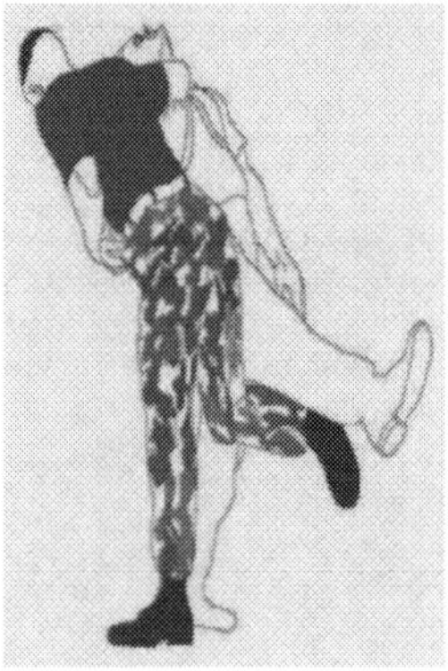

CHAPTER 6

CHOKES AND HOLDS

This chapter describes all techniques for a right-handed person. However, all techniques can be executed from either side.

The Marine is depicted in camouflage utilities. The opponent is depicted without camouflage.

WARNING

During training, never execute a choke at full force or full speed and never hold a choke for more than 5 seconds.

When Marines correctly perform a choke, they render an opponent unconscious in as little as 8 to 13 seconds. Chokes are easily performed regardless of size or gender. Marines must know how to apply chokes and how to counter a choke or a hold executed by an opponent.

1. Types of Chokes

There are two types of chokes: an air choke and a blood choke. An air choke closes off the airway to the lungs, thereby preventing oxygen from reaching the heart. A blood choke cuts off the blood flow to the brain. Both types can result in unconsciousness and eventual death for an opponent.

Air Choke

An air choke is performed on the opponent's windpipe (or trachea), cutting off the air to the lungs and heart. If Marines execute the air choke properly, the opponent loses consciousness within 2 to 3 minutes. Due to the length of time it takes to immobilize the opponent, air chokes are not recommended.

Blood Choke

A blood choke is performed on the opponent's carotid artery, which carries oxygen-enriched blood from the heart to the brain. The carotid artery is located on both sides of the neck. If Marines execute a blood choke properly, the opponent will lose consciousness within 8 to 13 seconds. The blood choke is the preferred choke because its intended effect (i.e., the opponent losing consciousness) can be executed quickly, ending the fight.

2. Chokes

Front Choke

Marines execute a front choke when they are facing the opponent. Marines use the opponent's lapels or collar to execute a front choke. To execute the front choke, Marines-

Grab the opponent's right lapel with the right hand, making certain that the knuckles or the back of the hand are against the opponent's carotid artery on the right side of his neck.

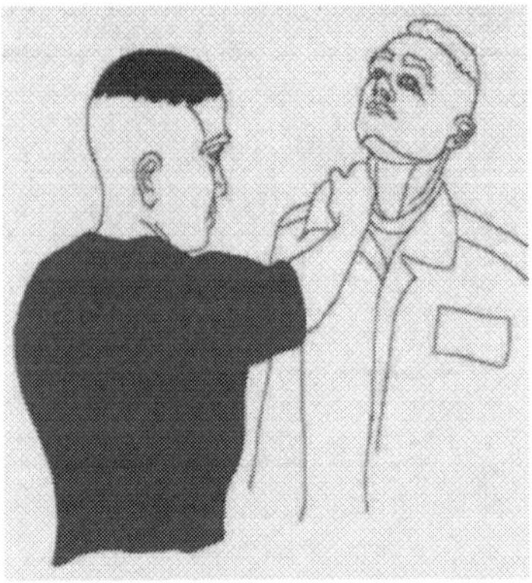

Keep the right hand against the opponent's neck, reach under the opponent's right arm with the left hand, grab the opponent's left lapel, and form an X with the wrists.

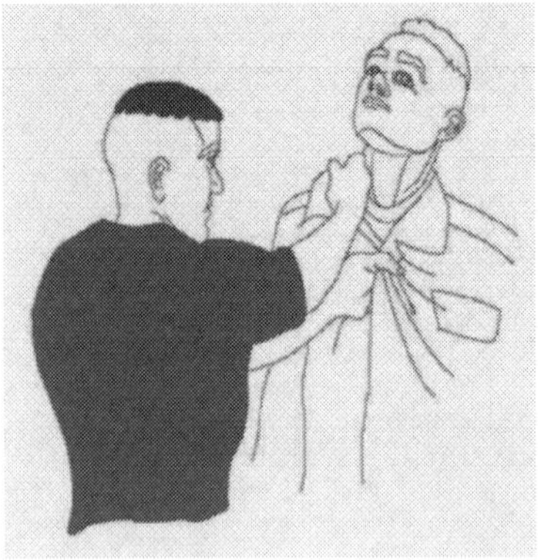

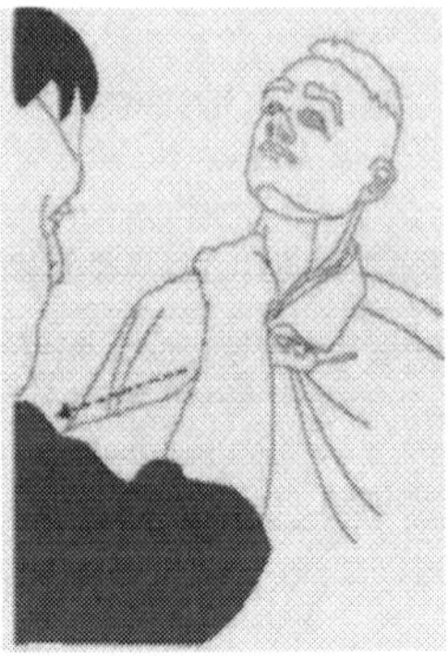

Press the right hand against the opponent's carotid artery.

Pull his lapel to the left with the left hand.

Side Choke

Marines execute a side choke when they are facing the opponent. The side choke is particularly effective when deflecting a punch thrown by an opponent. To execute the side choke, Marines—

Use the left hand to parry the opponent's arm inboard (to the inside of the opponent's reach).

Bring the right arm underneath the opponent's arm and up around the front of his neck.

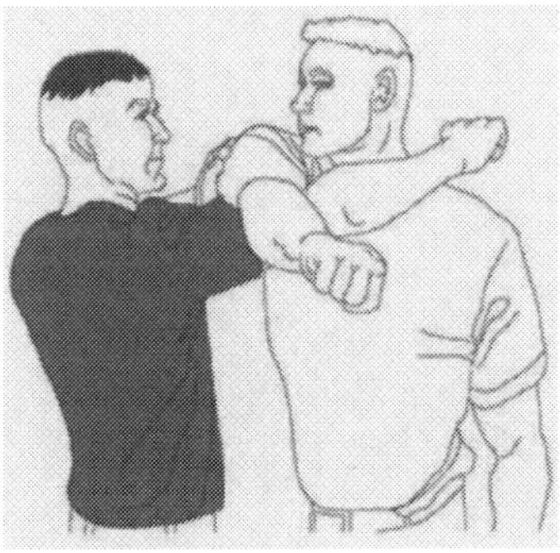

Extend the fingers, place the back of the forearm against the opponent's neck just below his ear, and press the carotid artery on his neck.

Note: If Marines are unable to place their fingers against the opponent's neck, they may use the back of the thumb or wrist.

Reach, with the left hand, around the back of the opponent's neck and clasp the hands together.

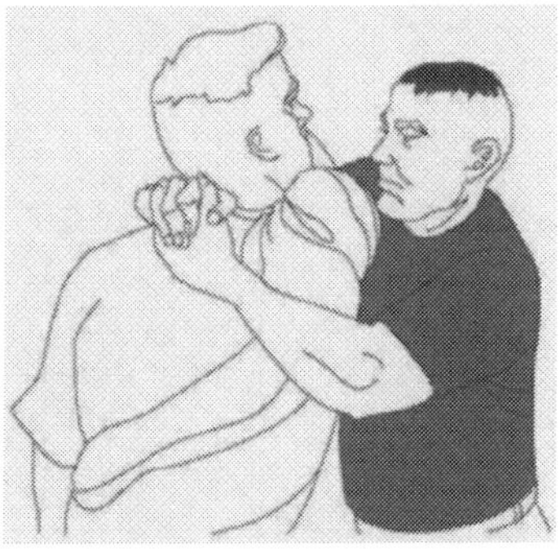

Pull the opponent toward the chest by pulling the clasped hands toward the chest.

Exert pressure on the side of the opponent's neck with the forearm.

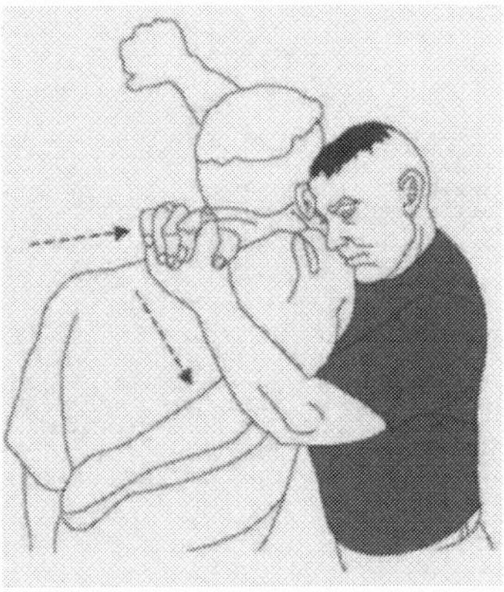

Rear Choke

Marines execute a rear choke when they are behind the opponent, the opponent is on the ground, or when they are taking the opponent to the ground. To execute the rear choke, Marines—

Reach, with the right arm, over the opponent's right shoulder and hook the bend of the arm around his neck.

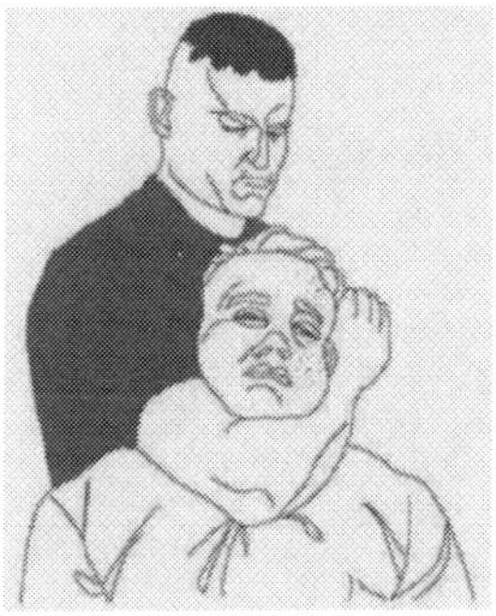

Clasp both hands together.

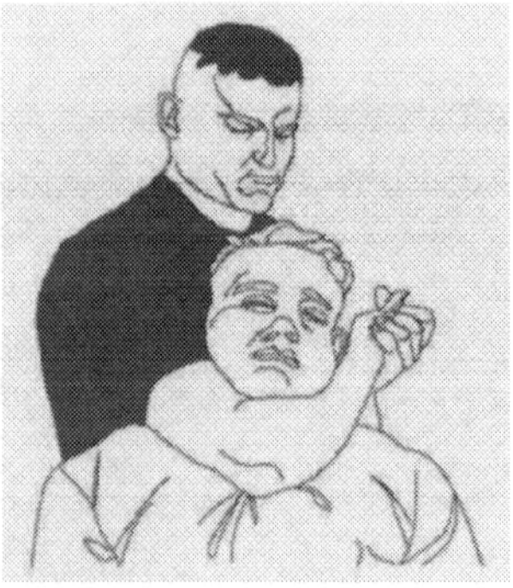

Exert pressure with the biceps and forearms on both sides of the opponent's neck on his carotid arteries.

Maintain pressure with the biceps and forearms on both sides of the neck and draw the opponent closer by drawing the right arm in.

Figure-Four Choke

A variation of the rear choke is the figure-four choke. The figure-four choke allows Marines to gain more leverage than the rear choke. If the rear choke cannot be secured, the figure-four variation is employed to increase the pressure of the choke on the opponent. To execute the figure-four variation of the rear choke, Marines—

Apply a rear choke. The Marine's body should be against the opponent's body.

Grasp the left biceps with the right hand and place the left hand against the back of the opponent's head.

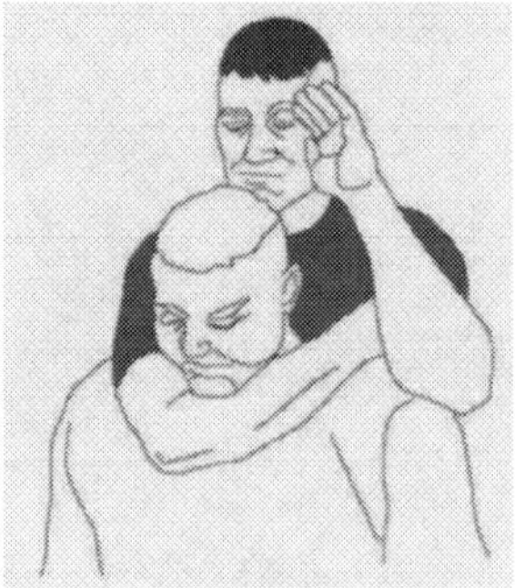

Push the opponent's head forward and down with the left hand.

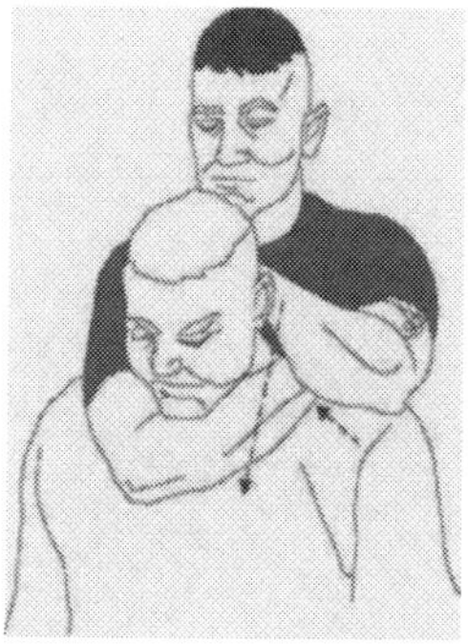

Draw the right arm in, maintaining pressure with the bicep and forearm on both sides of the opponent's neck.

3. Counters to Chokes and Holds

During a close combat situation, an opponent may apply a choke or hold on a Marine. If the opponent correctly applies a choke, a Marine quickly loses consciousness. If a choke is not executed properly, it often results in a hold, typically a bear hug or a headlock. A hold allows the opponent to control a Marine and removes the Marine's ability to attack. It is important for Marines to extract themselves from chokes and holds, regain the tactical advantage, and counter with strikes.

Although a choke causes unconsciousness in 8 to 13 seconds for a blood choke and 2 to 3 minutes for an air choke, the first movement in any counter to a choke is to clear the airway. Marines use softening techniques to loosen an opponent's grip and to clear their airway. Softening techniques are particularly effective if Marines lack the physical strength of their opponent. These techniques include groin strikes, eye gouges, foot stomps, etc. Softening techniques are not offensive; rather, they are used to loosen an opponent's hold.

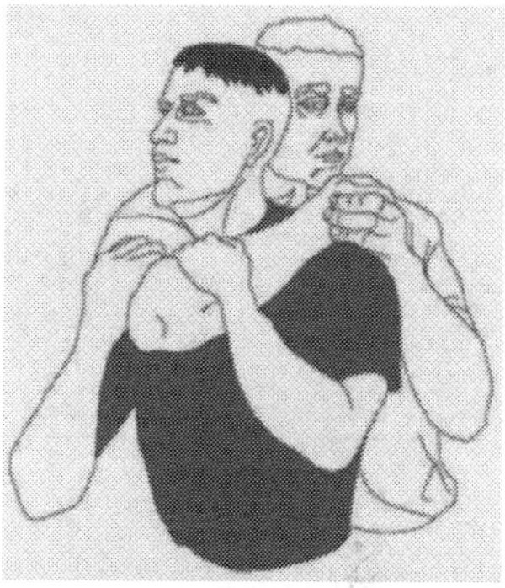

The second movement is to tuck the chin. Once the airway is clear, Marines tuck their chins to prevent the opponent from reapplying the choke.

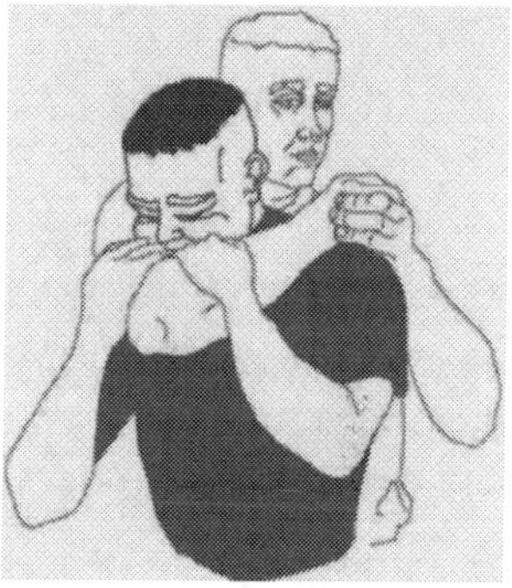

Counter to a Front Choke

Marines use a counter to a front choke when the opponent approaches from the front and uses both hands to choke a Marine around the throat. To execute the counter to the front choke, Marines—

Grasp the opponent's right forearm (where the elbow bends) with the left hand and apply downward pressure on the opponent's radial nerve with the fingers.

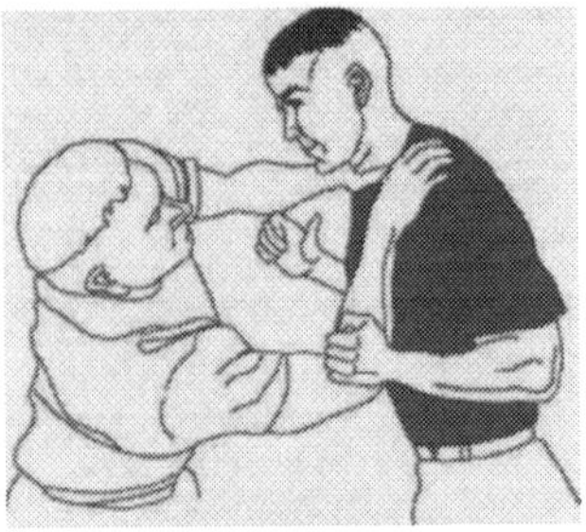

Execute a chin jab to the opponent's chin with the right hand. To generate power into the strike, bring the left foot to the outside of the opponent's right foot and rotate the hips into the strike.

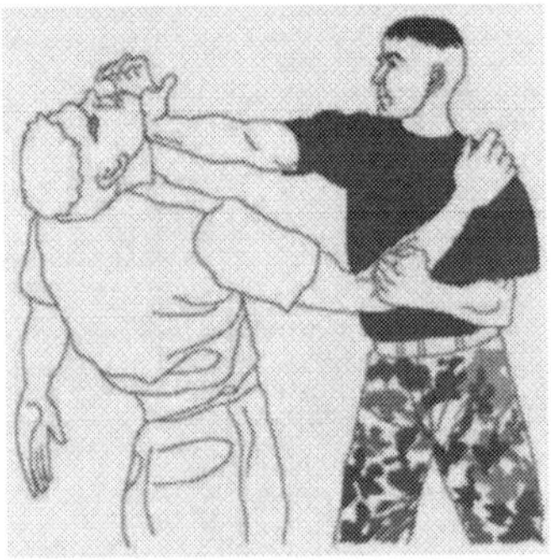

Counter to a Rear Choke

Marines execute a counter to a rear choke when the opponent approaches from the rear and puts his right arm around a Marine's throat. To execute the counter to the rear choke, Marines-

Grasp the opponent's forearm (at the radial nerve) and bicep with both hands and pull down just enough to clear the airway. Once the airway is clear, tuck the chin to protect the airway and to prevent the opponent from reapplying the choke.

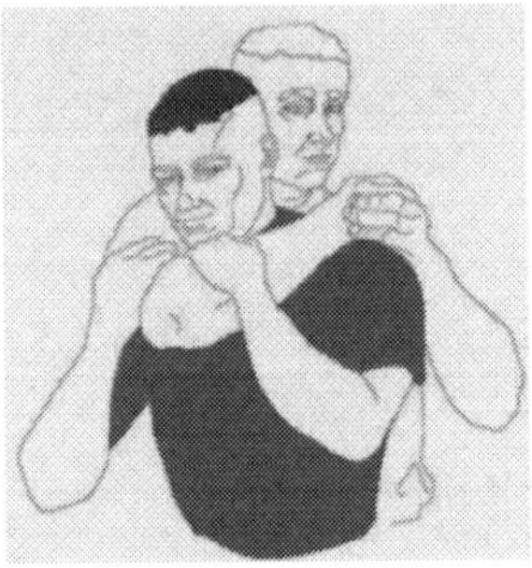

Step behind the opponent's right leg with the left foot, keeping both legs bent (almost in a squatting position).

Strike and drive the left elbow into the opponent's torso while rotating the hips and pivoting to the left, throwing the opponent back and to the ground.

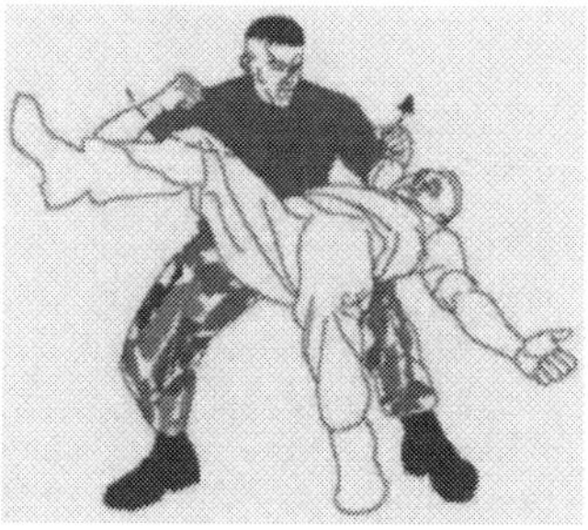

Counter to a Front Headlock

Marines use a counter to a front headlock when the opponent approaches from the front and puts his right arm around the Marine's neck, bends the Marine forward, and locks the Marine's head against his hip. To execute the counter to a front headlock, Marines—

Grasp the opponent's wrist and forearm with both hands and pull down to clear the airway. Maintain control of the opponent's wrist throughout the move. Once the airway is clear, tuck the chin to protect the airway and to prevent the opponent from reapplying the choke.

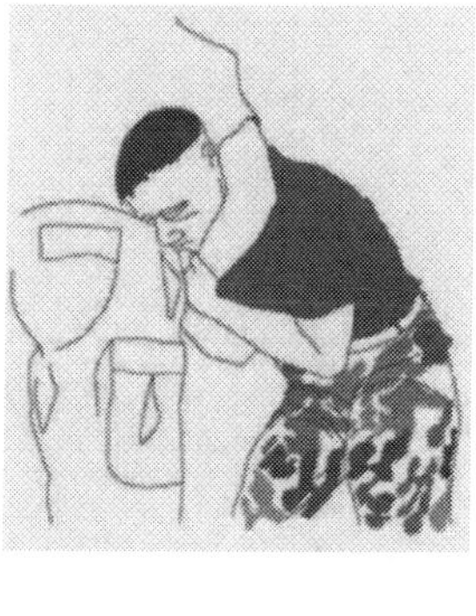

Move the right hand and arm across the opponent' s torso.

Step forward and to the left with the left foot at a 45-degree angle.

Execute a sweep with the right foot against the opponent's right leg. At the same time, push against the opponent's chest with the right arm and shoulder to generate power in the sweep.

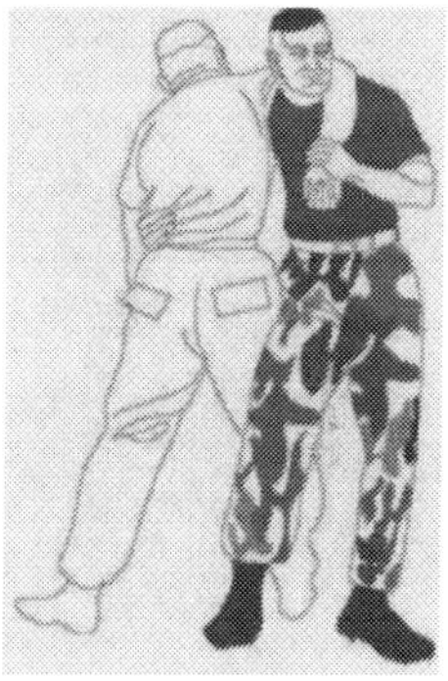

Counter to a Rear Headlock

Marines use a counter to a rear headlock when the opponent approaches from the rear and puts his right arm around the Marine's neck, bends the Marine forward, and locks the Marine's head against his hip. To execute the counter to the rear headlock, Marines-

Grasp the opponent's wrist and forearm with the right hand and pull down to clear the airway. Once the airway is clear, tuck the chin to protect the airway and to prevent the opponent from re-applying the choke.

Reach over the opponent's right shoulder with the left arm.

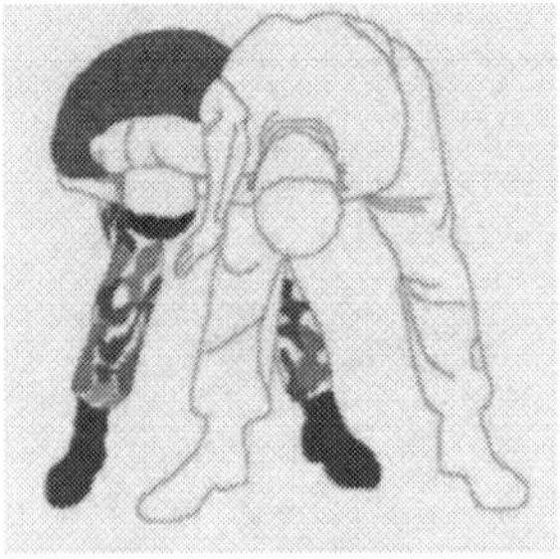

Grab any part of the opponent's face (chin, nose, eyes) and pull back while rising to a standing position.

Execute, with the right hand, a hammer fist strike to the opponent's exposed throat.

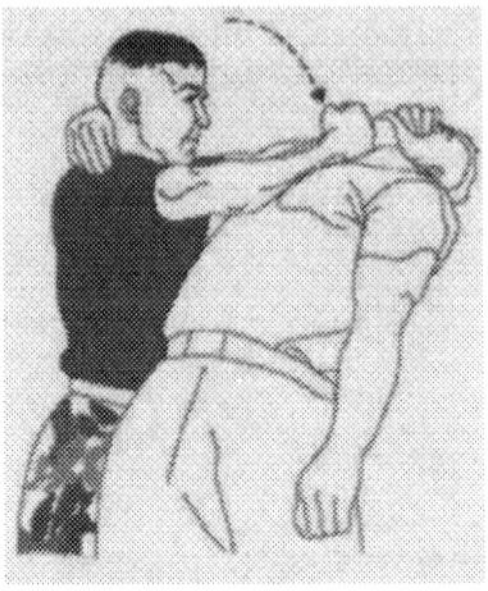

Counter to a Front Bear Hug

Marines execute a counter to a front bear hug when the opponent approaches from the front and puts both of his arms around the Marine's body, trapping the Marine's arms to the sides. To execute the counter to a front bear hug, Marines—

Step forward and to the left with the left foot at a 45-degree angle to the outside of the opponent's right leg, keeping the left leg bent.

Grasp the opponent's torso or arms to gain balance and to assist in throwing the opponent. It may be helpful to hook the opponent's right arm with the left arm.

Drive the right arm and shoulder forward and, at the same time, bring the right leg forward and sweep the opponent's right leg, bringing him to the ground.

Counter to a Rear Bear Hug

Marines use a counter to a rear bear hug when the opponent approaches from the rear and puts both of his arms around the Marine's body, trapping the Marine's arms to the sides. To execute the counter to a rear bear hug, Marines—

Step behind the opponent's right leg with the left foot, keeping both legs bent (almost in a squatting position). The left side of the body should be against the opponent's.

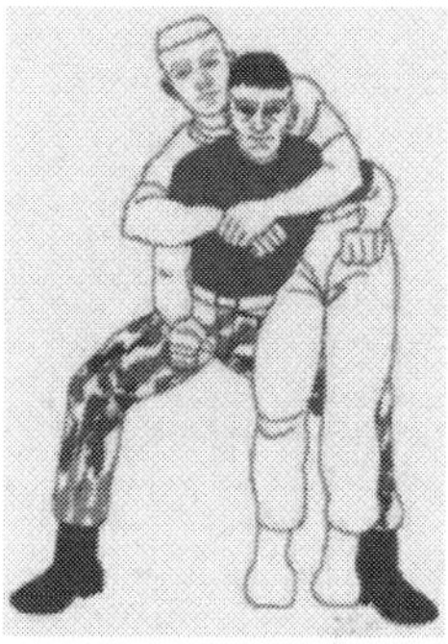

Pivot the hip, turning the body to the left and throwing the opponent back over the bent leg.

CHAPTER 7

GROUND FIGHTING

This chapter describes all techniques for a right-handed person. However, all techniques can be executed from either side.

In drawings, the Marine is depicted in woodland camouflage utilities; the opponent is depicted witlwut camouflage. In photographs, the Marine is depicted in woodland camouflage utilities; the opponent is depicted in desert camouflage utilities.

Marines should avoid being on the ground during a close combat situation because the battlefield may be covered with debris and there is an increased risk of injury. However, many close combat situations involve fighting on the ground. The priority in a ground fight is for Marines to get back on their feet as quickly as possible. In any ground fighting scenario, Marines will usually end up in one of four positions with the opponent. The offensive positions, in which Marines have a tactical advantage, are the guard and mount. The defensive positions, which are used as counters when the opponent has the tactical advantage, are the counter to the guard and the counter to the mount. Marines can also employ chokes during ground fighting to quickly end a fight.

1. Offensive Ground Fighting

Guard Position

Marines execute a guard position when the opponent is on top and the legs are wrapped around the opponent's legs. If the opponent is on top, he may try to choke the Marine, but the Marine still has the tactical advantage because the Marine is in a position to control the situation. To execute the guard position technique, Marines—

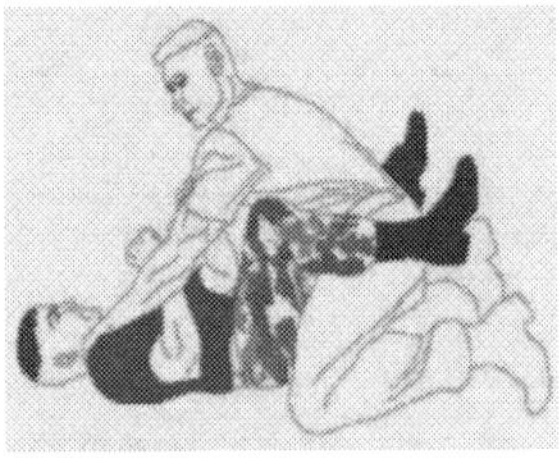

Trap the opponent's hands on the chest by crossing the hands on the chest so the forearms are resting on the opponent's forearms. Apply pressure with the elbows or forearms.

Strike the outside of the opponent's thigh with the cutting edge of the right heel. This causes the opponent to jerk to that side.

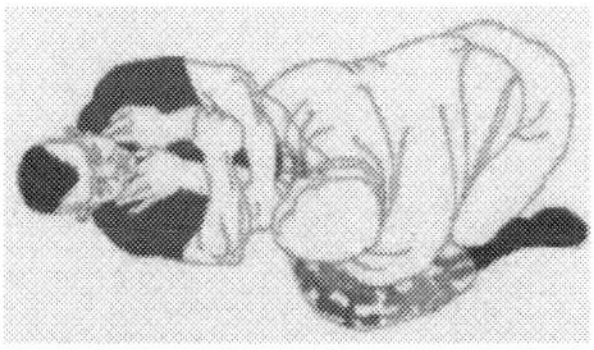

Move the head quickly to the left and swivel the hips to the right. At the same time, bring up both of the legs. Both legs are on the right side of the opponent's body.

Bring the right leg down, hooking the opponent's neck and head. Exert downward pressure to roll him over on his back. Grasp and maintain control of the opponent's left arm. Upon completion of the move, the Marine is sitting up with legs bent over the opponent while maintaining control of his arm.

Keep legs and knees bent. Maintain pressure against the opponent's neck with the back of the right foot and against his side with the left foot underneath his armpit. Squeeze the knees together, locking the opponent's arm.

Pull the opponent's arm straight up and fall back sharply, pulling his arm to the side in the direction of his little finger.

Return to standing.

Mount Position

Marines execute the mount if the opponent is lying on his back on the ground and the Marine is on top with legs wrapped around the opponent's body. This position is an offensive position because Marines are in a better position to control the opponent and to execute ground fighting techniques. To execute the mount position technique, Marines—

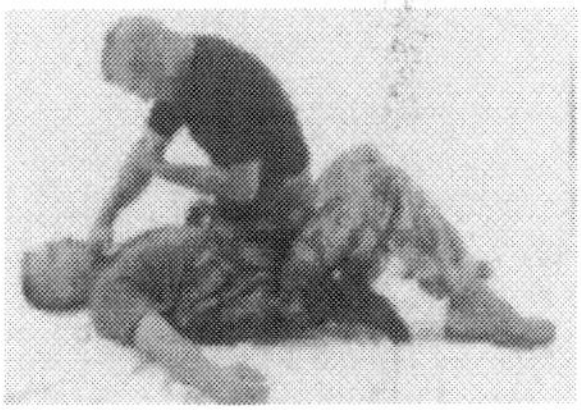

Grab the opponent's wrists or forearms with the hands. Hold them tightly against the torso.

Maintain control of the opponent's arms with the left arm and apply pressure to the opponent's brachial plexus (tie in) with the right hand.

Bring the left foot close to the opponent's right armpit and swing the right leg across the opponent' s head.

Keep the legs and knees bent. Maintain pressure against the opponent's neck with the back of the right foot and against his side with the left foot which is underneath his armpit. Squeeze the knees together, locking the opponent's arm.

Pull the opponent's arm straight up and fall back sharply, pulling his arm to the side in the direction of his little finger.

Return to standing.

2. Defensive Ground Fighting

Counter to the Guard

Marines use this technique if the opponent is lying on his back on the ground and the Marine is kneeling on the ground between the opponent's legs. To execute the counter to the guard position, Marines—

Strike the opponent's femoral nerve, located on the inside of the thigh, with the elbows. This forces the opponent to separate his legs.

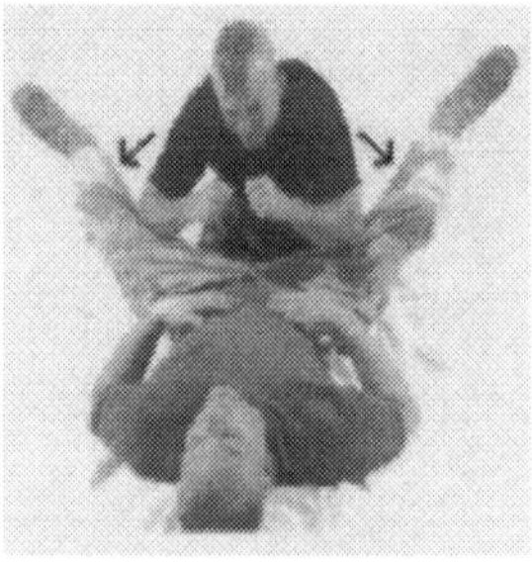

Strike the opponent's groin with the right fist.

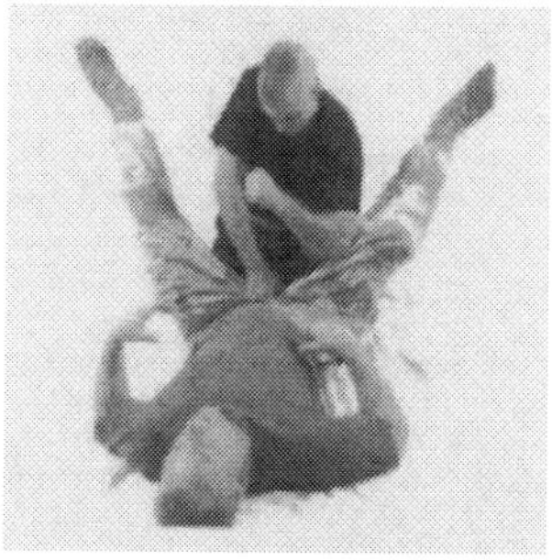

Hook the left arm underneath the opponent's right knee from the inside. Throw the opponent's leg over the head.

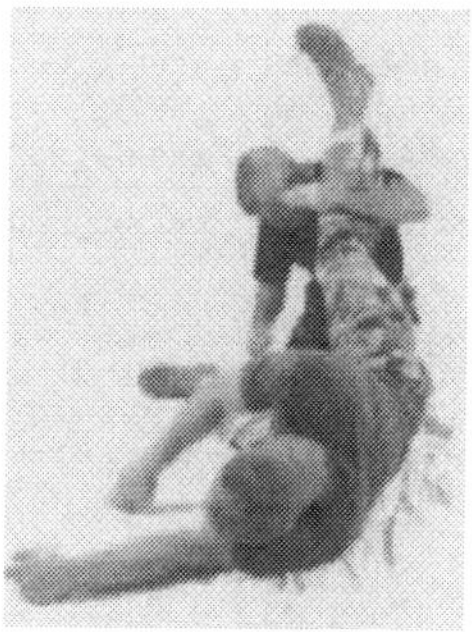

Duck the head quickly and move to the left.

Return to standing.

Counter to the Mount

Marines use this technique when they are lying on their back on the ground, the opponent is mounted on top of the Marine, and the opponent's legs are wrapped around the Marine. The opponent has the tactical advantage. To execute the counter to the mount position, Marines—

Grab the opponent's gear or clothing on his upper torso and pull him down close.

Use the right arm to hook the opponent's left arm, from the inside around the outside, above his elbow. With the right foot, hook the opponent's left leg or ankle.

Draw the elbow in to bend the opponent's elbow, bringing him down close. The arm must be hooked above the opponent's elbow in order to bend it.

Strike the opponent's side with the left hand.

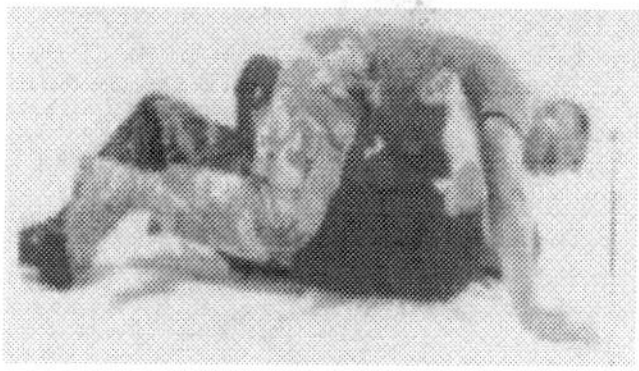

Push the opponent over and roll him off to the right side.

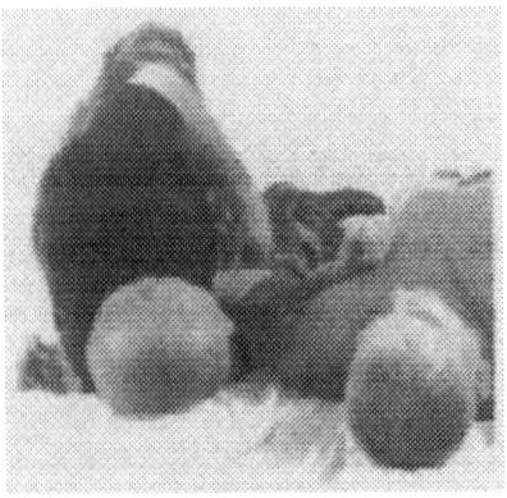

Maintain control of the opponent's hooked arm and move to a kneeling position.

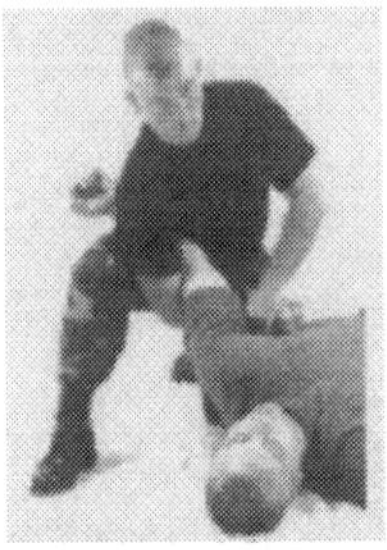

Move to a standing position while maintaining pressure on the opponent's arm and using the knee to apply pressure against the opponent's elbow.

3. Ground Fighting Chokes

WARNING

During training, never execute a choke at full force or full speed. Never hold a choke for more than 5 seconds.

The priority in ground fighting is for Marines to get back on their feet as quickly as possible. Sometimes, Marines can quickly end a ground fight by executing a choke on the opponent. When performed correctly, a choke can render an opponent unconscious in as little as 8 to 13 seconds. Chokes are easily performed regardless of size or gender. The chokes performed during ground fighting are the same as those performed while standing (see chap. 6).

Ground Fighting Front Choke

The ground fighting front choke is a blood choke performed most effectively from the mount position. The front choke employs the opponent's lapels or collar to execute the choke. To execute the ground fighting front choke, Marines—

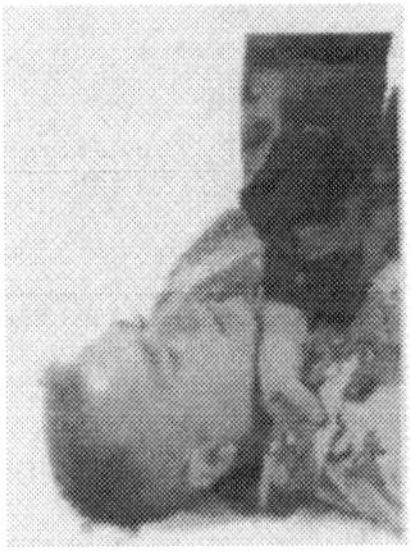

Grab the opponent's right lapel with the right hand, making certain that the knuckles or the back of the hand are against the carotid artery on the right side of the opponent's neck.

Keep the right hand pressed against the opponent's carotid artery, reach under the right arm with the left hand, and grab the opponent's left lapel, forming an X with the wrists.

Keep the right hand pressed against the opponent's carotid artery and pull the opponent's left lapel to the left with the left hand.

Ground Fighting Side Choke

The ground fighting side choke is a blood choke performed from the mount position. The ground fighting side choke is particularly effective when the opponent raises his arms and places them on the Marine's chest or throat. To execute the ground fighting side choke, Marines-

Use the left hand to parry the opponent's right arm inboard (to the inside of the opponent's reach). Bring the right arm underneath the opponent's arm and up around the front of his neck.

Keep the fingers extended, place the back of the forearm against the opponent's neck just below his ear and press the carotid artery on his neck. Depending on the Marine's or the opponent's body configuration, the back of the thumb or wrist may press against the opponent's neck.

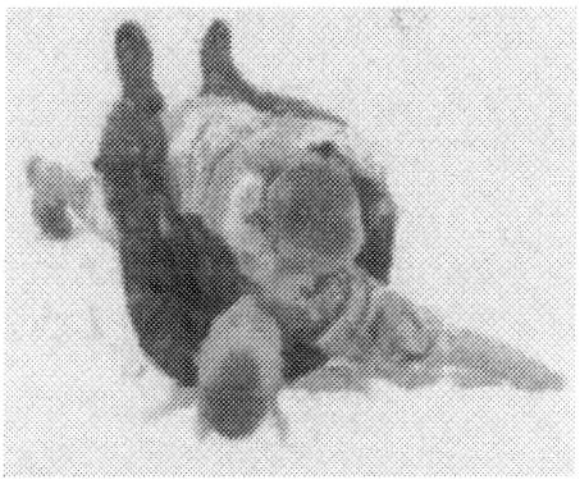

Use the left hand to reach around the back of the opponent's neck and clasp the hands together.

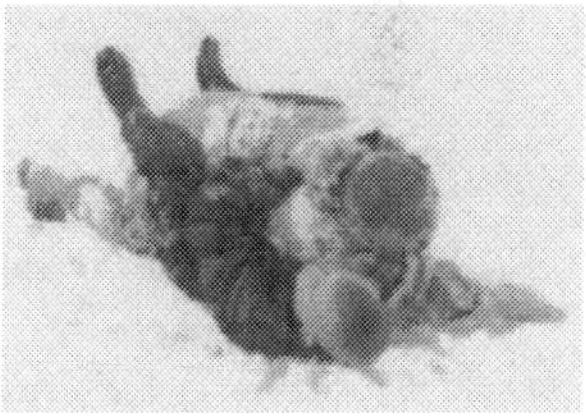

Pull the opponent toward the chest. Use the forearm to exert pressure on the side of his neck. This is done by pulling the clasped hands toward the chest.

Ground Fighting Rear Choke

The ground fighting rear choke is a blood choke performed when Marines are behind the opponent. To execute the ground fighting rear choke, Marines—

Wrap the legs around the opponent with the left foot against the inside of his left thigh and the right foot against the inside of his right thigh.

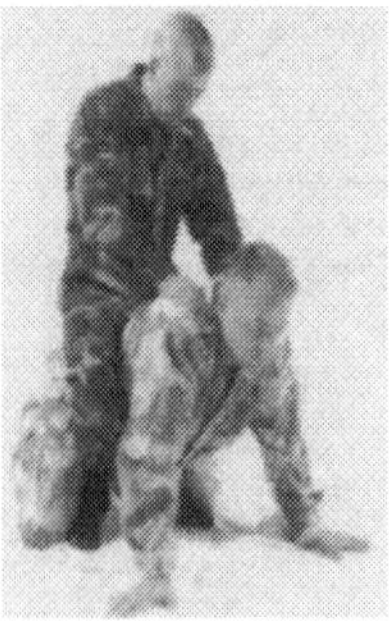

Use the right arm to reach over the opponent's right shoulder and hook the bend of the arm around his neck.

Use the left hand to clasp both hands together.

Exert pressure with the biceps and forearms on both sides of the opponent's neck. This applies pressure to the opponent's carotid arteries. While maintaining pressure with the biceps and forearms, draw the opponent closer by drawing the right arm in. To increase the effectiveness of the choke, lean backward by arching the back and pulling the opponent back. At the same time, push the feet against the opponent's thighs.

Figure-Four Choke

A variation of the ground fighting rear choke is the figure-four choke. The figure-four choke allows Marines to gain more leverage on the rear choke. If the rear choke cannot be secured, the figure-four variation may be applied to increase the pressure of the choke on the opponent. To execute the figure-four variation of the rear choke, Marines—

Apply the ground fighting rear choke. The chest should be against the opponent's back.

Grasp the left biceps with the right hand and place the left hand against the back of the opponent's head.

Use the left hand to push the opponent's head forward and down.

Draw the right arm in while maintaining pressure with the biceps and forearm on both sides of the opponent's neck.

Increase the effectiveness of the choke by arching backward and pushing the feet against the opponent's thighs. Continue to pull the opponent back with the right arm while exerting pressure forward with the left hand against the opponent's head.

CHAPTER 8

NONLETHAL TECHNIQUES

This chapter describes all techniques for a right-handed person. However, all techniques can be executed from either side.

In drawings, the Marine is depicted in woodland camouflage utilities; the opponent is depicted without camouflage. In photographs, the Marine is depicted in woodland camouflage utilities; the opponent is depicted in desert camouflage utilities.

The Marine Corps' involvement in military operations other than war—e.g., humanitarian, peacekeeping, or evacuation missions—has greatly increased. These missions require skills that span the spectrum of conflict and support operations within a continuum of force. But the Corps' day-to-day existence also demands a responsible use of force. Nonlethal techniques are among the skills Marines use to apply a responsible use of force.

1. Unarmed Restraints and Manipulation

Marines operate within a continuum of force, particularly in support of peacekeeping or humanitarian types of missions. In these situations, Marines must act responsibly to handle a situation without resorting to deadly force. Unarmed restraints and manipulation techniques including joint manipulation, come-alongs, and takedowns can be used to control a subject without resorting to deadly force. Marines must train to become proficient in nonlethal techniques and to respond in a responsible manner. These techniques are referred to as compliance techniques, and they are applied in the third level in the continuum of force.

WARNING

During training, never apply the techniques for unarmed restraints and manipulation at full force or full speed. Use a slow and steady pressure to avoid injury.

Compliance Techniques

Compliance techniques are unarmed restraint and manipulation techniques used to physically force a subject or opponent to comply. Compliance can be achieved through the close combat techniques of-

Pain compliance using joint manipulation and pressure points. (Pain compliance is the initiation of pain to get compliance on the part of the subject.)
Come-along holds.

Principles of Joint Manipulation

Joint manipulation is used to initiate pain compliance and gain control of a subject. It involves the application of pressure on the joints (elbow, wrist, shoulder, knee, ankle, and fingers). Pressure is applied in two ways:

In the direction in which the joint will not bend. For example, joints such as the knees

and elbows only bend in one direction and when pressure is applied in the opposite direction, pain compliance can be achieved. Beyond the point where the joint stops naturally in its range of movement (i.e., it no longer bends).

Since each joint has a breaking point, Marines should apply slow steady pressure only until pain compliance is reached. Continued pressure will break the joint and may escalate the violence of the situation.

Joint manipulation also uses the principle of off-balancing. A subject can be better controlled when he is knocked off balance.

Wristlocks

A wristlock is a joint manipulation that can be applied in a number of ways to achieve pain compliance. The wrist rotates in a number of directions and will bend in a single direction until its movement stops naturally. In a wristlock, pressure is exerted beyond that point by bending or twisting the joint. A wristlock is executed when an opponent tries to grab Marines or is successful in grabbing Marines or their equipment. A wristlock can also be performed by Marines if they wish to initiate control of an opponent.

Basic Wristlock. A basic wristlock is executed when Marines grab the opponent's left hand with the right hand. To execute the basic wristlock, Marines—

Use the right hand to grab the opponent's left hand by placing the thumb on the back of the opponent's hand so that the Marine's knuckles are facing to the left.

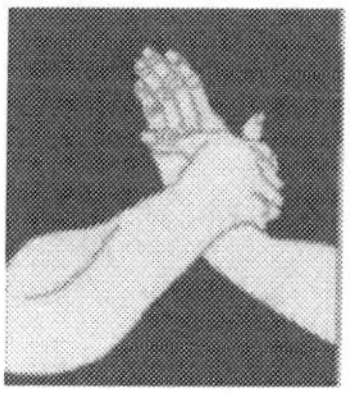

Hook the fingers across the fleshy part of the opponent's palm below the thumb. The fingers are used to anchor the hand so leverage can be applied to twist and bend the joint.

Exert downward pressure with the thumb to bend the opponent's joint. Rotate the opponent's hand to the right to twist the joint.

Step in to the opponent to keep the opponent's hand in close to the body to control him and provide more leverage on the wristlock.

Note: When executing the basic wristlock with the left hand, the Marine grabs the opponent's hand so that the Marine's knuckles are facing to the right, and then rotates and twists the opponent's hand to the left.

Use the left hand to further control the opponent.

Reverse Wristlock. A reverse wristlock is executed when Marines grab the opponent's right hand with the right hand. To execute the reverse wristlock, Marines—

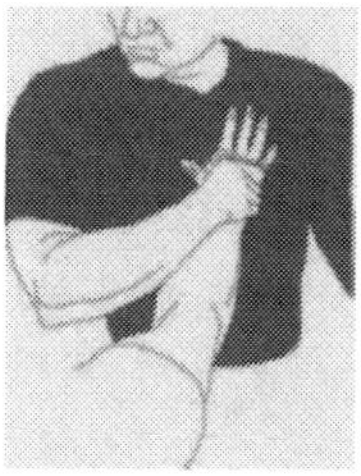

Place the right palm on the back of the opponent's right hand and wrap the fingers across the fleshy part of his palm below his little finger.

Twist the opponent's hand to the right while stepping in to place his hand against the chest. Apply downward pressure on the opponent's hand against the chest. Leave the opponent's hand on the chest to fully control the subject and to gain leverage.

Lean forward to use body weight to add additional pressure to the joint.

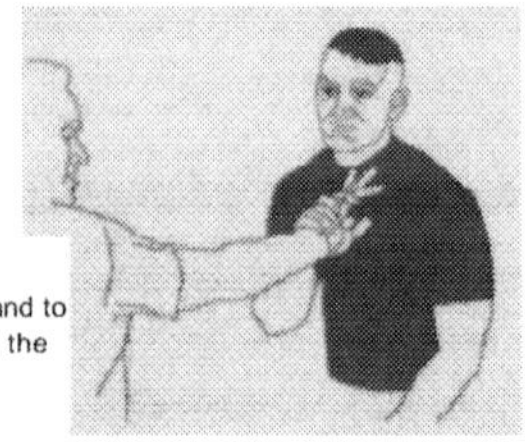

Use the left hand to further control the opponent.

Two-Handed Wristlock. Both hands can be used in the wristlock to maximize the leverage and pressure needed to bend and twist the joint. To execute the two-handed wristlock, Marines—

Place both thumbs on the back of the opponent's hand, thumbs crossed.

Hook the fingers of both hands around the fleshy part of the opponent's palm on both sides of his hand.

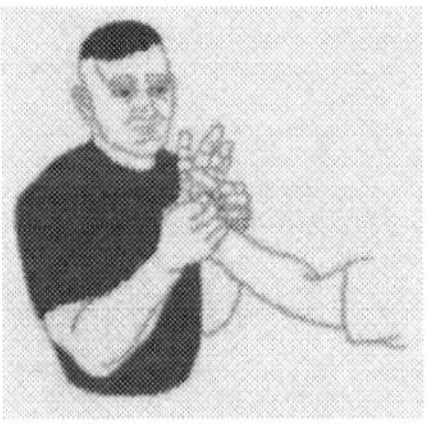

Step into the opponent and apply pressure downward on the back of his hand to bend the joint and rotate his wrist away from the body to twist the joint.

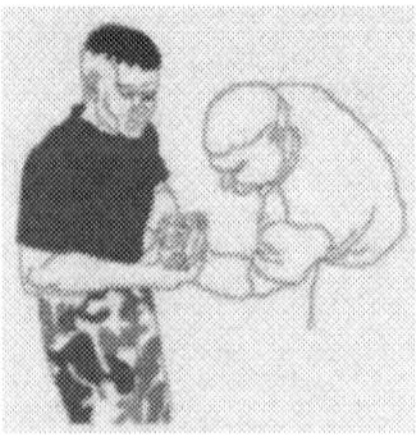

Enhanced Pain Compliance on Wristlock. Enhanced pain compliance techniques are applied in the third and fourth levels in the continuum of force. Additional pain be applied to a wristlock by-

Adding downward pressure to the elbow with the other hand or elbow by using the fingers to pull in on the opponent's radial nerve located on the inside of the forearm. When pressure is added to the opponent's radial nerve, his direction can be controlled.

Applying pressure against the opponent's finger joint to bend it in a direction it cannot bend (i.e., splitting the fingers).

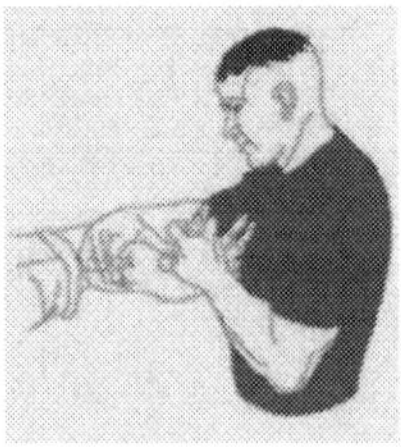

Come-Along Holds

Marines use a come-along hold to control and move an opponent.

Escort Position. A common come-along hold is the escort position. To execute the escort position, Marines—

Face the opponent. Use the left foot to step forward at a 45-degree angle. Turn to face the right side of the opponent.

Use the right hand to firmly grasp the opponent's right wrist. With the left hand, firmly grasp the opponent's right triceps.

Position the opponent's controlled arm diagonally across the torso, keeping his wrist against the right hip. The Marine should be standing to the right of and behind the opponent.

Note: This technique works well when escorting an opponent on either the right or left side. Take caution when escorting an opponent by ensuring his controlled hand is not in a position to grab the holstered weapon. The preferred escort position is from the Marine's left side, so that the opponent is kept further away from the weapon.

Wristlock Come-Along. To execute the wristlock come-along, Marines-

Use the left hand to execute a basic wristlock. Incorporate the right hand in a two-handed wristlock for more control.

Maintain pressure on the opponent's wrist with the right hand, step forward, and pivot around to stand next to the opponent.

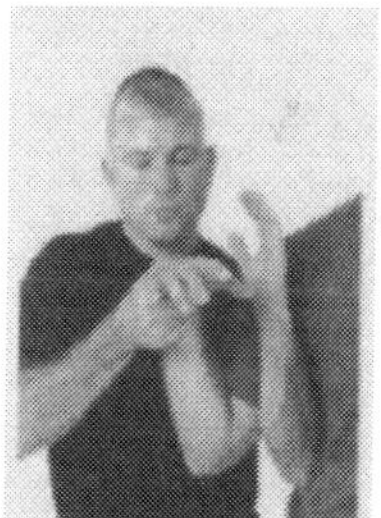

Release the left hand, quickly reach under the opponent's arm from behind, and grab his hand.

Use the left hand and apply downward pressure on the opponent's wrist.

Controlling Technique. The following controlling technique is used when an opponent grabs the Marine's wrist. To execute the technique, Marines—

Trap the opponent's hand with the palm of the other hand.

Rotate the opponent's trapped hand up and on his forearm while maintaining downward pressure on his trapped hand.

Apply downward pressure with both hands until the opponent is taken to the ground.

Armbars

An armbar is a joint manipulation in which pressure is applied on a locked elbow, just above the joint, in the direction the joint will not bend. An armbar has to be locked in quickly, but still requires a slow, steady pressure to gain compliance.

Basic Armbar. To execute a basic armbar, Marines-

Use the right hand to grab the opponent's right wrist.

Bring the left hand down on or above the opponent's elbow joint. To gain additional leverage, pivot to face the opponent.

Use the left hand to apply downward pressure on or above the opponent's elbow joint while pulling up on his wrist.

Armbar from a Wristlock. To execute an armbar from a wristlock, Marines-

Use the right hand to grab the opponent's right hand and execute a reverse wristlock.

Bring the left hand down on or above the opponent's elbow joint.

Use the left hand to apply downward pressure on or above the opponent's elbow joint while pulling up on his wrist.

Takedowns

A takedown is used to bring an opponent to the ground to further control him.

Takedown From a Wristlock Come-Along. To take the opponent to the ground from a wristlock come-along, Marines—

Use the right foot to push down on the opponent's calf or Achilles tendon.

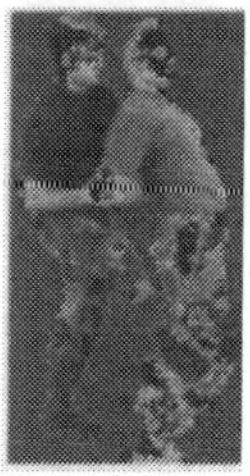

Maintain control of the opponent's wrist and elbow and apply a slow, steady pressure to bring him to the ground.

Armbar to a Takedown. This technique is used to take a noncompliant opponent to the ground from an armbar. To execute the armbar takedown, Marines—

Use the right hand to execute a reverse wristlock.

Bring the left hand or forearm down on or above the opponent's elbow joint.

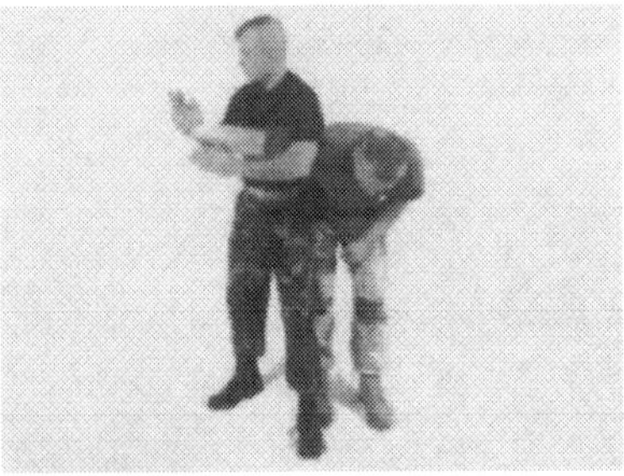

Pivot so the back is facing the opponent and, at the same time, lift the left elbow and slide the body so it is against the opponent, placing the armpit high above the opponent's elbow joint.

Lean back, placing the body weight on the opponent's arm until he complies or is taken to the ground.

Note: This technique may break the opponent's arm. Therefore, this technique should not be employed if the objective is a nonlethal takedown.

Wristlock Takedown. This technique is used to take a noncompliant opponent to the ground from a basic wristlock and to put him in a position where he can be handcuffed, if necessary. To execute the takedown, Marines-

Use the right hand to execute a basic wristlock.

Incorporate the left hand in a two-handed wristlock.

Apply downward pressure on the wristlock, pivot on the ball of the right foot, and quickly turn to the right to take the opponent to the ground.

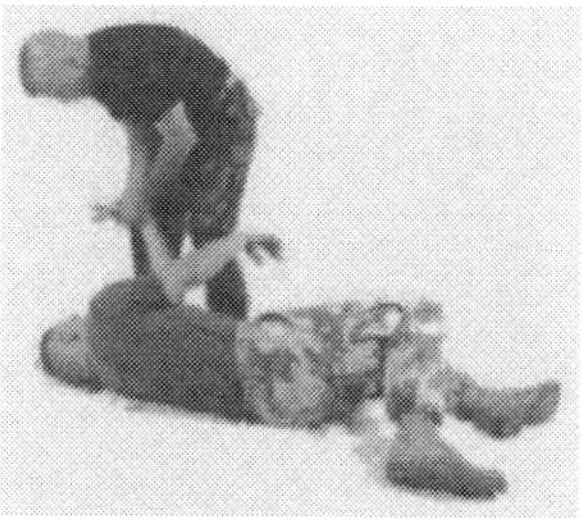

Continue to apply pressure on the wristjoint as the opponent lands on his back with his arm straight in the air.

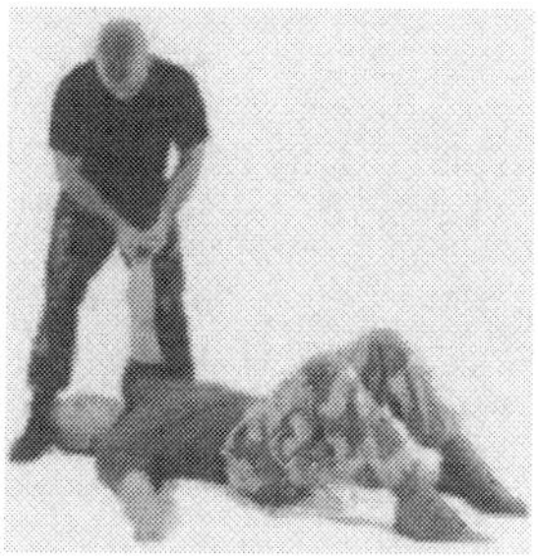

Slide the left foot under the opponent's back, underneath his armpit.

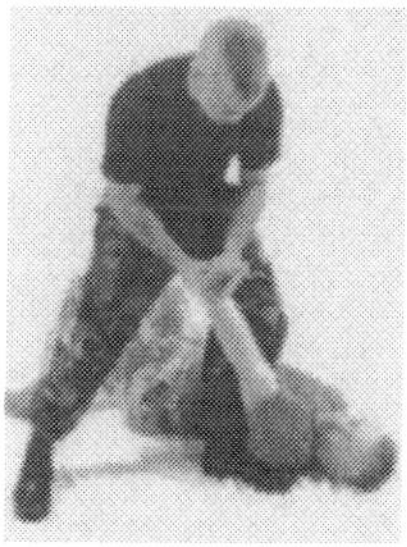

Apply pressure with the knee against the opponent's triceps while pulling back on his arm.

Continue applying pressure with the knee on the opponent's arm. Pivot and step around the opponent's arm to roll him on his stomach.

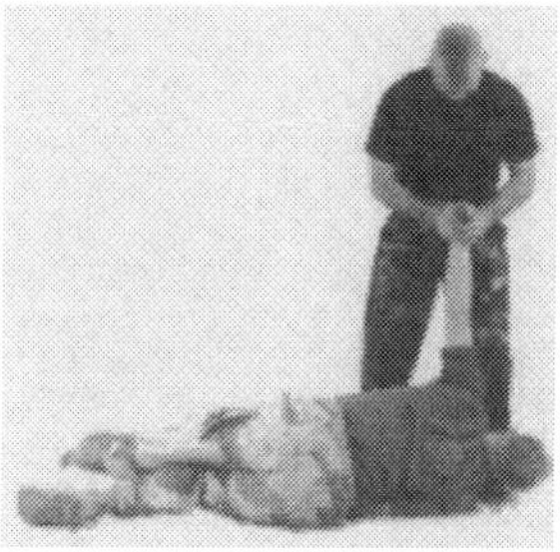

Kneel down with one knee on the opponent's back. The other knee is placed on the opponent's neck and shoulder on either side of his arm. Apply inward pressure with the knees to lock his arm in place.

Tell the opponent to put his other hand in the middle of his back. Bring the opponent's controlled hand to the center of his back.

Escort Position Takedown. This technique can be used to control a noncompliant opponent from the escort position. To execute the escort position takedown, Marines—

Lock the opponent's arm straight across the body while rotating his wrist away from the body.

Use the left hand or forearm and apply downward pressure above the opponent's elbow where the triceps meet.

Step back with the right foot and, keeping the opponent's hand controlled against the hip, pivot to the right while continuing to apply downward pressure on his arm to bring him to the ground.

Note: This technique works well when escorting a subject on the Marine's right or left side. When taking down a subject from the right side, step back and pivot to the left.

2. Nonlethal Baton

A baton or nightstick can be an effective compliance tool when used correctly. Batons can be used defensively (blocking), offensively (striking), and as a restraining device when needed. In the fourth level of the continuum of force (assaultive [bodily harm]), defensive tactics include baton or nightstick blocks and blows. Blows to the head or other bony parts of the body are considered deadly force. When deadly force is not authorized, Marines must be able to employ blocks, strikes, and restraints with the baton with the minimum of force.

Grips

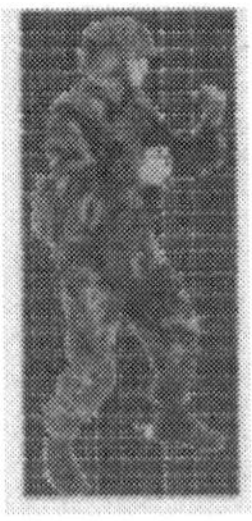

One-Handed Grip. To execute the one-handed grip, Marines—

Use the right hand to grasp the lower end of the baton, about 2 inches from the end.

Wrap the thumb and index finger around the baton so they are touching one another. The grip on the baton should be firm, but natural.

Two-Handed Grip. To execute the two-handed grip, Marines-

Use the right hand to grasp the lower end of the baton, about 2 inches from the end. Wrap the thumb and index finger around the baton so they touch one another.

Use the left hand to grasp the upper end of the baton, palm down, about 2 inches from the end. The hands should be approximately 10 to 12 inches apart.

Stance and Method of Carry

The basic warrior stance serves as the foundation for initiating nonlethal baton techniques. The method of cany provides effective defensive positions with a wide range of options to control a combative opponent.

One-Handed Carry. To execute the one-handed carry, Marines—

Grip the baton using the one-handed grip.

Elevate the baton, with the gripping hand at a level between the belt and shoulder.

Keep the left hand in the position of the basic warrior stance.

Two-Handed Carry. This carry is effective for blocks. To execute the two-handed carry, Marines-

Grip the baton using the two-handed grip.

Elevate the baton, with the left hand higher than the right hand.

Orient the weapon toward the subject.

Movement

In a nonlethal confrontation, movement may be made to create distance between Marines and the opponent or to close the gap to control him. When facing an opponent, Marines move in a 45-degree angle to either side of the opponent. Moving at a 45-degree angle is the best way to both avoid an opponent's strike and to put Marines in the best position to control the opponent.

Blocking Techniques

One-handed blocks are used when carrying the baton in the one-handed carry. The same one-handed blocks used in combative stick techniques (see page 3-6) apply in nonlethal baton engagements. The blocks include blocks for a vertical strike, a forward strike, and a reverse strike. Because the baton is often carried with two hands, there are also two-handed blocks that are effectively used from this carry. Two-handed blocks are discussed in the following subparagraphs.

High Block. Marines execute a high block to deter a downward vertical attack directed at the head and shoulders. To execute the high block, Marines—

Raise the baton up to a level even with or above the head. The baton should be in a horizontal position to block the blow. The fingers of the left hand should be open and behind the baton.

Place the baton perpendicular to the opponent's striking surface to absorb the impact of the blow.

Bend the elbows slightly to help absorb the impact of the blow. The arms should give with the strike of the blow.

Low Block. Marines execute a low block to deter an upward vertical attack directed at the abdomen, groin, or torso. The opponent's blow can be delivered by a foot, knee, or fist. To execute the low block, Marines—

Lower the baton to a level even with or below the groin. The baton should be in a horizontal position to block the blow. The fingers of the left hand should be open and behind the baton.

Place the baton perpendicular to the opponent's striking surface to absorb the impact of the blow.

Bend the elbows slightly to help absorb the impact of the blow. The arms should give with the strike of the blow.

Right Block. Marines execute a right block to deter an attack directed at the head, neck, flank, or hip. The opponent's blow can be delivered by a foot, knee, fist, or elbow. To execute the right block, Marines-

Thrust the baton in a vertical position to the right side.

Pivot to the right by stepping with the left foot and pivoting off the ball of the right foot. Rotate the hips and shoulders into the direction of the block. The fingers of the left hand should be open and behind the baton.

Place the baton perpendicular to the opponent's striking surface to absorb the impact of the blow.

Bend the elbows slightly to help absorb the impact of the blow. The arms should give with the strike of the blow.

Left Block. Marines execute a left block to deter an attack directed at the head, neck, flank, or hip. The opponent's blow can be delivered by a foot, knee, fist, or elbow. To execute the left block, Marines—

Thrust the baton in a vertical position to the left side.

Pivot to the left by stepping with the right foot and pivoting off the ball of the left foot. Rotate the hips and shoulders into the direction of the block. The fingers of the left hand should be open and behind the baton.

Place the baton perpendicular to the opponent's striking surface to absorb the impact of the blow.

Bend the elbows slightly to help absorb the impact of the blow. The arms should give with the strike of the blow.

Middle Block. Marines execute either a left or right block to deter an attack directed at the face, throat, chest, or abdomen. To execute the middle block, Marines—

Thrust the baton in a vertical position straight out in front of the body. The fingers of the left hand should be open and behind the baton.

Place the baton perpendicular to the opponent's striking surface to absorb the impact of the blow. The baton should be held with the left hand forward of the right.

Bend the elbows slightly to help absorb the impact of the blow. The arms should give with the strike of the blow.

Restraining Technique

The strong-side armlock is used to restrain an opponent who is not compliant. To execute the strong-side armlock, Marines—

Use the right hand to run the baton up under the opponent's left armpit, parallel to the ground.

Use the right foot to step forward at a 45-degree angle to the left side of the opponent. The baton is across his forearm.

Use the right hand to drive the baton forward and up so the action bends the opponent's arm behind his back. At the same time, continue moving around him to get behind him.

Place the baton on the opponent's forearm with the thumb and/or fingers and apply pressure to his forearm. At the same time, grasp the other end of the baton with the left hand.

Pull up on the low end of the baton with the right hand. At the same time, push down on the top end of the baton with the left forearm, reaching around with the left hand to grasp the opponent's biceps or shoulder.

Continue exerting downward pressure with the left forearm while pulling back on the opponent's biceps with the left hand. This places the opponent in a position where he is controlled and can be moved.

Apply pressure with the foot against the bend in the opponent's leg above his calf. This lowers the opponent to the ground rather than throwing him to the ground and risking severe injury.

Striking Target Areas

Marines must avoid striking an opponent in the head, neck, or other bony parts with the baton because this is considered deadly force and can lead to serious bodily injury or death. Instead, the legs, arms, and buttocks are target areas that are considered nonlethal.

Legs. Primary targets are the thighs and lower legs. Avoid striking the knee and ankle joints because this can cause permanent damage.

Arms. Primary targets are the upper arms. Avoid striking the shoulder and elbow and wrist joints because this can cause permanent damage.

Buttocks. Primary targets are the buttocks. Avoid striking any other part of the torso, including the chest, rib cage, spine, tail bone, and groin because strikes to these areas can cause permanent damage or death.

One-Handed Striking Techniques

One-Handed Forward Strike. A forward strike follows either a horizontal line or a downward diagonal line using a forehand stroke. To execute the one-handed forward strike, Marines-

Stand facing the opponent with the baton carried in a one-handed cany.

Place the right hand palm up, swing the baton from right to left, and make contact with the opponent.

One-Handed Reverse Strike. A reverse strike follows either a horizontal line or a downward diagonal line using a backhand stroke. To execute the one-handed reverse strike, Marines—

Stand facing the opponent with the baton carried in a one-handed carry.

Bend the right arm and cross the arm to the left side of the body. The baton should be close to or over the left shoulder.

Place the right hand palm down, swing the baton from left to right, and make contact with the opponent.

Two-Handed Striking Techniques

Two-Handed Forward Strike. This strike is an effective follow-up to a middle block or left block. To execute the two-handed forward strike, Marines—

Pull back with the left hand while driving the right hand forward toward the opponent. The baton should be horizontal to the ground. Power is generated by stepping forward with the right foot and rotating the right hip and shoulder into the strike.

Contact the opponent with the end of the baton.

Two-Handed Reverse Strike. This strike is an effective follow-up to a middle block or left block. To execute the two-handed reverse strike, Marines—

Pull back with the right hand while driving the left hand forward toward the opponent. The baton should be horizontal to the ground. Power is generated by stepping forward slightly with the left foot and rotating the left hip and shoulder into the strike.

Contact the subject with the end of the baton.

Front Jab. This strike is effective for countering a frontal attack. It can also be executed as a quick poke to keep a subject away. To execute the front jab, Marines thrust both hands forward in a quick jab. The baton is held either horizontal to the ground or at a slight downward angle.

Rear Jab. This strike is effective for countering a bear hug from the rear. To execute the rear jab, Marines thrust both hands rearward in a quick jab. The baton is held either horizontal to the ground or at a slight downward angle.

APPENDIX A

PUGIL STICK TRAINING

A pugil stick is a training device used to simulate a rifle bayonet so that effective, but safe, training can be conducted to build proficiency of rifle bayonet techniques. Pugil stick training builds on the techniques used to throw punches. Pugil stick training is the only "full contact" training provided to Marines in the Close Combat Program. Pugil stick training teaches Marines to function when faced with stress and violence, and it prepares them to deliver a blow and take a blow. It also provides them with the physical and mental skills vital to success on the battlefield.

1. Pugil Stick Training

Design

A pugil stick consists of a stick wrapped in padding at both ends that can be gripped like a rifle. The pugil stick is approximately the same weight and length of an unloaded rifle with a bayonet attached.

Holding the Pugil Stick

The pugil stick is held in the same manner as the sen ice rifle. All movements come from the basic warrior stance. To hold the pugil stick correctly, Marines—

- Use the right hand to grasp the lower end of the pugil stick overhanded.
- Use the left hand to grasp the upper end of the pugil stick underhanded.
- Use the right forearm to lock the lower end of the pugil stick against the hip.
- Orient the blade end of the pugil stick toward the opponent.

Safety Equipment

The following safety equipment must be worn during any pugil stick bout.

Groin Protection. Groin protection protects the groin from an accidental blow. It should be pulled high around the waist to protect the groin area, with the concave portion against the body.

Flak Jacket. The flak jacket provides protection to the body. It is worn completely fastened.

Neck Roll. The neck roll prevents whiplash if Marines receive a blow to the head. The neck roll further supports the head and protects the neck from blows. The neck roll is worn above the flak jacket and below the helmet. The tied end of the neck roll faces the front.

Helmet. A regulation football helmet protects the face and head. The helmet must fit snugly and the chin straps must be adjusted and snapped.

Mouthpiece. The mouthpiece is worn on the upper teeth to protect the teeth.

Gloves. Marines wear gloves to protect the hands if pugil sticks do not have gloves built-in.

Pugil Stick Screening

Prior to pugil stick training, instructors must ask participating Marines the following questions. If a Marine answers "yes" to any question, he must see the corpsman or a competent medical authority who evaluates him and determines whether or not the he can participate in the training.

- Are you on light duty?
- Are you restricted to running shoes by a corpsman or a doctor?
- Have you fought in a pugil stick bout within the last 7 days?
- Have you received a blow to the head within the last 7 days?
- Have you had a concussion within the last 6 months?
- Have you had dental surgery within the last 24 hours?
- Do you have stitches or staples on your body?
- Have you had a shoulder or head injury within the last 5 years?
- Are you taking a prescription drug?
- Do you have an ear infection or current sinus infection?
- Have you had a broken bone within the last 6 months?

Safety Personnel

The following safety personnel are required to conduct pugil stick training:

- One close combat instructor must officiate the bout.
- One close combat instructor trainer, commissioned officer, or staff noncommissioned officer must be in the training area to serve as range safety officer.
- One corpsman must be in the training area.

Training Area

To prevent injury, Marines train on areas with soft footing (i.e., sand or grass). Training mats are not recommended because feet can stick to the mats, prohibiting movement or causing joint injuries by twisting a knee. Bouts should not take place on a hard surface area; e.g., a flight deck or parking lot. A boxing ring may be used to conduct pugil stick bouts; ring dimensions can vary as long as there is ample room to execute the techniques in the training area.

Second Impact Syndrome

Second Impact Syndrome occurs when a second blow to the head produces a second concussion that occurs within 1 week following a previous concussion (before recovery from the first concussion). Second Impact Syndrome causes rapid brain swelling and can cause death. Therefore, there must be 7 days between pugil stick bouts to reduce the risk of severe injury resulting from Second Impact Syndrome. The 7 day separation between pugil stick bouts significantly reduces the possibility of injury, particularly in someone who may have suffered a brain injury or concussion but shows no symptoms.

Any Marine who experiences headaches or the following symptoms after training must be examined by appropriate medical personnel:

- Blurred vision.
- Ringing in the ears.
- Dilation of the pupils.
- Slurred speech.
- Bleeding from ears or mouth.
- Swelling in head or neck area.
- Any unnatural discoloration in head or neck area.

The Marine should not be allowed to participate in pugil stick training or any other activity where a heavy blow might be sustained for a minimum of 7 days after the headache or other symptoms have subsided.

Safety Measures

The following safety measures must be followed:

- Gear must be worn properly throughout training.
- Contact lenses or glasses will not be worn.
- False teeth will be removed from the mouth.
- Nothing will be worn around the neck except the neck roll.

Competition among groups of Marines is authorized as long as it does not overshadow training objectives or compromise safety procedures.

Safety and proper techniques are paramount.

Safety is more important than competition.

2. General Rules and Regulations Governing Pugil Stick Bouts

Instructors and Support Personnel Requirements

At a minimum, one close combat instructor observes a pugil stick bout. For safety purposes, it is better to have two instructors judging a bout because each instructor can fully observe each of the fighters. The best position for observation is to the right of a fighter. This allows the instructor to see the fighter's facial expression and body movement. The instructor's position must not interfere with the fight.

In addition, one commissioned or staff noncommissioned officer (to serve as range safety officer) and one corpsman will be in the training area.

Prior to the Bout

Prior to the bout, Marines—

Are paired according to height, weight, and gender.

Wear the proper safety equipment.

Once Marines have properly donned the safety equipment, they wait for instructions from the close combat instructor. On the command of the close combat instructor, a pair of Marines will enter the ring. Once in the ring, the close combat instructor inspects each Marine for the proper safety equipment.

Note: There should be enough gear so that when two Marines are training, two other Marines can be donning safety equipment for the next bout.

During the Bout

The bout begins when the close combat instructor blows the whistle. All strikes are directed above the waist. Upon hearing a whistle blast, all fighting immediately ceases.

Stopping a Bout

There are two reasons for stopping a bout: delivery of a scoring blow or an unsafe condition. The close combat instructor trainer, close combat instructor, or the range safety officer may stop the bout at any time an unsafe condition is observed.

Scoring Blow. A scoring blow is an offensive technique delivered to a vulnerable area of an opponent with sufficient force and precision to be considered as a disabling or killing blow. Scoring blows are not judged solely on the degree of force with which a blow is delivered, but on the accuracy and techniques employed. A scoring blow is defined as—

A straight thrust with the blade end of the weapon (red end of the pugil stick) to the opponent's face mask or throat.

A slash to the side of the opponent's helmet (below the ear) or neck with the red end of the pugil stick.

A heavy blow to the opponent's head with an authorized technique (i.e., buttstroke, smash) using the butt (black end) of the pugil stick.

When a scoring blow is delivered, the close combat instructor blows the whistle to stop the bout.

Unsafe Condition. The bout will be stopped as soon as an unsafe condition exists. An unsafe condition exists when a Marine is unable to defend himself, loses his balance and falls down on one or two knees or falls down completely, shows instability (e.g., buckling at the knees), loses muscular tension in his neck and his head snaps back or to one side, or appears disoriented. An unsafe condition also exists if a Marine lets go of one end of the pugil stick, the equipment (e.g., helmet, neck roll) falls off, or a Marine fails to use the proper techniques.

If any of these conditions occur—

The close combat instructor trainer, close combat instructor, range safety officer, or anyone supervising training stops the bout and separates the two Marines.

The corpsman evaluates the possibly injured Marine to determine if the fight can continue. The corpsman—

- Ensures that he is alert and responsive, both verbally and physically.
- Talks to him to see if he is coherent.
- Makes certain he comprehends and replies to verbal questions.
- Makes certain his speech is not slurred.
- Checks his physical signs.
- Ensures his eyes are focused and not dazed or glazed.
- Ensures his legs are not wobbly or shaky.

The pugil stick is not used as a baseball bat. The use of unauthorized techniques will result in expulsion from the bout. Only techniques taught in bayonet training are authorized. These techniques include-

Straight thrust.

Buttstroke (horizontal and vertical).

Smash.

Slash.

Parry.

Blocks (low, high, left, right).

After the Bout

A third whistle blast is used prior to identifying the Marine who delivered the killing blow.

A fourth whistle blast is used prior to demonstrating the killing technique used by the Marine who won the bout.

3. Directions for Making Pugil Sticks

Close combat instructor trainers are taught to make pugil sticks to standard specifications at school. It is recommended that all pugil sticks be made under the supervision of a close combat instructor trainer. Until the Marine Corps acquires a standardized pugil stick that units can order through the supply system, instructors use the following information and instructions to make pugil sticks locally.

Supplies

The following supplies are needed to make one pugil stick:

One 45" by 2" circular, oak, wood dowel.

Four 9" wide by 27" long foam pad pieces to make two inner handguards of double thickness.

Four 11" wide by 19" long foam pad pieces to make two outer handguards of double thickness.

Canvas material to cover inner and outer handguards.

One foam pad cut 1/4" thick by 7" wide by 13" long to make the centerguard.

One 8" wide by 14"long piece of canvas material to make a pugil stick sock.

Canvas material to make two pugil stick end caps. Ideally, two different color fabrics should be used to simulate the bayonet and buttstock ends of a rifle. Common colors for end caps are black and red. Each end cap is a different color to help the officiating certified close combat instructor determine the scoring blow during the pugil stick bout.

Two 1/4" thick by 11" wide by 54" long foam pad pieces to make two ends.

Two foam pads cut 1/4" thick by 3" wide by 14 1/2" long to make two end plug inserts.

One can of 3M Photo MountQ spray adhesive (NSN 8040-01-120-4009)to adhere foam padding to the stick. (Spray adhesive is available from self-service.)

One roll of duct tape or riggers tape (NSN 7510-00-266-5016) to protect foam pad and canvas from wear and to reinforce gluing action of spray adhesive. (Duct tape or riggers tape is available from self-service.)

One roll para-cord or 5-50 cord to tie both the inner and outer handguards together.

One roll of thread strong enough to sew canvas material together.

Sixteen grommets, one for each of the four corners of the inner and outer handguards.

One foam sleeping mat (NSN 8465-01-109-3369). Use unserviceable foam sleeping mats obtained from Defense Reutilization and Marketing Office.

Tools

The following tools are needed to make a pugil stick:

Tape measure, ruler, or yardstick.
Scissors.
Grommet machine.
Knife.
Sewing machine.

Directions

To make a pugil stick, Marines perform the following steps.

Preparing Materials

Step 1. Cut foam sleeping mat into the dimensions listed above. Cut mats to double thickness.

Step 2. Use spray adhesive to adhere the two 9" by 27" pieces of foam together. Use the spray adhesive to adhere the two 11" by 19" pieces of foam together. Let adhesive dry before using the double-thick foam for handguards.

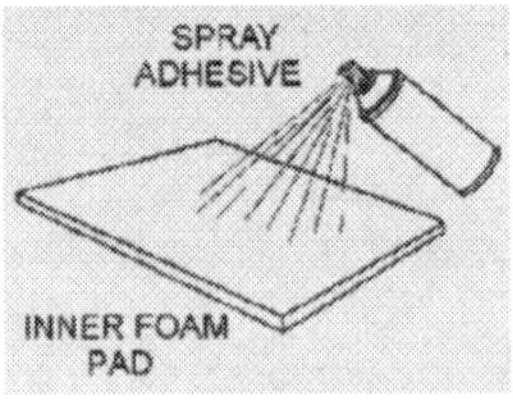

Step 3. Cut canvas material to make two coverings for the inner handguards. Cut material to make two coverings for the outer handguards. Two inner handguards and two outer handguards are required for each pugil stick.

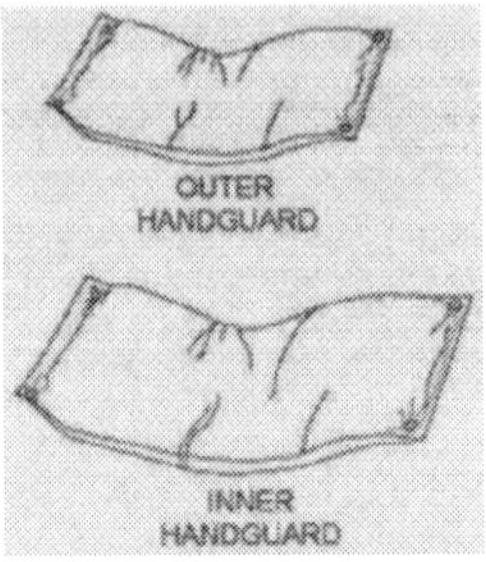

Step 4. Use a sewing machine to make a sleeve out of the 9" by 27" canvas. Sew down three sides. Insert foam pad inside the canvas sleeve and then sew down the fourth side. Use the grommet machine to create holes reinforced with metal grommets in each of the four corners. Once finished, the inner handguards are complete.

Note: It is recommended that the Fabric Repair Shop be used to sew canvas sleeves and apply grommets.

Step 5. Repeat step 4 to make the other inner handguard and both outer handguards.

Constructing Ends, End Plugs, and End Caps

Step 1. Apply spray adhesive on approximately 8 inches of one end of the dowel. This adhesive serves to glue the foam pad to the wooden dowel.

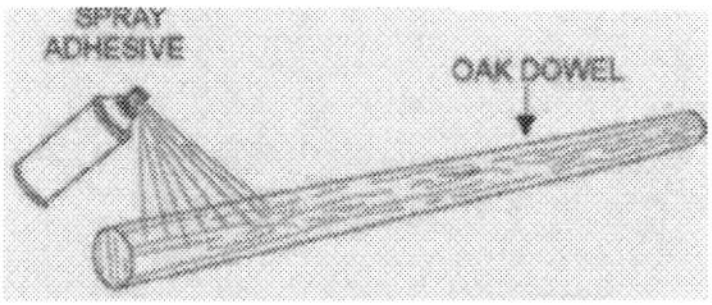

Step 2. Apply spray adhesive on one side of the 11" by 54" foam pad. This causes the foam pad to stick to itself when it is rolled around the end of the wooden dowel. Ensure 3 inches of the rolled foam pad extends past the wooden dowel for stability.

Step 3. Apply spray adhesive on one side of 3" by 14 1/2" foam pad. Roll pad into a tight cylinder.

Step 4. Apply spray adhesive on the exposed end of the dowel and fill the hole in the end with the end plug. The end plug prevents the wooden dowel from protruding beyond the edge of the rolled foam padding. This is an important safety precaution to prevent injury.

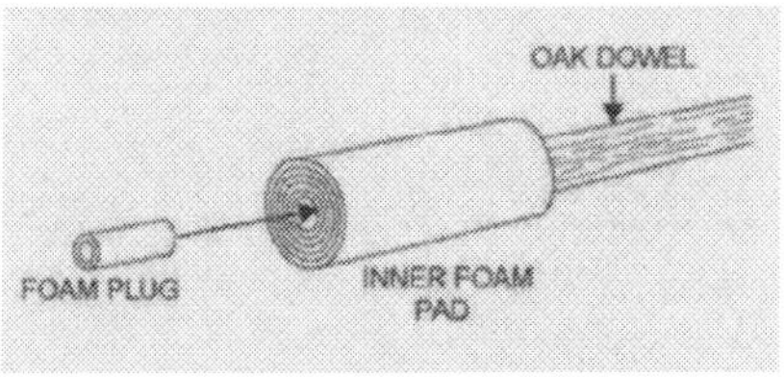

Step 5. Use duct tape or riggers tape to secure and reinforce the foam pad once the end plug is inserted in the end.

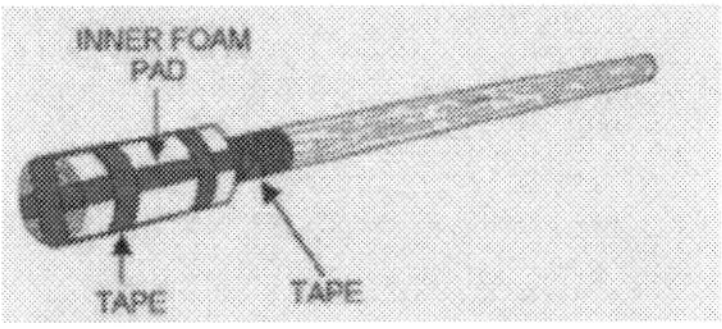

Step 6. Measure the dimensions of the ends (typically, the length of the cylinder should be 12 inches long to allow for taping). Cut and sew canvas material in the shape of a cylinder with one open end. Once complete, this becomes the end cap. The end cap is a colored sleeve that covers the rolled and taped foam pad at each end of the pugil stick.

Note: It is recommended that the Fabric Repair Shop be used to make the end cap.

Step 7. Slide one of the end caps over the rolled foam pad and tape it to the dowel.

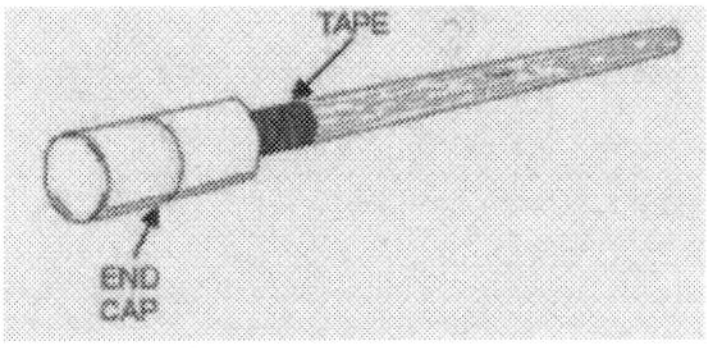

Building the Centerguard

Step 1. Spray one side of the 7" by 13" foam pad with spray adhesive and wrap it tightly around the center of the dowel.

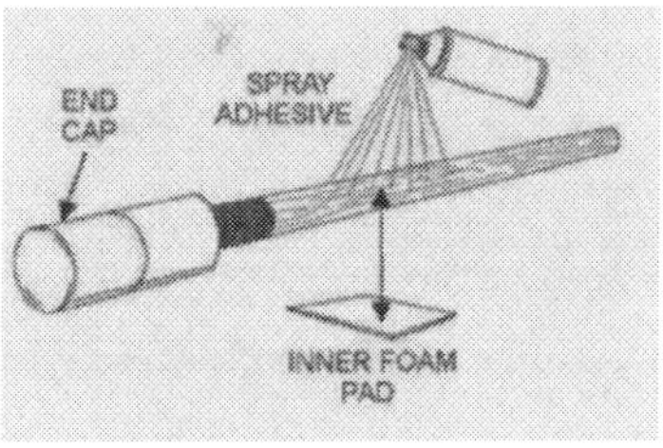

Step 2. Wrap duct tape or riggers tape around the foam pad to secure.

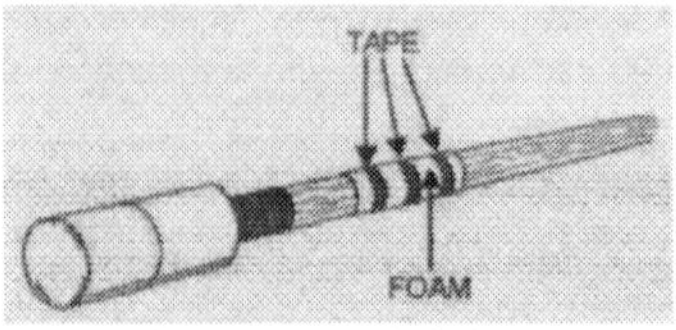

Step 3. Use a sewing machine to make a tubular pugil stick sock out of the 8" by 14" piece of canvas. Slide the pugil stick sock over the rolled foam pad and secure it with the duct tape or riggers tape. Duct tape or riggers tape protects the edges of the centerguard and reinforces the gluing effects of the spray adhesive.

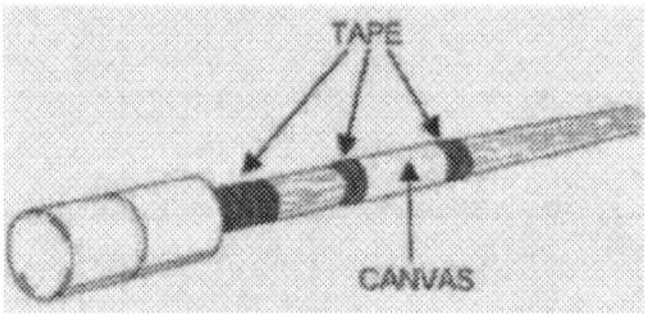

Note: It is recommended that the Fabric Repair Shop be used to make the pugil stick sock.

Finishing the Opposite End Cap

After one end cap is complete and the centerguard is finished, then repeat steps 1 through 7 in Constructing Ends, End Plugs, and End Caps for the other end of the pugil stick. Ensure that the opposite color end cap is used.

Fastening Handguards

Step 1. Place the inner handguards (longer piece) around the dowel. The inner handguard is placed on first because it protects the wrist. One inner handguard is needed for the left hand and one for the right hand.

Step 2. Fold the inner handguard around the pugil stick and run the para-cord or 5-50 cord through the metal grommets.

Step 3. Tighten down the corner ends of the inner handguards by pulling on the para-cord or 5-50 cord.

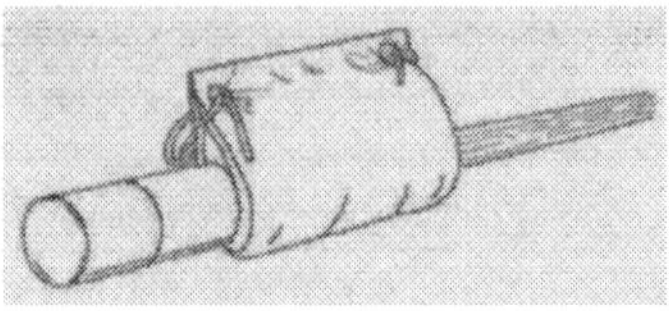

Step 4. Tie off the cord with a square knot above the metal grommets.

Step 5. Place the outer handguards (shorter piece) over the inner handguards (longer piece). Repeat steps 1-4 for the other handguard. The outer handguards provide additional protection to the knuckles and fingers and provides additional stability for the inner handguards.

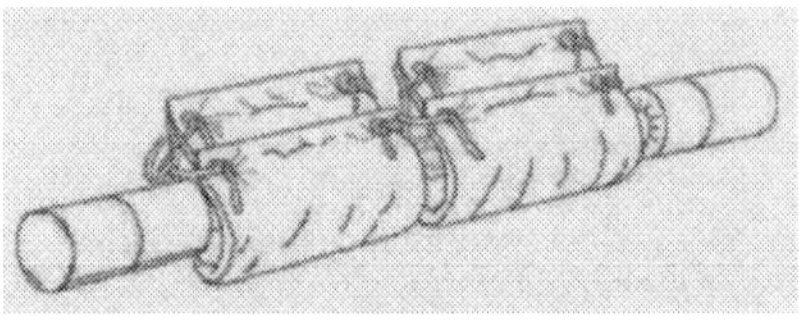

Serviceability Inspection

Prior to any training, all pugil sticks are inspected to ensure they are fit for use. If any of the following conditions exist, the pugil stick is unsafe and will not be used:

Stick protrudes through the end caps.
Frayed or worn edges exist on canvas surfaces.
The stick can be felt through the pads or the padding is too soft.
End caps or handguards are not securely attached.

APPENDIX B

SAFETY PRECAUTIONS DURING TRAINING

1. General Safety Precautions

When training close combat techniques, certain safety precautions must be adhered to to prevent injuries.

Most training should be conducted on a training area with soft footing such as a sandy or grassy area. If training mats are available, they should be used. A hard surface area is not appropriate for close combat training.

All techniques should be executed slowly at first. Marines can increase the speed of execution as they become more proficient. Marine-on-Marine training that requires contact (chokes, throws, ground fighting, and unarmed restraints and manipulation) should not be executed at full force or full speed.

If a technique is applied to the point that a Marine is uncomfortable, the Marine must "tap out." This indicates immediate release of the pressure being applied or to immediately stop the technique. The Marine "taps out" by firmly tapping his hand several times on any part of the opponent's body that will get his attention or by saying *stop*.

Second Impact Syndrome occurs when a second concussion develops within hours, days, or weeks following a prior concussion (and before recovery from the first concussion). Second Impact Syndrome causes rapid brain swelling and can cause death. Marines who experience headaches or other symptoms following training must be examined by appropriate medical personnel. These symptoms can include, but are not limited to, blurred vision, ringing in the ears, dialation of the pupils, bleeding from the ears or mouth, slurred speech, swelling in head or neck area, or any unnatural discoloration of head or neck. They should not be allowed to participate in pugil stick training or any other activity where a heavy blow might be sustained for a minimum of 7 days after the symptoms have subsided.

2. Safety Precautions for Individual Techniques

Falls

When training Marines to fall, they should progress from the ground, to a kneeling or squatting position, and then to a standing position. This ensures they are comfortable and understand the technique before progressing to executing falls from a higher profile. This instructional technique vastly reduces the risk of injury.

Strikes and Punches

When training Marines to strike and punch, they begin by executing the techniques "in the air." As they become more proficient, they execute strikes on equipment (when available) such as an air shield or a heavy bag. At no time should they be permitted to execute strikes on another student.

Chokes and Ground Fighting Chokes

When training Marines to execute chokes, they will not apply pressure to the opponent's throat during training because the trachea and windpipe can be crushed. During training, Marines should practice the proper procedures for blood chokes. They should never execute air chokes. At no time during training should a choke be applied at full force or full speed. No choke should be held for longer than 5 seconds.

Unarmed Restraints and Manipulation

When training Marines to execute unarmed restraints and manipulation techniques, they utilize slow and steady pressure. Never apply these techniques at full force or at full speed.

Pugil Stick Training

A blow to the head during training will bruise the brain. A second blow to the head can cause death. This is known as the Second Impact Syndrome. There must be 7 days between pugil stick training to prevent injury or death.

TECHNICAL MANUAL
ARMY NO. 9-1005-319-23&P
AIR FORCE TO 11W3-5-5-42

***ARMY TM 9-1005-319-23&P**
AIR FORCE TO 11W3-5-5-42

DEPARTMENTS OF THE ARMY AND AIR FORCE
Washington, DC, 1 May 1991

Technical Manual for 5.56MM,

M16A2, M4 and M4A1 Carbine

DISTRIBUTION STATEMENT. APPROVED FOR PUBLIC RELEASE; DISTRIBUTION IS UNLIMITED.

REPORTING ERRORS AND RECOMMENDING IMPROVEMENTS

You can help improve this manual. If you find any mistakes or if you know of a way to improve the procedures please let us know.

Army users mail your letter, DA Form 2028 (recommended Changes to Publications and Blank Forms), or DA Form 2028-2 located in the back of this manual direct to: Director, Armament and Chemical Acquisition and Logistics Activity, ATTN: AMSTA-AC-NML, Rock Island, IL 61299-7630.

Air Force users submit AFTO Form 22, Technical Order System Publication Improvement Report and Reply, to: WR-ALC/MMDET, Robins AFB, GA 31098-5609.

A reply will be furnished to you.

			Page
		HOW TO USE THIS MANUAL	iii
CHAPTER	1	**INTRODUCTION**	1-1
		External View of 5.56mm Rifle, M16A2	1-0
		External View of 5.56mm Carbine, M4	1-0.1
		External View of 5.56mm Carbine, M4A1	1-0.1
		Chapter Overview	1-1
Section	I	General Information	1-1
Section	II	Equipment Description and Data	1-3
Section	III	Principles of Operation	1-6

*This manual supersedes ARMY TM 9-1005-319-23&P dated 28 August 1987, including all changes.

			Page	Illus Figure
CHAPTER	**2**	**UNIT MAINTENANCE INSTRUCTIONS**	2-1	
		Chapter Overview	2-1	
Section	I	Repair Parts, Special Tools, TMDE, and Support Equipment	2-1	
Section	II	Service Upon Receipt	2-1	
Section	III	Preventive Maintenance Checks and Services (PMCS)	2-3	
Section	IV	Troubleshooting	2-20	
Section	V	Maintenance Procedures	2-33	
Section	VI	Preparation for Storage or Shipment	2-70	
CHAPTER	**3**	**DIRECT SUPPORT MAINTENANCE INSTRUCTIONS**	3-1	
		Chapter Overview	3-1	
Section	I	Repair Parts, Special Tools, TMDE, and Support Equipment	3-1	
Section	II	Troubleshooting	3-2	
Section	III	Maintenance Procedures for the M16A2 Rifle, M4 and M4A1 Carbine	3-15	
Section	IV	Preembarkation Inspection of Materiel in Units Alerted for Overseas Movement (A.F. Only)	3-95	
CHAPTER	**4**	**MAINTENANCE OF AUXILIARY EQUIPMENT**	4-1	
		Chapter Overview	4-1	
Section	I	Auxiliary Equipment Repair	4-1	
APPENDIX	**A.**	**REFERENCES**	A-1	
APPENDIX	**B.**	**MAINTENANCE ALLOCATION CHART**	B-1	
APPENDIX	**C.**	**REPAIR PARTS AND SPECIAL TOOLS LIST**	C-1	
Section	I	Introduction	C-1	
Section	II	Repair Parts List	C-1-1	
Group	00	5.56MM Rifle, M16A2, Carbine, M4 and M4A1	C-1-1	C-1
	00.1	M4 and M4A1 Carrying Handle Assembly	C-1A-1	C-1A
	00.101	M4 and M4A1 Rear Sight Assembly	C-1B-1	C-1B
Group	01	Bolt carrier assembly	C-2-1	C-2
	0101	Bolt assembly	C-3-1	C-3
	0102	Key and bolt carrier assembly	C-4-1	C-4
Group	02	Handle assembly	C-5-1	C-5
Group	03	Upper receiver and barrel assembly	C-6-1	C-6
	0301	Barrel assembly, M16A2	C-7-1	C-7
	0301	Replacement Barrel & Front Sight Assembly, M4/M4A1	C-7-1	C-7
	0302	Upper receiver assembly	C-8-1	C-8
	030201	Forward assist assembly	C-9-1	C-9
	030202	Rear sight assembly, M16A2	C-10-1	C-10
Group	04	Lower receiver and buttstock assembly	C-11-1	C-11
	0401	Buttstock assembly, M16A2	C-12-1	C-12
	0401A	Buttstock assembly, M4/M4A1	C12A-1	C12A
	0402	Hammer assembly	C-13-1	C-13
	0403	Trigger assembly	C-14-1	C-14
	040301	Trigger subassembly,M16A2 and M4	C-15-1	C-15
	0404	Lower receiver and receiver extension assembly, M16A2	C-16-1	C-16
	0404A	Lower receiver and receiver extension assembly, M4 and M4A1	C16A-1	C16A

			Page	Illus Figure
APPENDIX	**C.**	**REPAIR PARTS AND SPECIAL TOOLS LIST--Cont.**		
Section	III	Special Tools List	C-17-1	C-17
Section	IV	National Stock Number and Part Number Index	I-1	
APPENDIX	**D.**	**EXPENDABLE/DURABLE SUPPLIES AND MATERIALS LIST**	D-1	
APPENDIX	**E.**	**ILLUSTRATED LIST OF MANUFACTURED ITEMS**	E-1	
		ALPHABETICAL INDEX	Index-1	

WARNING

ALL WARNINGS in this technical manual pertain to both the rifle and the carbines unless otherwise specified.

Before starting an inspection, be sure to clear the rifle. Do not pull the trigger until the rifle has been cleared. Inspect the chamber to ensure that it is empty and no ammunition is in position to be chambered.

Do not keep live ammunition near work area.

To avoid injury to your eyes, use care when removing and installing spring-loaded parts.

All M1 6A2 rifles and M4 carbines must be inspected and gaged at least once annually for safety and serviceability. Initial gaging is required 1 year from receipt of the weapons. Air Force users refer to inspection requirements in Air Force Manual (AFM) 36-2227, Volume 1.

All Army Reserve and Army National Guard M16A2 rifles and M4 carbines must be inspected and gaged at least once every 2 years, after the initial inspection/gaging procedures have been accomplished. This initial gaging procedure is required 1 year from receipt of the weapons. This 2 year interval may be maintained unless preventive maintenance checks and services (PMCS) or other physical evidence indicates that an individual unit's M16A2 rifles and M4 carbines require inspection/ gaging at a more frequent interval. If it is determined that a yearly inspection is necessary for an individual unit, only that unit will be affected. This will not affect other units in regard to the interval of inspection.

It is recommended that training units inspect/gage all rifles and carbines at the end of each training cycle. Training units will inspect/gage all rifles and carbines at least once annually.

Below direct support maintenance, **DO NOT** interchange bolt assemblies from one rifle/carbine to another. Doing so may result in injury to, or death of, personnel.

Bolt cam pin must be installed or rifle/carbine will blow up while firing the first round. If the bolt cam pin is not installed, injury to or death of, personnel may result.

Dry cleaning solvent is flammable and toxic and should be used in a well ventilated area. The use of rubber gloves is necessary to protect the skin when washing rifle parts.

When using solid film lubricant or dichloromethane, be sure the area is well ventilated.

When using carbon removing compound (item 8, app D), avoid skin contact. If carbon removing compound comes In contact with the skin, wash thoroughly with running water. using a good lanolin base cream after exposure to the compound is helpful. Using gloves and protective equipment is required.

The lock plate prevents the selector lever from being placed in BURST and will be installed at the discretion of the unit commander. It is mandatory for use in civil disturbance (riot control).

Only blank cartridge M200 is to be used when the blank firing attachment is attached to the rifle/carbine.

WARNING (CONT)

Do not fire blank ammunition at a representative enemy at distances of less than 20 feet (6.10 Om). The unburned propellant grains can cause injury within this distance.

For further information on safety, care, and handling of ammunition: Army and Air Force users refer to M16A2 Rifle Operator's Manual.

For additional first aid data, see Field Manual (FM) 21-11.

HOW TO USE THIS MANUAL

Read this manual carefully before performing required maintenance. This manual will be referred to for Inspection/Maintenance and Repair procedures.

GENERAL

There are several things you need to know to use this manual efficiently.

1. All references in the manual are to pages only. Reference to maintenance procedures is to the page where the respective initial setup appears.

2. Illustrations for the maintenance procedures show only those parts affected by the operation being performed.

3. Whenever the male gender is mentioned in the manual (i.e., crewman, repairman), it also pertains to females.

4. When the term "evacuate to support maintenance" is used, the entire rifle must be evacuated.

5. When a procedure is common to M16A2 rifle, and M4/M4A1 carbine, **ONLY** the M16A2 configuration will be depicted. If a procedure is not common to both weapons, the procedure will be incorporated.

6. When the word rifle is referenced in text, it will reference the rifle and the carbines.

INDEXES

This manual is organized to help you find the information you need quickly. There are several useful indexes.

1. **Table of Contents**. Lists in order all chapters, sections, and appendixes. Gives page references.

2. **Nomenclature Cross-References List.**

3. **Chapter Overviews**. Summarize material covered in the chapter. Are located at the beginning of each chapter.

4. **Symptom Index**. Located just before the troubleshooting table in each maintenance chapter. Lists, in alphabetical order, parts of the rifle with possible malfunctions. References pages of the troubleshooting table.

5. **Alphabetical Index**. Located at the end of the manual. An extensive subject index for everything in the manual. Gives page references.

MAINTENANCE PROCEDURES

There are two maintenance chapters:

Army personnel use chapter two for unit maintenance procedures and chapter three for direct support maintenance procedures.

Air Force personnel: Only Air Force Specialty Code 753XX Combat Arms Training and Maintenance (CATM) specialists, technicians, and gunsmiths are authorized to perform maintenance procedures contained in this manual.

Each maintenance task has an initial setup containing a list of the following things you will need in order to do your maintenance task.

1. **Tools and Special Tools**. For standard and special tools, see appendixes B and C. Army users are to use the Tool Set, Gage Set, and/or Shop Set listed in the initial setup.

2. **Materials/Parts**. Lists expendable materials and 100 percent replaceable parts. Each material or part is followed by a part number or appendix reference.

3. **References**. Lists other publications containing necessary information.

4. **Equipment Condition**. Lists conditions to be met before starting the procedure. The reference on the left of the condition is a page reference to instructions for setting up the condition.

5. **General Safety Instructions**. Lists safety instructions to follow before performing maintenance procedures.

EXTERNAL VIEW OF 5.56 MM RIFLE M16A2

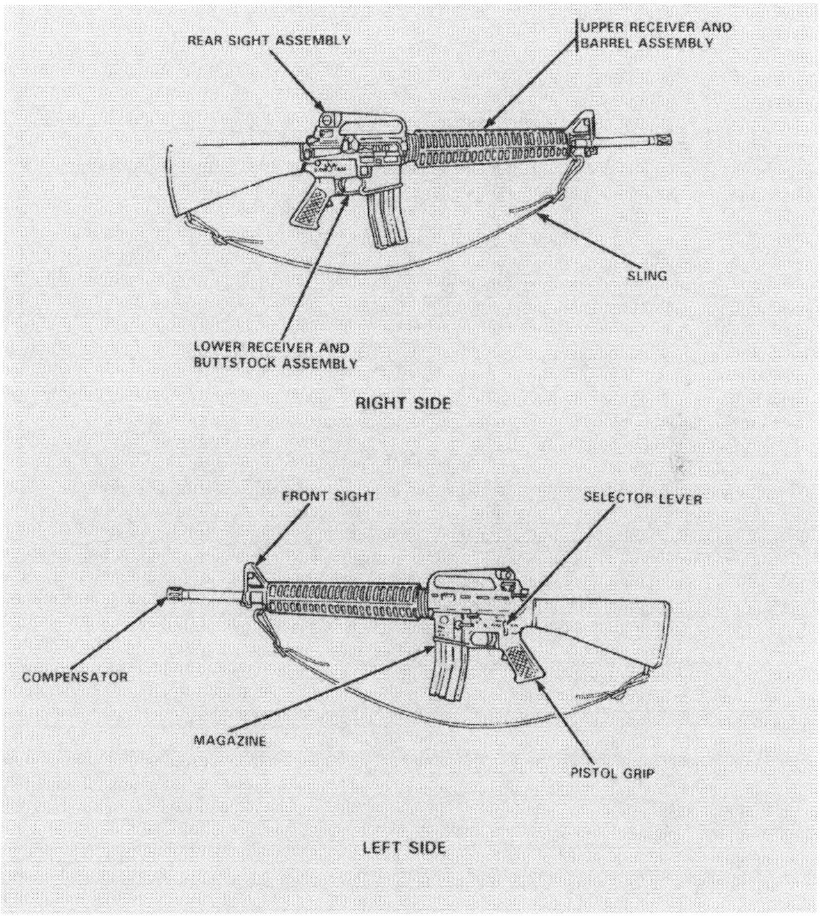

EXTERNAL VIEW OF 5.56MM CARBINE, M4/M4A1

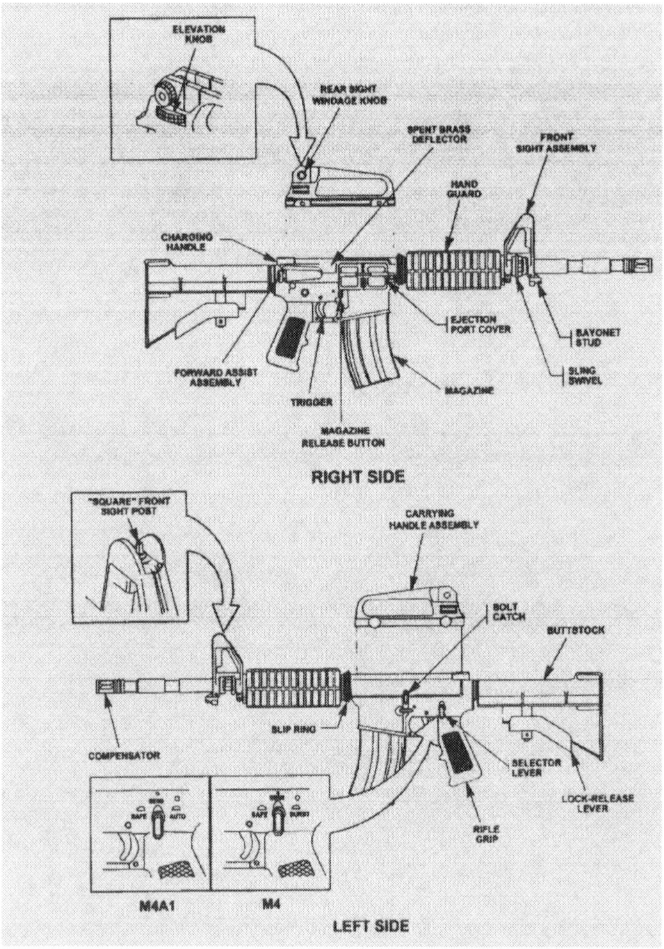

CHAPTER 1
INTRODUCTION

CHAPTER OVERVIEW

This chapter contains general information, equipment description and data, and principles of operation for the M16A2 rifle and M4/M4A1 carbines.

Section I. GENERAL INFORMATION

1-1. SCOPE.

a. Type of Manual: Unit and Direct Support Maintenance.

b. Model Number and Equipment Name: 5.56mm Rifle M16A2, M4 and M4A1 Carbines.

c. Purpose of Equipment. Provides personnel an offensive/defensive capability to engage targets with small arms fire.

1-2. MAINTENANCE FORMS, RECORDS, AND REPORTS. Department of the Army forms and procedures used for equipment maintenance will be those prescribed by DA PAM 738-750, The Army Maintenance Management System.

Air Force users refer to TO 11W-1-10 for applicable forms and records.

1-3. DESTRUCTION OF ARMY MATERIEL TO PREVENT ENEMY USE. See TM 750-244-7.

1-4. PREPARATION FOR STORAGE OR SHIPMENT. Refer to page 2-70.

Air Force users refer to Special Package Instruction (SPI) 00-856-6885.

1-5. OFFICIAL NOMENCLATURE, NAMES AND DESIGNATIONS.

NOMENCLATURE CROSS-REFERENCE LIST

Common Name	Official Nomenclature
Action Spring	Compression Helical Spring
Ball Bearing	Bearing Ball
Bolt Catch Spring	Compression Helical Spring
Bolt Carrier Key Tool	Machine Key
Burst Disconnector	Lock-Release Lever
Cam Clutch Spring	Helical Spring
Carbine	M4/M4A1 Carbine
Charging Handle Assembly	Handle Assembly

1-5. OFFICIAL NOMENCLATURE, NAMES AND DESIGNATIONS (CONT).

NOMENCLATURE CROSS-REFERENCE LIST

Common Name	Official Nomenclature
Disconnector Springs	Compression Helical Spring
Ejector Spring	Helical Spring
Extractor Spring Assembly	Spring Assembly
Hammer Spring	Torsion Helical Spring
Lower Receiver Extension	Spring Receiver Holder
Magazine	Cartridge Magazine
Magazine Catch Spring	Compression Helical Spring
Peel Washer	Shim
Pistol Grip	Rifle Grip
Pivot Pin Detent	Takedown Pin Detent
Rifle	Rifle, 5.56mm, M16A2
Rifle Barrel Assembly	Barrel Assembly
Selector Lever	Fire Control Selector
Semiautomatic Disconnector	Lock-Release Lever
Sling	Small Arms Sling
Trigger Spring	Torsion Helical Spring
Upper Receiver	Upper Cartridge Receiver

1-6. REPORTING EQUIPMENT IMPROVEMENT RECOMMENDATIONS (EIR). If your M16A2 rifle needs improvement, let us know. Send us an EIR. You, the user, are the only one who can tell us what you don't like about your equipment. Let us know why you don't like the design.

Army users submit SF 368 (Product Quality Deficiency Report) to: Commander, U.S. Army Armament Research, Development and Engineering Center, ATTN: AMSTA-AR-QAW (R)/Customer Feedback Center, Rock Island, IL 61299-7300.

Air Force users submit Materiel Deficiency Report (MDR) to: DIR MAT MGT ROBINS AFB GA//MMIBTC// and Product Quality Deficiency Report to: DIR MAT MGT ROBINS AFB GA//MMQA// IAW Technical Order 00-35D-54.

A reply will be sent to you.

1-7. CORROSION PREVENTION AND CONTROL (CPC). CPC of Army materiel is a continuing concern. It is important that any corrosion problems with this item be reported so that the problem can be corrected and improvements can be made to prevent the problem in future items,

While corrosion is typically associated with rusting of metals, it can also include deterioration of other materials such as rubber and plastic. Unusual cracking, softening, swelling, or breaking of these materials may be a corrosion problem.

If a corrosion problem is identified, it can be reported using Standard Form 368, Product Quality Deficiency Report. Use of key words such as "corrosion", "rust", " deterioration", or "cracking" will assure that the information is identified as a CPC problem.

Army users submit Product Quality Deficiency Report (SF 368) to:

Commander
U.S. Army Armament Research, Development and Engineering Center
ATTN: AMSTA-AR-QAW (R)
Rock Island, IL 61299-7300

Air Force users submit Materiel Deficiency Report (MDR) to:

DIR MAT MGT
ATTN: MMIBTC
Robins AFB, GA

and Product Quality Deficiency Report to:

DIR MAT MGT
ATTN: MMQA
Robins AFB, GA

Section II. EQUIPMENT DESCRIPTION AND DATA

1-8. EQUIPMENT CHARACTERISTICS, CAPABILITIES, AND FEATURES.

a. Characteristics.

(1) Light weight
(2) Air-cooled
(3) Gas-operated
(4) Magazine-fed
(5) Semiautomatic or burst fire

b. Capabilities. Provides personnel an offensive/defensive capability to engage targets with direct small arms fire.

c. Features.

(1) Receivers are made of light-weight aluminum alloys; however, the safety , durability, and function of the rifles are in no way reduced. The portability and logistical values are greatly increased, particularly when air transport is used.

(2) The bolt locking action is one of the mechanical features of the rifle. The bolt assembly and barrel extension contain locking lugs which engage and lock the bolt assembly firmly in the barrel extension. The initial force of the explosion of the cartridge is absorbed by the barrel, barrel extension, and bolt assembly.

1-8. EQUIPMENT CHARACTERISTICS, CAPABILITIES, AND FEATURES (CONT).

(3) The trigger guard is easily adaptable to winter operations. A spring-loaded retaining pin is depressed to allow ready access to the trigger when wearing arctic mittens.

(4) The ejection port cover prevents dirt or sand from getting into the ejection port. The ejection port cover must be closed during periods when firing is not anticipated. It opens automatically by the forward or rearward movement of the bolt carrier assembly.

1-9. LOCATION AND DESCRIPTION OF MAJOR COMPONENTS.

(A) MAGAZINE. 30 cartridge capacity.

(B) SLING. The sling is adjustable and provides a means to carry the weapon.

(C) BOLT CARRIER ASSEMBLY. Carries bolt assembly to chamber and fires the weapon. Contains the firing pin, cartridge extractor, bolt assembly, cartridge ejector, and bolt cam pin.

(D) CHARGING HANDLE ASSEMBLY. Provides a means of charging the weapon.

(E) **M16A2 UPPER RECEIVER AND BARREL ASSEMBLY**. Upper receiver contains rear sight assembly, ejection port, ejection port cover, and a housing for the key and bolt carrier assembly and bolt assembly. Rifle barrel assembly is air-cooled, contains compensator and front sight assembly, and holds the two handguard assemblies and the sling swivel.

(F) LOWER RECEIVER AND BUTTSTOCK ASSEMBLY. Lower receiver contains the trigger assembly, sear, hammer assembly, selector lever, rifle grip, bolt catch, and buttstock assembly. The buttstock assembly houses the action spring, buffer assembly, and extension assembly.

(G) M4/M4A1 CARRYING HANDLE. Provides a means of carrying carbine.

(H) **M4/M4A1 UPPER RECEIVER AND BARREL ASSEMBLY**. Upper receiver contains, ejection port, ejection port cover, a housing for key and bolt carrier assembly and bolt assembly, and mounting surface for the carrying handle assembly. Carbine barrel assembly is air-cooled, contains compensator and front sight assembly, and holds the two handguard assemblies and the sling swivel.

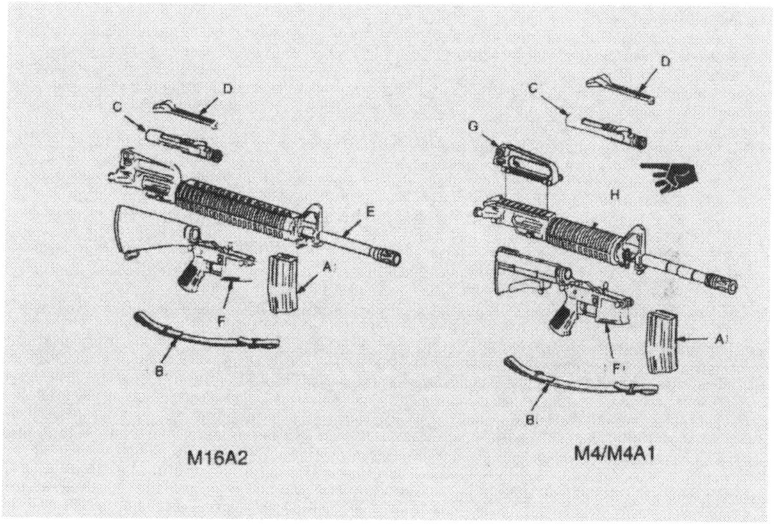
D
C
E
A
F
B
M16A2
D
C
G
H
A
F
B
M4/M4A1

1-10. EQUIPMENT DATA.

	US CUSTOMARY	METRIC
Weight:		
Carbine, M4/M4A1 without magazine and sling	6 lb 7 oz	2.91 kg
Rifle, MI 6A2 without magazine and sling	7 lb 8 oz	3.40 kg
Sling, adjustable	4 oz	0.11 kg
Empty magazine	4 oz	0.11 kg
Loaded magazine	1 lb 1 oz	0.48 kg
Carbine, M4/M4A1 w/sling and loaded magazine ...	7 lb 12 oz	3.51 kg
Rifle M16A2 w/sling and loaded magazine.............	8 lb 13 oz	4.00 kg
Bayonet-Knife M7	10.5 oz	0.30 kg
Scabbard M10 o	5 oz	0.14 kg
Length:		
Carbine with compensator, buttstock extended.......	33.0 in	83.82 cm
Carbine with compensator, buttstock closed	29.75 in	75.57 cm
Rifle with compensator	39.63 in	100.66 cm
Barrel (Carbine)	14.5 in	36.83 cm
Barrel (Rifle)...	20 in	50.8 cm
Barrel with compensator (Carbine)..........................	15.5 in	39.37 cm
Barrel with compensator (Rifle)	21 in	53.34 cm
Mechanical features:		
Rifling...	right-hand twist 6 grooves, 1 turn	
....	in 7 inches (17.78 cm)	
Method of operation...	direct gas	
Type of breech mechanism	rotating bolt	
Method of feeding	magazine	
Cooling..	air	
Trigger pull (M16A2 & M4).......................................	5.5 to 9.5 lb	2.49 to 4.31kg
Trigger pull (M4A1)..	5.5 to 8.5 lb	2.49 to 3.86 kg
Ammunition:		
Caliber	223	5.56mm
Type...	ball, blank, dummy, and tracer	
Firing characteristics:		
Muzzle velocity (Carbine) (approximate)	2,970 fps	905.85 mps
Muzzle velocity (Rifle) (approximate).......................	3,100 fps	945.5 mps
Chamber pressure	52,000 psi	358,540 kPa
Cyclic rate of fire (Carbine) (approximate)...............	700-970 rds/m	
Cyclic rate of fire (Rifle) (approximate)	700-900 rds/m	

1-10. EQUIPMENT DATA (CONT).

	US CUSTOMARY	METRIC
Maximum rate of fire:		
Semiautomatic	45 rds/m	
Burst	90 rds/m	
Sustained rate of fire	12/15 rds/m	
Maximum range	3,938 yards	Approximately 3,600 meters
Maximum effective range:		
Individual/point targets (Carbine)	547 yards	500 meters
Individual/point targets (Rifle)	602 yards	550 meters
Area targets (Carbine)	650 yards	600 meters
Area targets (Rifle)	875 yards	800 meters

Section III. PRINCIPLES OF OPERATION

1-11. GENERAL. The 5.56mm M16A2 and M4/M4A1 carbine:

a. Is gas-operated. It fires in either the semiautomatic or burst mode.

b. Has positive locking of the bolt. Firing pin is part of the bolt carrier assembly and cannot strike the primer until the bolt assembly is fully locked.

Section III. PRINCIPLES OF OPERATION (CONT).

1-12. PRINCIPLES OF OPERATION.

(A) **MAGAZINE**. Holds cartridges ready for feeding and provides a guide for positioning cartridges for stripping. Provides quick reload capabilities for sustained firing.

(B) **SLING**. Provides the means for carrying the weapon.

(C) **BOLT CARRIER ASSEMBLY**. Provides stripping, chambering, locking, firing, extraction, and ejection of cartridges using the drive springs and projectile propelling gases for power.

(D) **CHARGING HANDLE ASSEMBLY**. Provides initial charging of the weapon. The handle latch locks the charging handle assembly in the forward position during sustained fire to prevent injury to the operator.

(E) **M16A2 UPPER RECEIVER AND BARREL ASSEMBLY**. Provides support for the bolt carrier assembly. The barrel chambers the cartridge for firing and directs the projectile.

(F) **LOWER RECEIVER AND BUTTSTOCK ASSEMBLY**. Provides firing control for the rifle and carbine. M16A2 ONLY provides storage for basic cleaning materials.

(G) **M4/M4A1 CARRYING HANDLE ASSEMBLY**. Provides a means of carrying the carbine, contains rear sight assembly, and can be removed for mounting various optics.

(H) **M4/M4A1 UPPER RECEIVER AND BARREL ASSEMBLY**. Provides support for the bolt carrier assembly. The barrel chambers the cartridge for firing and directs the projectile.

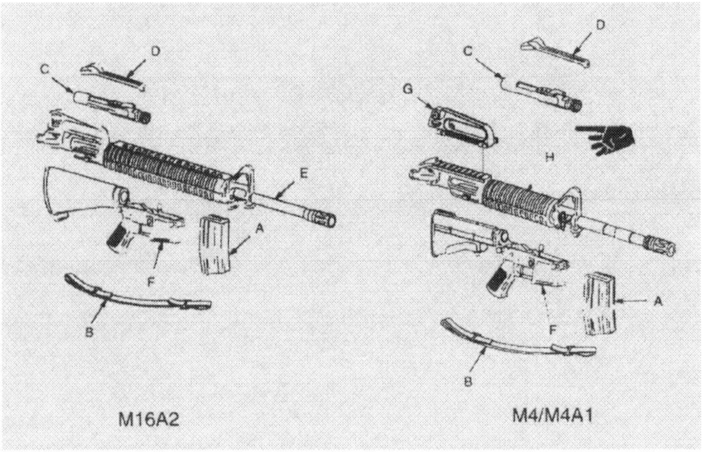

CHAPTER 2
UNIT MAINTENANCE INSTRUCTIONS

CHAPTER OVERVIEW

This chapter provides information and Instructions to help keep the rifle in good repair and contains the following sections:

a. Repair Parts, Special Tools, TMDE, and Support Equipment
b. Service Upon Receipt
c. Preventive Maintenance Checks and Services (PMCS)
d. Troubleshooting
e. Maintenance Procedures

Section I. REPAIR PARTS, SPECIAL TOOLS, TMDE, AND SUPPORT EQUIPMENT

2-1. COMMON TOOLS AND EQUIPMENT. For authorized common tools and equipment, refer to the Modified Table of Organization and Equipment (MTOE) applicable to your unit.

Air Force users must maintain the following common tools:

3-ounce soft-brass hammer
Vise
Flat tip screwdriver
Punch
Tweezers/round nose pliers
Hammer
Needle nose pliers

2-2. SPECIAL TOOLS, TMDE, AND SUPPORT EQUIPMENT. Special tools required for unit support are listed in appendixes B and C. Fabricated tools are listed and illustrated in appendix E.

2-3. REPAIR PARTS. Repair parts are listed and illustrated in appendix C of this manual.

Section II. SERVICE UPON RECEIPT

2-4. GENERAL.

a. Inspect the rifle for damage Incurred during shipment. If rifle has been damaged, report the damage on SF 364, Report of Discrepancy (ROD).

b. Check the rifle against the packing slip to see If shipment is complete. Army users report all discrepancies in accordance with DA PAM 738-750.

Air Force users submit Materiel Deficiency Report (MDR) to: DIR MAT MGT ROBINS AFB GA//MMIBTC// and Product Quality Deficiency Report to: DIR MAT MGT ROBINS AFB GA//MMQA//. IAW Technical Order 00-35D-54.

c. Check to see whether the equipment has been modified.

2-5. SERVICE UPON RECEIPT OF MATERIEL.

WARNING

Before starting an inspection, be sure to clear the rifle. Do not actuate the trigger before clearing the rifle. Inspect the chamber to make sure it is empty and free of obstructions. Check to see there are no obstructions in the barrel and no ammunition is in position to be chambered.

SERVICE UPON RECEIPT

LOCATION	ITEM	ACTION	REMARKS
1. Container	a. M16A2 rifle or M4/M4A1 carbine	a. Remove rifle from containers.	
		b. Inspect the equipment for damage incurred during shipment	If the equipment has been damaged, report the damage on SF Form 364, Report of Discrepancy (ROD).
		c. Check the equipment against the packing list to see if the shipment is complete	Report all discrepancies in accordance with the instructions of DA PAM 738-750.
	b. Basic issue items	Check for missing items	Refer to TM 9-1005-319-10 (operator's manual).
2. M16A2 rifle or M4/M4A1 carbine	a. Barrel assembly	If volatile corrosion inhibitor (VCI) is in barrel, remove and discard.	
	b. All parts	a. Field-strip rifle and inspect for missing, damaged, and rusted or corroded parts.	Refer to operator's manual.
		b. Clean and lubricate	Refer to operator's manual.
		c. Reassemble	Refer to operator's manual.
		d. Function check	Refer to page 2-69.

SERVICE UPON RECEIPT (CONT)

LOCATION	ITEM	ACTION	REMARKS
		e. Check to see whether the equipment has been modified	Refer to DA PAM 25-30.
	c. Magazine	Check for positive retention and functioning of bolt catch	Refer to operator's manual.

Section III. PREVENTIVE MAINTENANCE CHECKS AND SERVICES (PMCS)

2-6. GENERAL This section contains the procedures and instructions necessary to perform unit preventive maintenance checks and services. These services are to be performed by unit maintenance personnel with the assistance of the operator where practical.

2-7. PREVENTIVE MAINTENANCE CHECKS AND SERVICES

WARNING

Before starting an inspection, be sure to clear the rifle. Do not keep live ammunition near the work area.

a. General. The PMCS procedures are contained in the table following. They are arranged in logical sequence requiring a minimum amount of time and motion on the part of the persons performing them and are arranged so that there will be minimum interference between persons performing checks simultaneously on the same end item.

b. Item No. Column. Checks and services are numbered in disassembly sequence. This column shall be used as a source of item numbers for the "TM Number" column on DA Form 2404, Equipment Inspection and Maintenance Worksheet, in recording results of PMCS.

c. Interval Column. This column gives the designated interval when each check is to be performed.

d. Item To Be Checked Or Serviced Column. This column lists the items to be checked or serviced.

e. Procedure Column. This column contains a brief description of the procedure by which the check is to be performed. It contains all the information required to accomplish the checks and services. Information marked SH Indicates a specific equipment shortcoming and the procedure needed to correct the shortcoming.

NOTE

For the purpose of this technical manual, the following definition is supplied. This definition is not intended to apply to any other document

Shortcoming (SH): A fault that requires maintenance or supply action on a piece of equipment, but does not render equipment Not Mission Capable

f. Not Fully Mission Capable If: Column. This column contains a brief statement of the condition (e.g., malfunction, shortage) that would cause the covered equipment to be less than fully ready to perform its assigned mission.

2-7. PREVENTION MAINTENANCE CHECKS AND SERVICES (CONT).

PREVENTIVE MAINTENANCE CHECKS AND SERVICES FOR M16A2 RIFLE (CONT)

Item No.	Interval	Item to be Checked or Serviced	Procedure	Not Fully Mission Capable if:

WARNING

Before starting an inspection, be sure to clear the weapon. Do not pull the trigger until the weapon has been cleared. Inspect the chamber to ensure that it is empty and no ammunition is in position to be chambered. Do not keep live ammunition near work area.

NOTE

An inactive weapon is a weapon which has been stored in an arms room for a period of 90 days without use. The weapon may or may not have been assigned to an individual.

Inactive weapons shall receive quarterly PMCS unless inspection reveals more frequent servicing is necessary.

Normal cleaning (PMCS) of an inactive weapon will be performed every 90 days. Should the unit armorer detect corrosion on a weapon prior to the end of the 90-day period, the PMCS should be performed immediately.

Solid Film Lubricant (SFL) is the authorized touch up for the M16A2 Rifle and M4/M4A1 Carbine and may be used on up to one third of the exterior finish of the weapon.

FOR ARMY CONUS USE ONLY AND AIR FORCE TRAINING WEAPONS ONLY: Solid Film Lubricant may be used as a touch up without limitation on the upper receiver and barrel assembly. This is to say that units which **DO NOT** fall under the category of Divisional Combat Units or rapid deployment type units may have up to 100 percent of the exterior surface of the upper receiver and barrel assembly protected with SFL.. Prior to application of SFL, the surface must be thoroughly cleaned and inspected for corrosion and/or damage. If corroded or damaged, the part must be repaired or replaced prior to application of SFL. Continued use under combat conditions would result in an unprotected surface when the SFL wears off. This would result in a large light reflecting surface and accelerated deterioration of the unprotected surface. Therefore, Divisional Combat Units and units which fall under the definition of Rapid Deployment type must adhere to the limitation of **NOT** over one third of their exterior surface covered by SFL.

When determining mission capability, deadline if it is a deficiency.

PREVENTIVE MAINTENANCE CHECKS AND SERVICES FOR M16A2 RIFLE (CONT)

Item No.	Interval	Item to be Checked or Serviced	Procedure	Not Fully Mission Capable if:
1	Quarterly	Magazine (serviceability check)	Disassemble as In TM 9-1005-319-10 (operator's manual). Inspect tube (1) for bulges, dents, or damaged feeder lips (2). Inspect spring (3) and follower (4) for kinks or damage. SH-Replace the magazine if any of these conditions exist. Reassemble magazine and check for binding during operation of follower (4). SH-Replace the magazine if the follower binds. 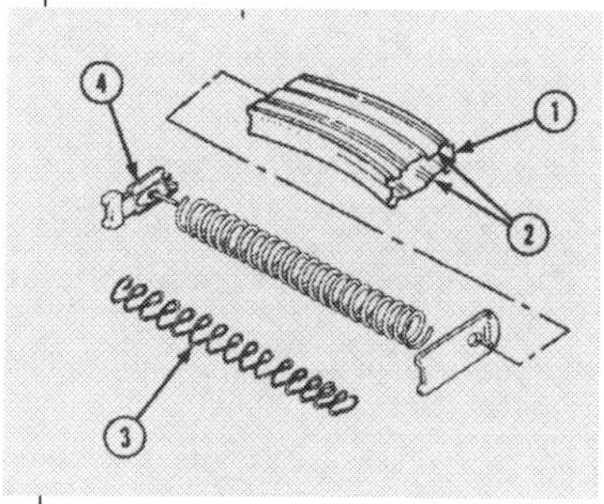	A magazine is not available fu; use with the rifle.
2	Quarterly	Charging handle assembly and selector lever	**WARNING** If the rifle falls any of the following selector lever tests, evacuate It to support maintenance. Continued use of the rifle could result In injury to, or death of, personnel.	

2-7. PREVENTIVE MAINTENANCE CHECKS AND SERVICES (CONT).

PREVENTIVE MAINTENANCE CHECKS AND SERVICES FOR M16A2 RIFLE (CONT)

Item No.	Interval	Item to be Checked or Serviced	Procedure	Not Fully Mission Capable if:
2	Quarterly (cont)	Charging handle assembly and selector lever (cont)	Pull charging handle (1) to rear. Check that chamber is clear. Let bolt carrier assembly (2) close. Leave hammer in cocked position. Do not pull trigger	Charging handle does not lock in place when in the forward position.
			Place selector lever (3) in SAFE position Pull trigger. Place selector lever (3) in SEMI position.	Hammer falls.
			NOTE For the purpose of the following test, "SLOW" is defined as 1/4 to 1/2 the normal rate of trigger release.	
			Pull trigger	Hammer does not fall.
			Hold trigger to the rear, charge weapon, and release the trigger with a slow, smooth motion, without hesitations or stops, until the trigger is fully forward (an audible click should be heard).	Hammer falls.
			Repeat the above SEMI position test five times	The weapon malfunctions during any of these five tests.

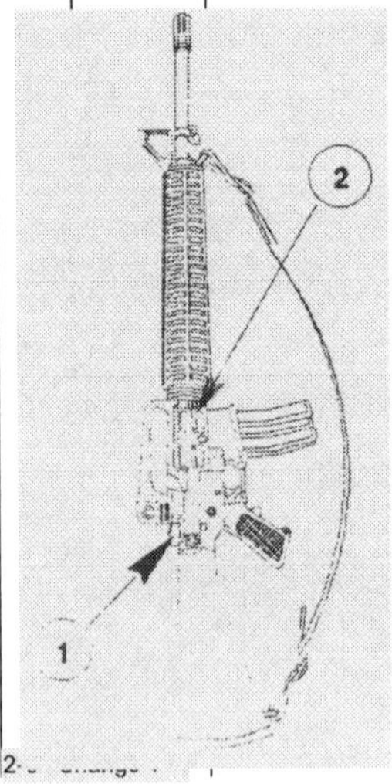

PREVENTIVE MAINTENANCE CHECKS AND SERVICES FOR M16A2 RIFLE (CONT)

Item No.	Interval	Item to be Checked or Serviced	Procedure	Not Fully Mission Capable if:
2	Quarterly (cont)	Charging handle assembly and selector lever (cont)	**M16A2 and M4 ONLY** Place selector lever (3) in BURST position. Charge weapon and squeeze trigger.	Hammer does not fall.
			Hold trigger to the rear, pull the charging handle to the rear and release it three times. Release trigger. Squeeze trigger.	Hammer does not fall.
			NOTE The burst disconnector should have held the hammer to the rear when it engaged the deep notch of the burst cam.	
			M4A1 ONLY Place selector lever (3) in AUTO position Charge carbine and squeeze trigger Hammer should fall.	Hammer does not fall.
			Hold trigger to the rear, charge carbine, and release trigger. Squeeze trigger. Hammer should not fall.	Hammer falls.

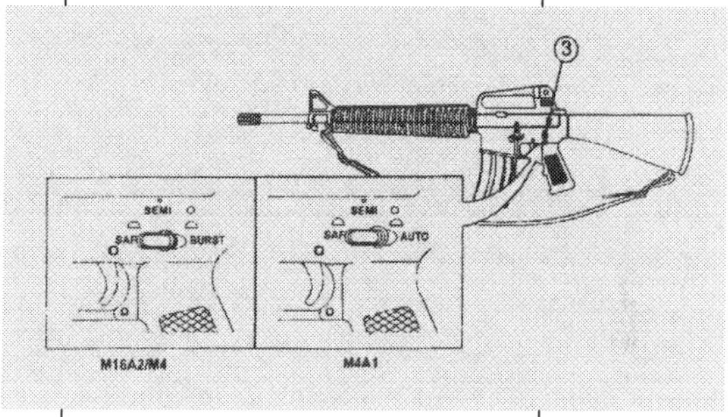

2-7. PREVENTIVE MAINTENANCE CHECKS AND SERVICES (CONT).

PREVENTIVE MAINTENANCE CHECKS AND SERVICES FOR M16A2 RIFLE (CONT)

Item No.	Interval	Item to be Checked or Serviced	Procedure	Not Fully Mission Capable if:
2	Quarterly (cont)	Charging handle assembly and selector lever (cont)	**NOTE** Automatic sear should have released hammer while holding trigger in the squeezed position before releasing and resqueezing the trigger. **All weapons** With hammer in forward position, using moderate finger/thumb pressure attempt to place the selector lever (3) in SAFE position	Moderate finger/thumb pressure moves selector lever to SAFE position.

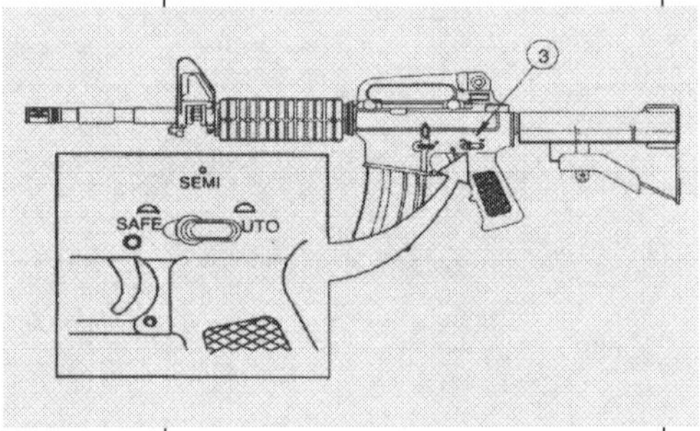

2-7. PREVENTIVE MAINTENANCE CHECKS AND SERVICES (CONT).

PREVENTIVE MAINTENANCE CHECKS AND SERVICES FOR M16A2 RIFLE (CONT)

Item No.	Interval	Item to be Checked or Serviced	Procedure	Not Fully Mission Capable if:
3	Quarterly	Upper receiver and barrel assembly (handguard assemblies) 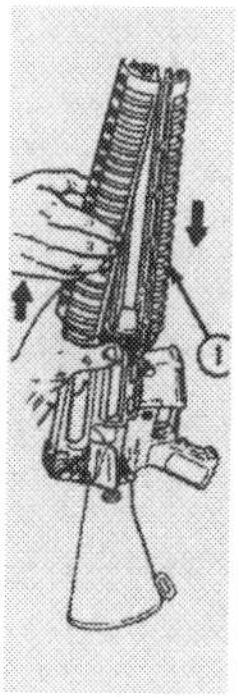	**CAUTION** Do not use screwdriver or any other tool when removing the handguard assemblies, doing so may damage the handguard assemblies and/or slip. **NOTE** Refer to operator's manual for "buddy system" procedure on removing handguard assemblies. Remove and Inspect handguard assemblies (1) internally and externally for cracks and/or damage. Cracks are acceptable providing they do not extend into the handguard retaining flange, or adversely affect rifle operation or operator safety or proper retention of handguard assembly. Discard and replace the handguard assembly (1) If the heatshield is loose enough to rattle when Installed on rifle.	Handguard missing or unserviceable.

PREVENTIVE MAINTENANCE CHECKS AND SERVICES FOR M16A2 RIFLE (CONT)

Item No.	Interval	Item to be Checked or Serviced	Procedure	Not Fully Mission Capable if:
4	Quarterly	Upper receiver and barrel assembly (serviceability check) 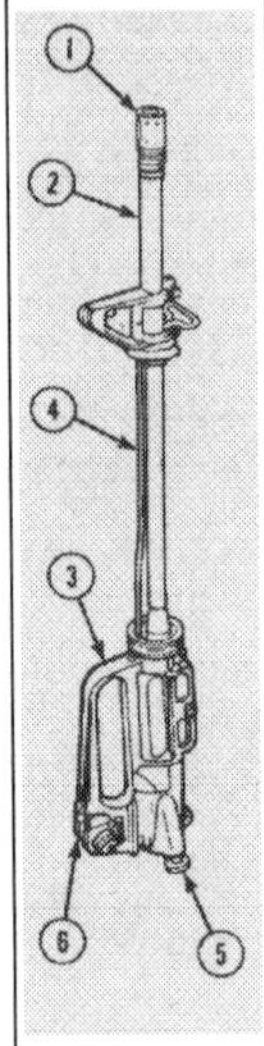 	**WARNING** Dry cleaning solvent is flammable and toxic and should be used in a well ventilated area The use of rubber gloves is necessary to protect the skin when washing rifle parts **CAUTION** Damage may occur if excessive force is used to release takedown pin or pivot pin. Use hand pressure ONLY. Release takedown pins and open and separate receivers Hand check compensator (1) for looseness on barrel (2), then hand check barrel (2) for looseness on upper receiver (3). Check center slot of compensator for alignment (p 2-50). If compensator or barrel is loose, evacuate to support maintenance Check gas tube (4), forward assist assembly (5), and rear sight assembly (6) for damage, The rear sight spring should retain the rear sight assembly (6) In either position with firmness SH - If damaged, evacuate to support maintenance.	Compensator or barrel is loose

2-7. PREVENTIVE MAINTENANCE CHECKS AND SERVICES (CONT).

PREVENTIVE MAINTENANCE CHECKS AND SERVICES FOR M16A2 RIFLE (CONT)

Item No.	Interval	Item to be Checked or Serviced	Procedure	Not Fully Mission Capable if:
4	Quarterly (cont)	Upper receiver and barrel as sembly (service ability check) (cont)	**NOTE** If front or rear sight is moved, return to original position Check front sight post, de tent, and spring (7) for dam age and corrosion Clean and lubricate Check charg ing handle (8) and election port cover (9) for defects and proper function Check sling swivel (10) and rivet (11) for damage and proper function SH Other corn ponents are defective, re place as necessary.	Charging handle (8) is defective.

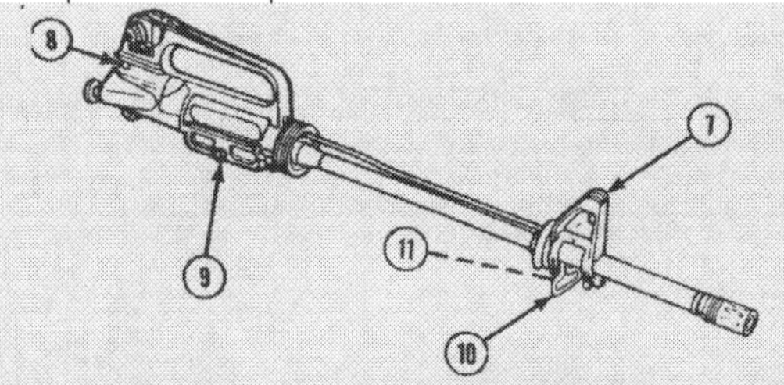

CAUTION

Do not use a wire brush to roughen surfaces Use a well ventilated area during cleaning and application of solid film lubricant. If solid film lubricant comes in contact with moving parts or functioning surfaces of the rifle, remove lubricant immediately by washing with dry cleaning solvent.

NOTE

Shiny metal exterior surfaces of the rifle should be recoated with solid film lubricant (Item 21, app D). Clean surface with dry cleaning solvent (item 16, app D); dry, roughen with abrasive cloth (Item 13, app D) and apply solid film lubricant.

PREVENTIVE MAINTENANCE CHECKS AND SERVICES FOR M16A2 RIFLE (CONT)

Item No.	Interval	Item to be Checked or Serviced	Procedure	Not Fully Mission Capable if:
4	Quarterly (cont)	Upper receiver and barrel assembly (serviceability check) (cont) 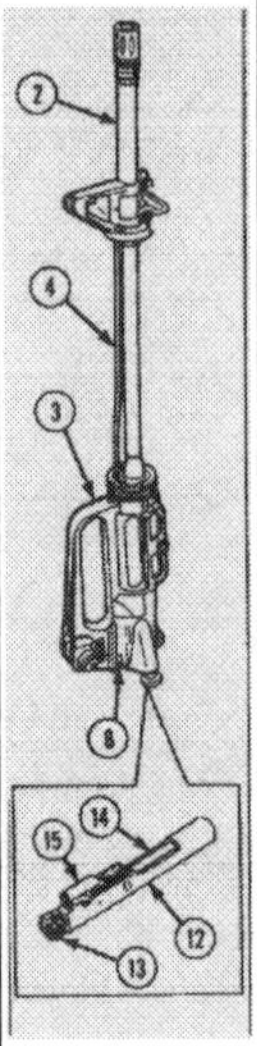	Inspect upper receiver (3) finish for scratches or worn shiny spots. If scratched or worn shiny in spots, disassemble and remove all lubricant from surface with dry cleaning solvent (item 16, app D). Wear rubber gloves (item 18, app D) and use a wash pan (item 24, app D) to apply solvent. Let parts dry thoroughly. Roughen the surface using abrasive cloth (item 1 3, app D) and apply solid film lubricant (item 16, app D). Allow 16 to 24 hours to dry before handling. Hold barrel (2) at 40-degree angle (muzzle down). Pull charging handle (8) to rear. Hold bolt carrier assembly (12) to rear and push charging handle forward. Release bolt carrier assembly (12). The bolt carrier assembly should close and lock under its own weight. If It does not, remove the bolt assembly (13) from the key and bolt carrier assembly (14) and slide the key and bolt carrier assembly (without bolt) back and forth In the upper receiver and barrel assembly. If the gas tube (4) hits the carrier key (1 5), or If the gas tube binds in the carrier key, try to correct the malfunction by adjusting (slightly bending) the gas tube In the area of the handguard assemblies. If unable to adjust, evacuate to support maintenance.	Adjustment does not correct the malfunction.

2-7. PREVENTIVE MAINTENANCE CHECKS AND SERVICES (CONT).

PREVENTIVE MAINTENANCE CHECKS AND SERVICES FOR M16A2 RIFLE (CONT)

Item No.	Interval	Item to be Checked or Serviced	Procedure	Not Fully Mission Capable if:
4	Quarterly (cont)	Upper receiver and barrel assembly (serviceability) check) (cont)	**M4/M4A1** Inspect carrying handle assembly (16) and mounting surface (17) of upper receiver for damage. If the carrying handle is missing or can not be correctly mounted, repair as authorized or evacuate to support maintenance. Inspect carrying handle assembly (16) to insure unit applied identification (ID) code matches unit applied ID code on carbine. If it doesn't match locate correct carrying handle assembly and match up to correct M4/ M4A1 carbine. If a match can not be found, the weapon should be re-zeroed by the operator.	

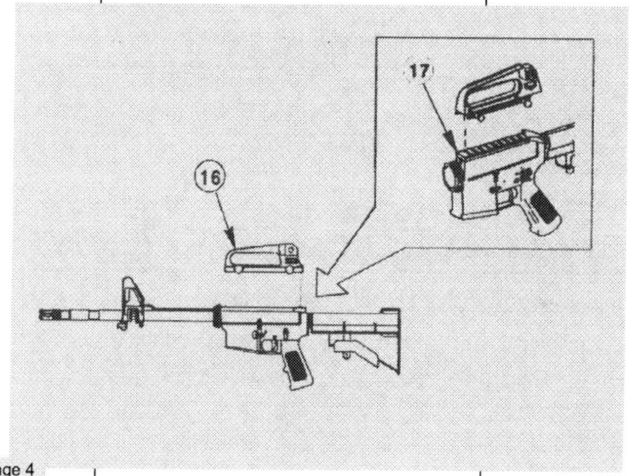

2-7. PREVENTIVE MAINTENANCE CHECKS AND SERVICES (CONT).

PREVENTIVE MAINTENANCE CHECKS AND SERVICES FOR M16A2 RIFLE (CONT)

Item No.	Interval	Item to be Checked or Serviced	Procedure	Not Fully Mission Capable if:
			WARNING Below direct support maintenance, do not interchange bolt assemblies from one rifle to another. Doing so may result in injury to, or death of, personnel.	
5	Quarterly	Key and bolt carrier assembly and bolt assembly (serviceability check)	Remove and disassemble. Visually inspect bolt assembly (1) for cracks, especially in the area of the bolt cam pin hole (2). Check for cracks on locking lugs (3), for a cluster of pits or chipped bolt face (4), and for an elongated firing pin hole (5). If cracked or broken, evacuate to support maintenance for repair.	Defects are found.
			Check for worn or missing bolt rings (6) Check for proper staggering of bolt rings Insert the bolt assembly (1) into the key and bolt carrier assembly (7). Turn key and bolt carrier assembly (7) so the bolt assembly (1) points down. The bolt assembly must not drop out. Remove bolt assembly (p 2-35). Check for broken or missing firing pin retaining pin (8) and bolt cam pin (9); replace as necessary.	The bolt assembly drops out of the key and bolt carrier assembly due to its own weight. Missing or broken firing pin retaining pin or bolt cam pin.

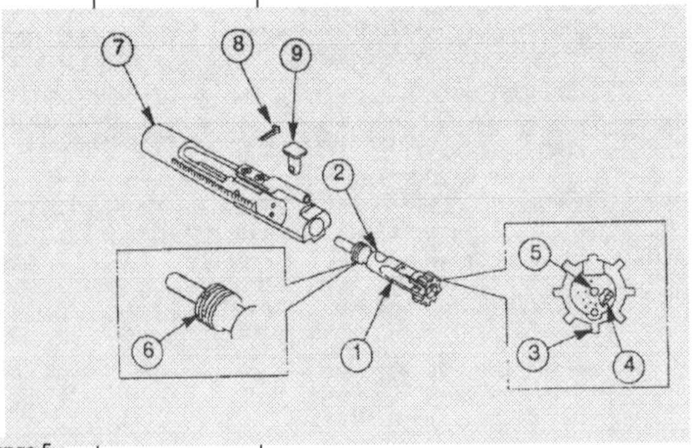

PREVENTIVE MAINTENANCE CHECKS AND SERVICES FOR M16A2 RIFLE (CONT)

Item No.	Interval	Item to be Checked or Serviced	Procedure	Not Fully Mission Capable if:
5	Quarterly (cont)	Key and bolt carrier assembly and bolt assembly (serviceability check) (cont)	Check cartridge extractor (10), extractor spring assembly (11), cartridge ejector (12), and ejector spring (13) for dirt and serviceability. If dirty, clean, lubricate and assemble. If unserviceable, replace as necessary.	Parts are missing or unserviceable
			Check key and bolt carrier assembly (7) and carrier key (14) for damage and looseness. If damaged or loose, evacuate to support maintenance	Key and bolt carrier assembly or carrier key is damaged, or carrier key is loose.
			NOTE If carrier key Is dented, evacuate to support maintenance.	
			Check firing pin 15) for chips or breaks If damaged, evacuate to support maintenance.	Firing pin is damaged.
			Pits or wear in area Illustrated (16) is permissable	

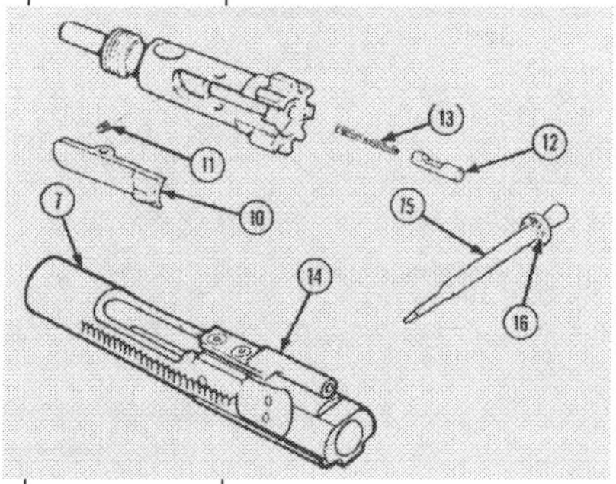

2-7. PREVENTIVE MAINTENANCE CHECKS AND SERVICES (CONT).

PREVENTIVE MAINTENANCE CHECKS AND SERVICES FOR M16A2 RIFLE (CONT)

Item No.	Interval	Item to be Checked or Serviced	Procedure	Not Fully Mission Capable if:
6	Quarterly	Lower receiver and buttstock assembly (serviceability check)	Remove buffer assembly (1) and action spring (2) Check buffer assembly for cracks SH - Buffer assembly is cracked Check action spring (2) for kinks and free length Free length should be **RIFLE**: 11 3/4 Inches (29 85 cm) minimum to 13 1./2 inches (34 29 cm) maximum SH - If action spring is kinked or does not meet free length requirements **CARBINE**: 10 1/16 inches (25.56 cm) minimum to 11 1/4 inches (28.58 cm) maximum. Do not attempt to adjust spring length SH - If action spring is kinked or does not meet free length requirements	

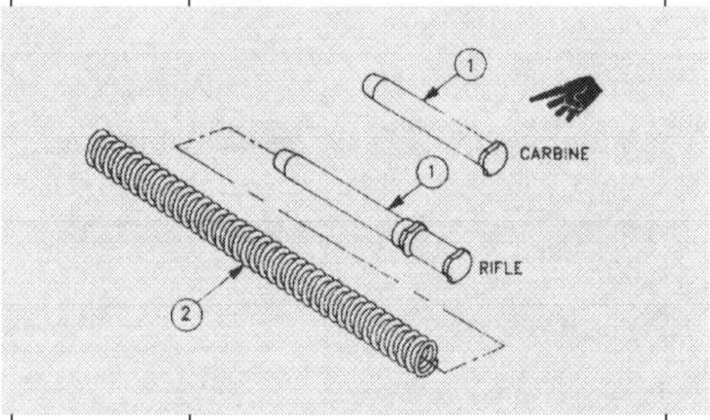

PREVENTIVE MAINTENANCE CHECKS AND SERVICES FOR M16A2 RIFLE (CONT)

Item No.	Interval	Item to be Checked or Serviced	Procedure	Not Fully Mission Capable if:
6	Quarterly (cont)	Lower receiver and buttstock assembly (serviceability check) (cont)	Remove pistol grip screw (3), lockwasher (4), pistol grip (5), helical spring (6), safety detent (7), pivot pin (8), pivot pin detent (9), and helical spring (10). Clean and lubricate metal components. Also clean and generously lubricate pivot pin holes and spring/detent holes. Replace defective/damaged components as necessary.	Components are defective/damaged.
			Disengage takedown pin (11) and pull out, push back in to re-engage takedown pin (an audible click should be heard). If an audible click is not heard, see page 2-57 for repair.	Components are defective/damaged.

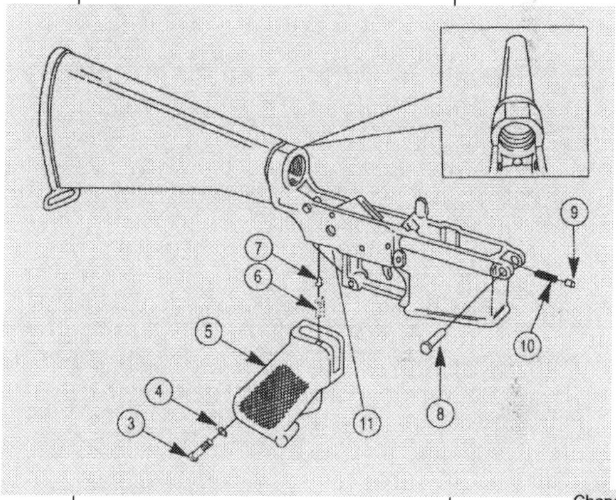

2-7. PREVENTIVE MAINTENANCE CHECKS AND SERVICES (CONT).

PREVENTIVE MAINTENANCE CHECKS AND SERVICES FOR M16A2 RIFLE (CONT)

Item No.	Interval	Item to be Checked or Serviced	Procedure	Not Fully Mission Capable if:
6	Quarterly (cont)	Lower receiver and buttstock assembly (serviceability check) (cont)	Lubricate helical compression spring and takedown pin detent (11) by placing one drop of lubricant on takedown pin detent and lowering the buttstock assembly (12) to vertical position. Allow the lubricant to work Its way around the helical compression spring and takedown pin detent (11). Check buttstock assembly(12) components for damage. 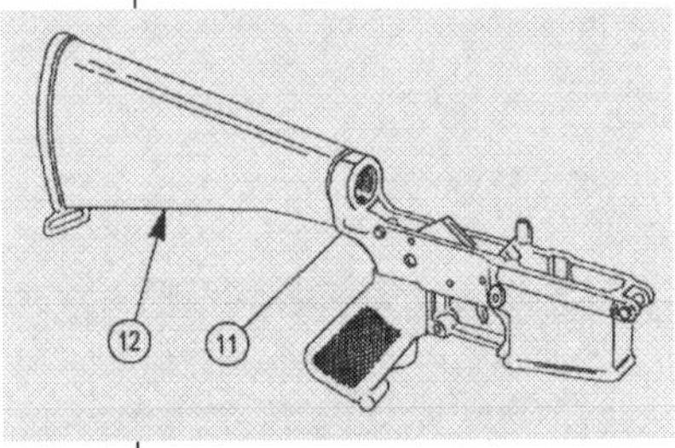 	
			RIFLE ONLY Under the following conditions, hairline cracks (no chipped away material allowed) originating from the buttplate end of the buttstock are acceptable.	Components are damaged.

PREVENTIVE MAINTENANCE CHECKS AND SERVICES FOR M16A2 RIFLE (CONT)

Item No.	Interval	Item to be Checked or Serviced	Procedure	Not Fully Mission Capable if:
6	Quarterly (cont)	Lower receiver and buttstock assembly (serviceability check) (cont)	**RIFLE ONLY** a. One hairline crack, not to exceed 1 in. (2.54 cm) in length, per side of buttstock	The buttstock is cracked in the critical area or does not meet the crack criteria.
			b. Two additional hairline cracks up to 0.25 in. (0.64 cm) in length, per side of buttstock.	
			c. A total of three cracks per side of the buttstock, originating from the buttplate end, are allowable.	
			Cracks in the critical area at the front end of the buttstock are not acceptable.	
			Check buttstock assembly (12) for forward to rear movement and/or a 1/32 in. (0.079 cm) gap between the buttstock assembly (12) and the lower receiver (13). If forward to rear movement and/or a 1/32 in. (0.079 cm) gap appears, tighten self-locking screw. If still not tight, remove buttstock assembly and check for loose lower receiver extension. If loose evacuate to support maintenance. If not loose, replace buttplate (14).	Lower receiver extension cannot be tightened.

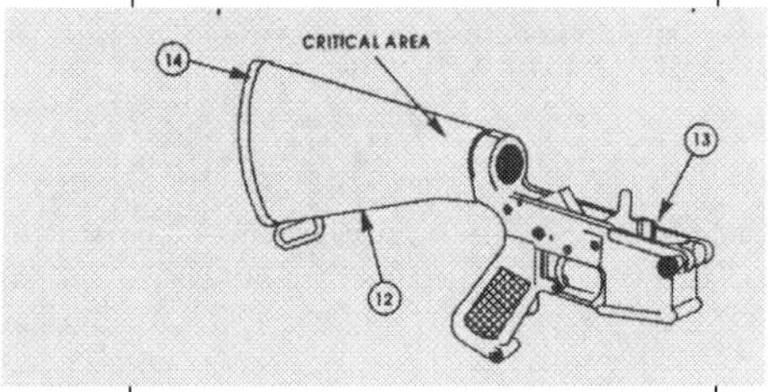

2-7. PREVENTIVE MAINTENANCE CHECKS AND SERVICES (CONS).

PREVENTIVE MAINTENANCE CHECKS AND SERVICES FOR M16A2 RIFLE (CONT)

Item No.	Interval	Item to be Checked or Serviced	Procedure	Not Fully Mission Capable if:
6	Quarterly (cont)	Lower receiver and buttstock assembly (serviceability check) (cont)	**RIFLE ONLY** Small amounts of side-to-side, up-and-down or rotational movement of the buttstock assembly is acceptable. (1) Cracks visible around the butt-plate mounting holes while screws are mounted SH - Cracks are visible around mounting holes when installed on rifle. (2) Cracks or separations around the door assembly are visible when the door assembly is closed SH - Cracks are visible when door assembly is closed. (3) If buttplate is cracked in excess of 0.25 in (O 64 cm) in length and extends through the buttplate (14). see pg. 2-64 for repair SH - If cracked in excess of 0 25 in. extends thru buttplate (4) The buttplate (14) should not be removed other than for repair or replacement of parts at which time a new self-locking screw, NSN 5305-01-147-8585, must be used.	

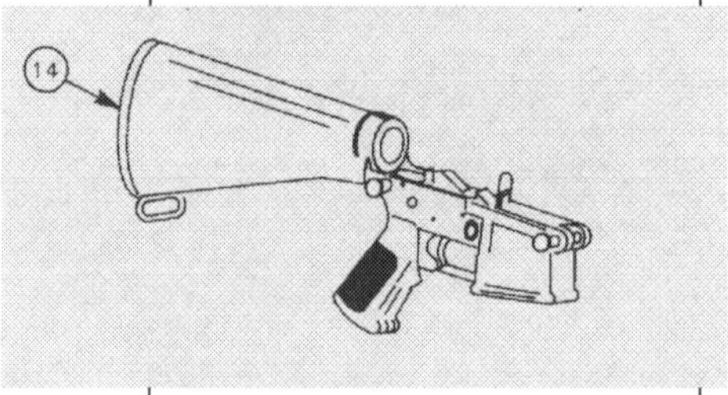

PREVENTIVE MAINTENANCE CHECKS AND SERVICES FOR M16A2 RIFLE (CONT)

Item No.	Interval	Item to be Checked or Serviced	Procedure	Not Fully Mission Capable if:
6	Quarterly (cont)	Lower receiver and buttstock assembly (serviceability check) (cont)	**NOTE** If a weapons lower receiver is missing one third or more of its exterior protective finish, resulting in an unprotected/light reflecting surface, it is candidate for overhaul. This missing finish will be considered a shortcoming. This shortcoming requires action to obtain a replacement weapon. Once a replacement has been received, evacuate the original weapon to depot for overhaul. If scratched or worn shiny in spots, repair in the same manner as outlined for upper receiver (see item 4 above).	
7	Quarterly	M16A2 Rifle	Assemble as in TM 9-1005-319-10 (operator's manual). Check sling for damage. If damaged, replace. Check for improperly assembled, broken, missing, or damaged parts. Check over all general appearance. Replace parts as required and authorize evacuation to support maintenance for repair.	

2-7. PREVENTIVE MAINTENANCE CHECKS AND SERVICES (CONT).

~~PREVENTIVE MAINTENANCE CHECKS AND SERVICES FOR M16A2 RIFLE (CONT)~~

Item No.	Interval	Item to be Checked or Serviced	Procedure	Not Fully Mission Capable if:
8	Quarterly	Annual DS safety and serviceability inspection and gaging	Check to ensure annual DS safety and serviceability inspection and gaging has been done and that the next gaging and inspection is scheduled. If annual gaging has not been performed within the last year, notify support maintenance.	Annual gaging has not been performed.

Section IV. TROUBLESHOOTING

2-8. GENERAL.

a. This section contains unit level troubleshooting information for locating and correcting most of the operating troubles which may develop in the M16A2 rifle, M4/M4A1 carbine. Each malfunction for the individual part or assembly is followed by a list of tests or inspections which will help you to determine the corrective actions in the order listed.

b. This manual cannot list all malfunctions that may occur, nor all tests or inspections and corrective actions. If a malfunction is not listed or is not corrected by listed corrective actions, see individual repair sections in the maintenance procedures on each major assembly.

2-9. TROUBLESHOOTING PROCEDURES. Refer to troubleshooting table for malfunctions, tests, and corrective actions. The symptom index is provided for a quick reference of the malfunctions covered in the table.

SYMPTOM INDEX

	Troubleshooting Procedures Page
Failure of magazine to lock in rifle	2-21
Failure to feed	2-22
Failure to chamber	2-22
Failure to lock	2-23
Failure to fire	2-24
Failure to unlock	2-26
Failure to extract	2-26
Failure to eject	2-27
Failure to cock	2-28
Short recoil	2-28
Rifle cannot be zeroed	2-30
Failure to cycle with selector lever set on BURST	2-31
Fires two rounds with one pull of trigger with selector lever set on SEMI (double firing)	2-31
Fires with selector lever on SAFE or when trigger is released with selector lever on SEMI	2-31
Bolt assembly falls to lock to rear after firing last round	2-32

TROUBLESHOOTING

MALFUNCTION
TEST OR INSPECTION
CORRECTIVE ACTION

1. FAILURE OF MAGAZINE TO LOCK IN RIFLE.

Step 1. Dirty or corroded magazine catch (1).

Disassemble and clean.

Step 2 Defective magazine catch spring (2)

Evacuate to support maintenance.

Step 3 Worn or broken magazine catch (1).

Evacuate to support maintenance

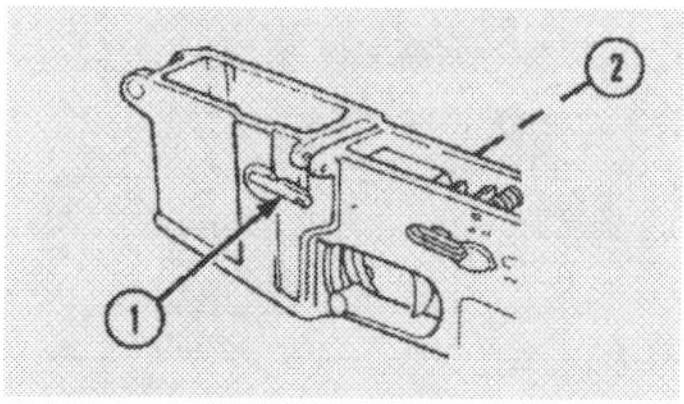

2-9. TROUBLESHOOTING PROCEDURES (CONT).

TROUBLESHOOTING (CONT)

MALFUNCTION
TEST OR INSPECTION
CORRECTIVE ACTION

2. FAILURE TO FEED

Step 1. Magazine catch spring weak or broken

Evacuate to support maintenance

Step 2. Magazine catch (1) defective.

Evacuate to support maintenance

Step 3. Magazine catch (1) out of adjustment (will not retain magazine)

Refer to operator's manual.

Step 4. Short recoil.

Refer to page 2-28

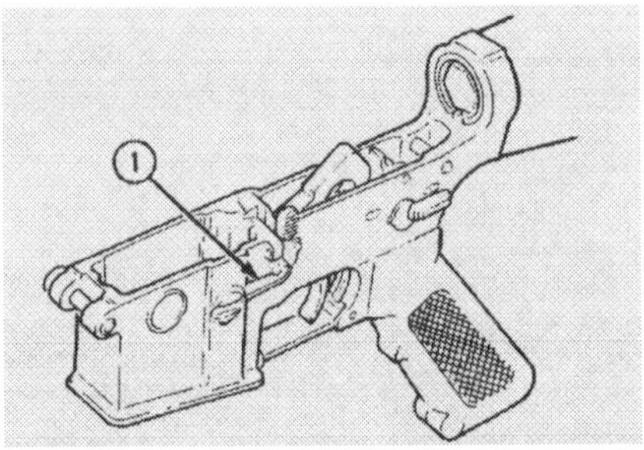

3. FAILURE TO CHAMBER.

Step 1. Weak or broken action spring (1), RIFLE ONLY (free length 11 3/4 inches (29.85 cm) minimum to 13 1.'2 Inches (34 29 cm) maximum) CARBINE ONLY (10 1 16 inches (2S 56 cm) minimum to 11 1,4 inches (28 58 cm) maximum)

Replace action spring (p 2-57)

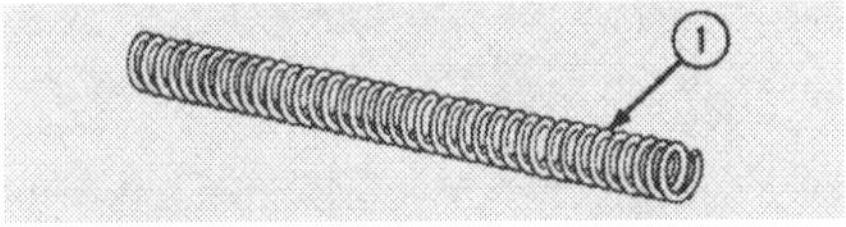

TROUBLESHOOTING (CONT)

MALFUNCTION
TEST OR INSPECTION
CORRECTIVE ACTION

Step 2. Short recoil

Refer to page 2-28

4. FAILURE TO LOCK

Step 1. Bolt cam pin (1) missing

Replace (p 2-35)

Step 2. Loose or damaged bolt carrier key (2).

a. Evacuate to support maintenance.

b. Dented bolt carrier key may be repaired (p 2 35).

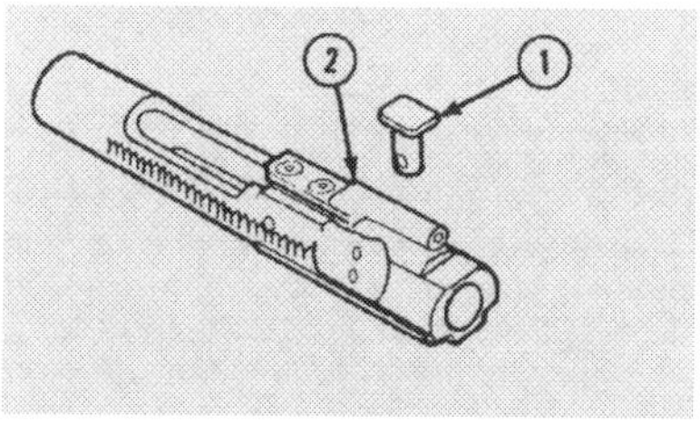

Step 3. Improperly assembled extractor spring assembly (3)

Assemble correctly (p 2-38).

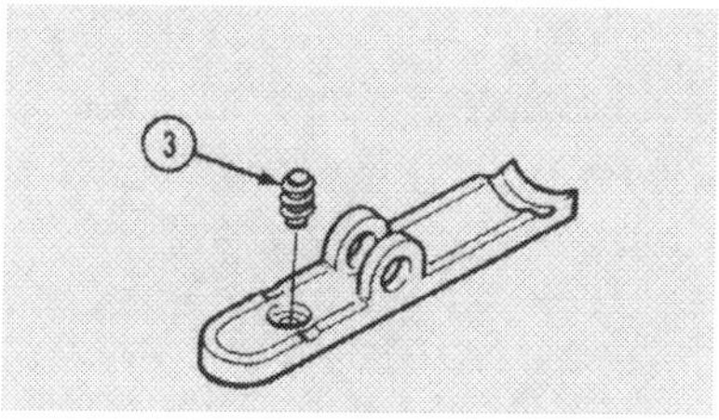

2-9. TROUBLESHOOTING PROCEDURES (CONT).

TROUBLESHOOTING (CONT)

MALFUNCTION
TEST OR INSPECTION
CORRECTIVE ACTION

4. FAILURE TO LOCK (CONT)

Step 4. Bent gas tube 14J

a Adjust to Its original configuration by bending in area of hand-guard assembly.

b If the gas tube cannot be returned to Its original configuration, evacuate the rifle to support maintenance

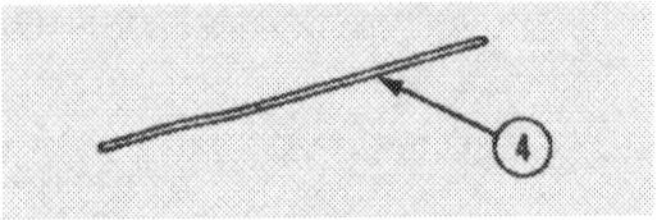

Step 5 Weak or broken action spring (5); RIFLE ONLY: (free length 11 3/4 inches (29,85 cm) minimum to 13 1,/2 Inches (34 29 cm) maximum). CARBINE ONLY (10 1/16 Inches (25 56 cm) minimum to 11 1,4 inches (28 58 cm) maximum

Replace action spring (p 2-571

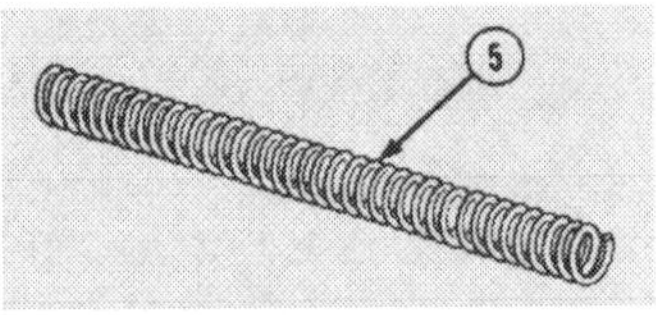

5. FAILURE TO FIRE.

Step 1 Broken or chipped firing pin (1).

Evacuate to support maintenance.

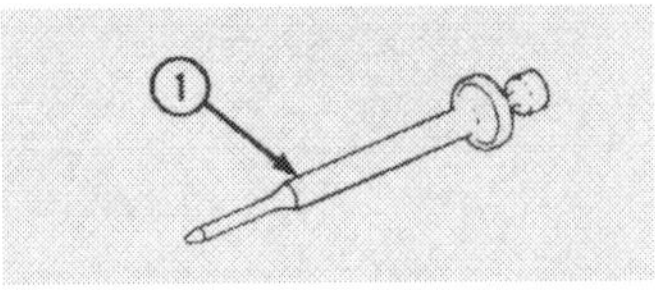

TROUBLESHOOTING (CONT)

MALFUNCTION
TEST OR INSPECTION
CORRECTIVE ACTION

Step 2 Carbon buildup in firing pin recess inside bolt assembly

Remove cartridge extractor and clean recess with pipe cleaner (Item 11, app D), refer to operator's manual.

Step 3. Firing mechanism (2) and or lower receiver assembly (31 improperly assembled or has worn, broken, or missing parts.

Evacuate to support maintenance

Step 4. Broken, defective, or missing firing pin retaining pin (4).

Replace (p 2-35)

Step 5. Selector lever (5) frozen on SAFE position.

Evacuate to support maintenance

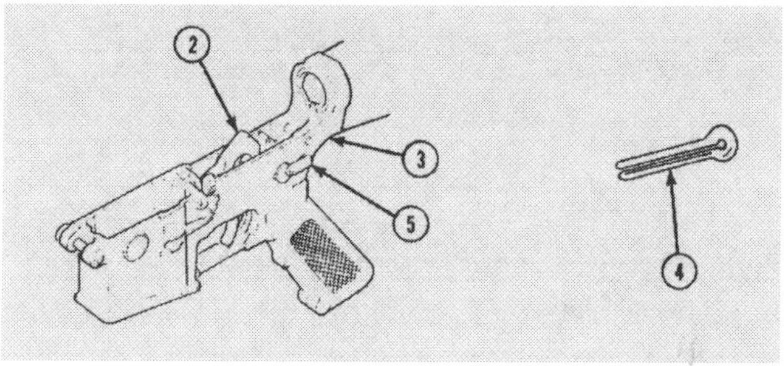

2-9. TROUBLESHOOTING PROCEDURES (CONT).

TROUBLESHOOTING (CONT)

MALFUNCTION
TEST OR INSPECTION
CORRECTIVE ACTION

6. FAILURE TO UNLOCK.

Step 1. Burred locking lugs (1) on bolt assembly.

Remove burrs.

Step 2. Burred lugs (2) on barrel extension

Remove burrs.

Step 3 Short recoil.

Refer to page 2-28.

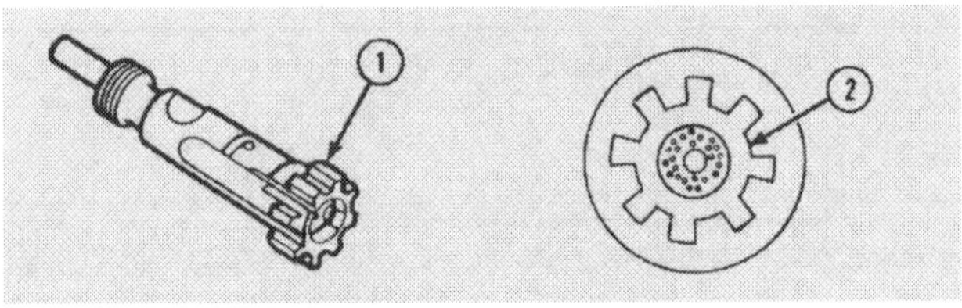

7. FAILURE TO EXTRACT.

Step 1. Defective extractor pin (1), cartridge extractor (2), and or extractor spring assembly (3).

Replace extractor pin (1), cartridge extractor (2), and or extra(:tor spring assembly (3) (p 2-381

Step 2. Short recoil

Refer to page 2-28

NOTE

Rubber insert and spring are an assembly Illustration shows Insert out of assembly for clarification only Do not remove the rubber Insert from the extractor spring assembly.

TROUBLESHOOTING (CONT)

MALFUNCTION
TEST OR INSPECTION
CORRECTIVE ACTION

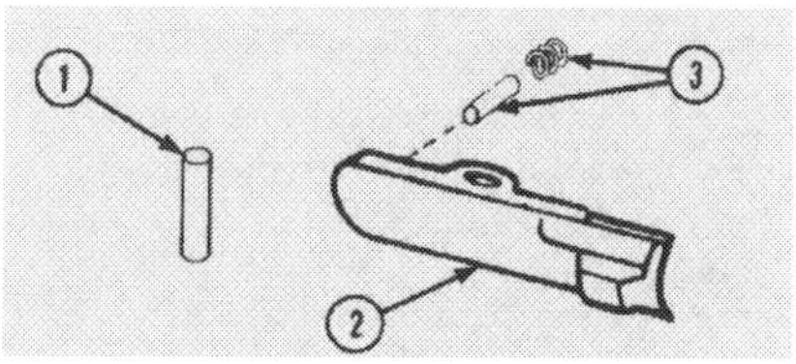

8. FAILURE TO EJECT.

Step 1. Broken cartridge ejector (1).

Replace (p 2-38).

Step 2. Cartridge ejector (1) stuck in bolt body (2).

Disassemble and clean (p 2-38).

Step 3. Weak or broken ejector spring (3).

Replace (p 2-38)

Step 4. Short recoil

Refer to page 2-28.

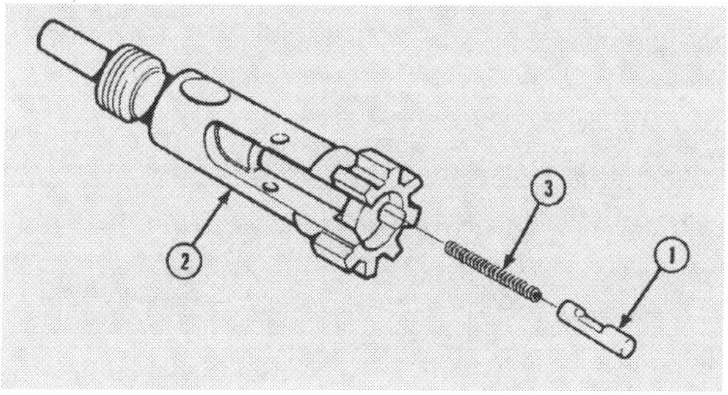

2-9. TROUBLESHOOTING PROCEDURES (CONT).

TROUBLESHOOTING (CONT)

MALFUNCTION
TEST OR INSPECTION
CORRECTIVE ACTION

9. FAILURE TO COCK

Step 1 Worn, broken, or missing parts of firing mechanism

Evacuate to support maintenance.

Step 2 Short recoil

Refer to below

10 SHORT RECOIL

Step 1 Broken or damaged action spring (1).

Replace action spring (p 2-57).

Step 2 Unlubricated or dirty action spring and receiver extension

Clean and lubricate

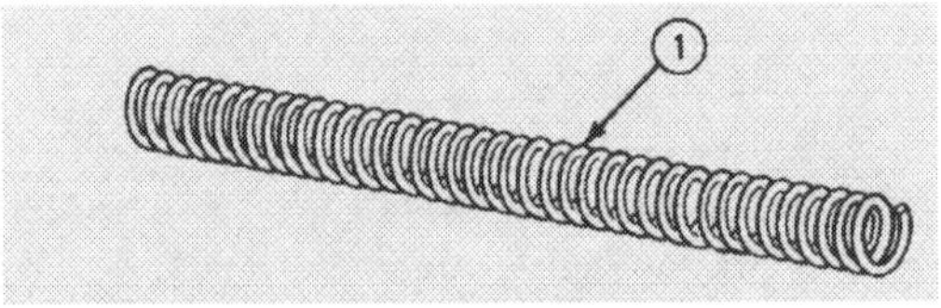

Step 3 Improper gap space or worn, missing, or broken bolt rings (2)

a. Stagger bolt ring gaps (approximately 13 turn apart}

b. Evacuate to support maintenance if bolt rings are worn, broken, or missing.

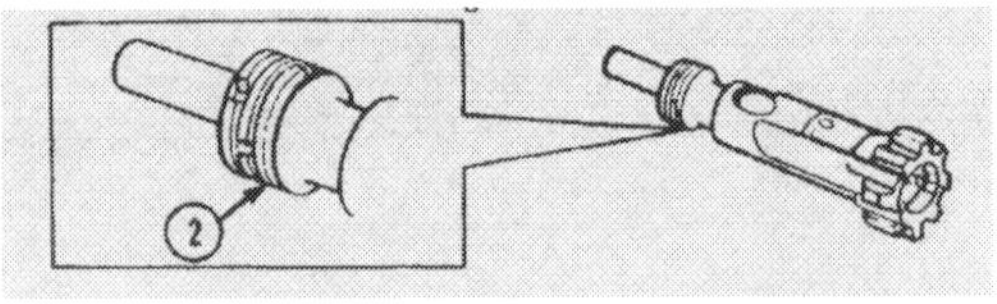

TROUBLESHOOTING (CONT)

MALFUNCTION
TEST OR INSPECTION
CORRECTIVE ACTION

Step 4. Carbon build-up or foreign matter In the narrow passage of the bolt carrier key (3).

Clean with CLP (Item 9, app D) and a pipe cleaner (Item 11, app D).

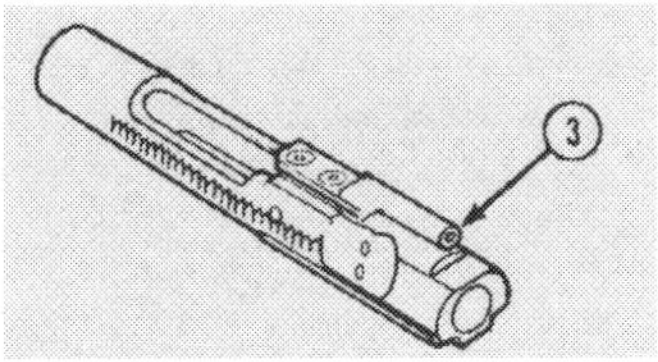

Step 5. Gas leakage caused by broken or loose gas tube (4) around front sight base.

Evacuate to support maintenance.

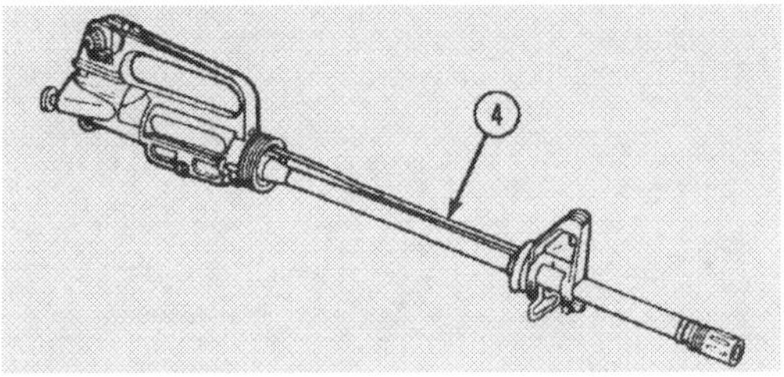

Step 6. Improper alignment of gas tube and .carrier key.

a. Adjust gas tube alignment by bending in area of handguard assembly to Its original configuration

b. If gas tube cannot be returned to Its original configuration, evacuate the rifle to support maintenance.

2-9. TROUBLESHOOTING PROCEDURES (CONT).

TROUBLESHOOTING (CONT)

MALFUNCTION
TEST OR INSPECTION
CORRECTIVE ACTION

11. RIFLE CANNOT BE ZEROED.

Step 1. Defective barrel assembly (1).

Evacuate to support maintenance

Step 2. Barrel assembly out of alignment with rear sight assembly (2) on upper receiver.

Evacuate to support maintenance

Step 3. Defective front sight (3).

Remove front sight post (4), front sight detent (5), and helical spring (6). If damaged, replace.

Step 4. Defective -ear sight assembly (2).
Evacuate to support maintenance.

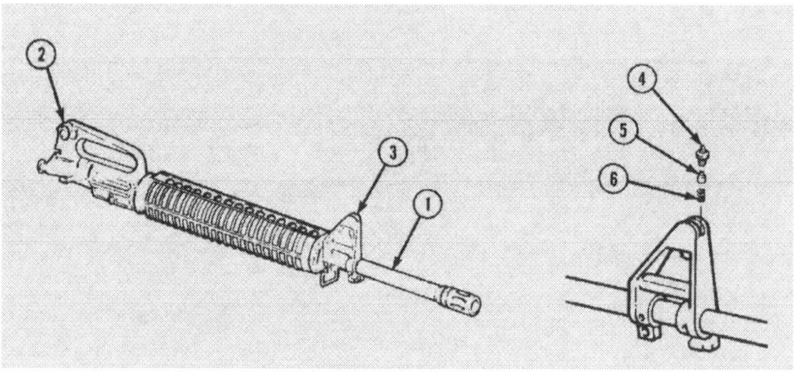

TROUBLESHOOTING (CONT)

MALFUNCTION
TEST OR INSPECTION
CORRECTIVE ACTION

12. FAILURE TO CYCLE WITH SELECTOR LEVER SET ON BURST (M16A2 AND M4).

Faulty selector lever or broken cam, cam clutch spring, or burst disconnector.

Evacuate to support maintenance.

12.1. FAILURE TO CYCLE WITH SELECTOR LEVER SET ON AUTO (M4A1).

Faulty selector lever.

Evacuate to support maintenance.

13. FIRES TWO ROUNDS WITH ONE PULL OF TRIGGER WITH SELECTOR LEVER SET ON SEMI (DOUBLE FIRING).

Perform function test.

If any part of function test (p 2-68) fails, evacuate to support maintenance.

14. FIRES WITH SELECTOR LEVER ON SAFE OR WHEN TRIGGER IS RELEASED WITH SELECTOR LEVER ON SEMI.

Worn, broken, or missing parts of firing mechanism.

Evacuate to support maintenance.

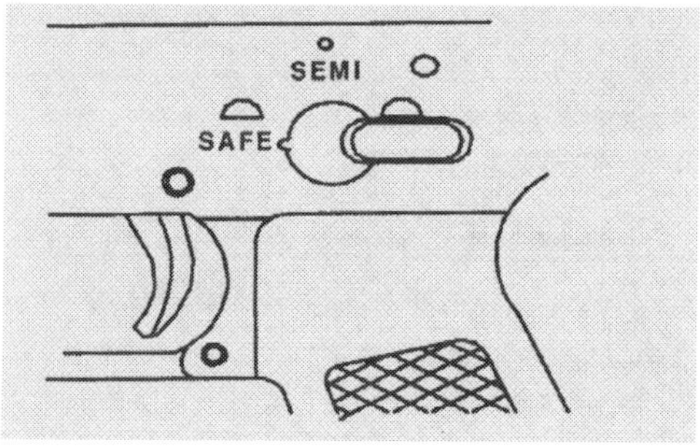

2-9. TROUBLESHOOTING PROCEDURES (CONT).

TROUBLESHOOTING (CONT)

MALFUNCTION
TEST OR INSPECTION
CORRECTIVE ACTION

15 BOLT ASSEMBLY FAILS TO LOCK TO REAR AFTER FIRING LAST ROUND.

Step 1. Magazine follower (1) worn or broken.

Replace magazine.

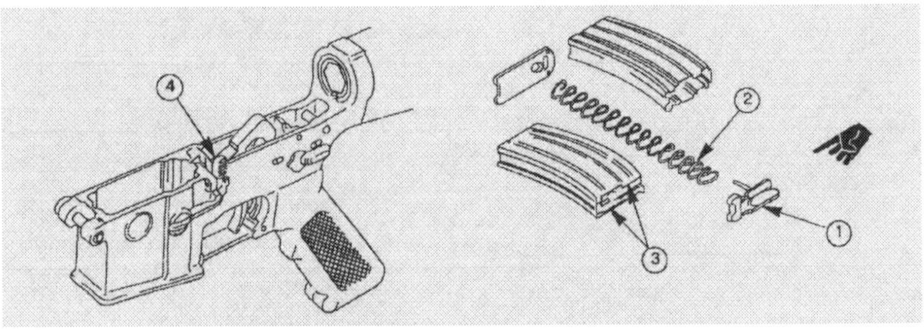

Step 2. Magazine catch spring (2) weak or broken.

Replace magazine.

Step 3. Magazine feeder lips (3) bent or broken.

Replace magazine.

Step 4. Magazine follower (1) binds during operation.

Replace magazine.

Step 5. Broken bolt catch (4) and/or spring.

Evacuate to support maintenance.

Section V. MAINTENANCE PROCEDURES

2-10. INITIAL SETUP. The following information will reduce the space required for the initial setup portion of the maintenance procedures.

a. Materials/Parts required are not listed unless they apply to the procedure

b. Personnel Required is listed only if the task requires more than one person If Personnel Required Is not listed, It means one person can do the job.

c. The normal standard equipment condition Is that the item is removed from the en(1 Item or next higher assembly and is in the assembled condition. Equipment Condition is not listed unless some other condition is required.

d. The approximate time required Is listed on the applicable Maintenance Allocation Chart (MAC).

e. When the term evacuate to support maintenance is used, the entire rifle must be evacuated.

2-11. LUBRICATION GENERAL.

a. Whenever the term "cleaner, lubricant, and preservative (CLP)" or the words "lubricant", "lube", "LSA", or "LAW" are cited In this TM, they are to be interpreted to mean CLP (Item 9, app D), weapons lubricating oil (LSA) (Item 23, app D), or weapons lubricating oil (LAW) (Item 22, app D) can be utilized as applicable The following constraints must be adhered to:

b. Under all but the coldest arctic conditions, LSA or CLP are the lubricants to use on the rifle. Either may be used at 100F (230C) and above. However, do not use both on the same rifle at the same time.

c. LAW is the lubricant to use during cold arctic conditions, + 100F (12C) and below

d. Any of the lubricants may be used from 10°F to + 100F (230C to 12°C)

e. ARMY ONLY: Do not mix lubricants on the same rifle The rifle must be thoroughly cleaned during change from one lubricant to another Dry Cleaning Solvent (SD) (Item 16, app D) is recommended for cleaning during change from one lubricant to another.

f. Rifle Bore Cleaner (RBC) (item 12, app D) may be used to remove carbon buildup In the bore and other portions of the rifle.

2-12. MAJOR COMPONENTS OF M16A2 RIFLE AND CARBINES.

This task covers disassembly

INITIAL SETUP:

References
TM 9-1005-319-10 (operator's manual)

Equipment Conditions
Rifle and Carbines assembled

General Safety Instructions
Before starting an inspection, be sure to clear the weapon. Do not keep live ammunition near the work area.

To avoid injury to your eyes, use care when removing and installing spring-loaded parts.

Below direct support maintenance, do not interchange bolt assemblies from one weapon to another. Doing so may result in injury to, or death of, personnel.

DISASSEMBLY

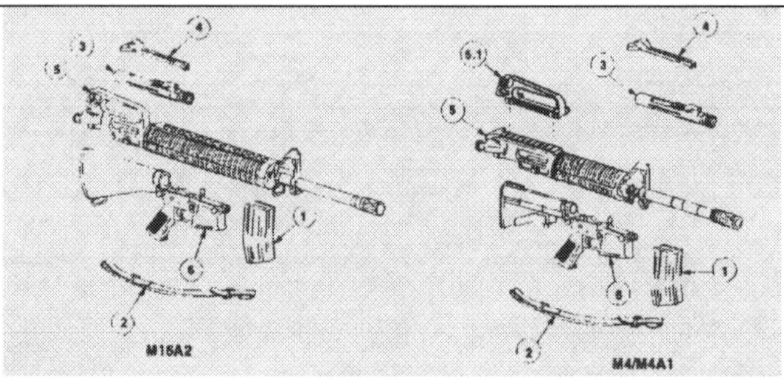

a. Refer to operator's manual.

b. Remove magazine (1), sling (2), bolt carrier assembly (3), charging handle assembly (4), and upper receiver and barrel assembly (5), carrying handle assembly (5.1), from lower receiver and buttstock assembly (6).

NOTE

Solid Film Lubricant (SFL) is the only authorized touchup for the MI 16A2 rifle and M4/M4A1 carbines and may be used on up to one third of the exterior finish of the weapon. **FOR ARMY CONUS USE ONLY AND AIR FORCE TRAINING WEAPONS ONLY**: SFL may be used as touchup without limitation on the upper receiver and barrel assembly. This is to say that units which DO NOT fall under the category of Divisional Combat Units or rapid deployment type units may have up to 100 percent of the exterior surface of the upper receiver and barrel assembly protected with SFL if necessary.

12-13. BOLT CARRIER ASSEMBLY.

This task covers:

a. **Disassembly**
b. **Cleaning**
c. **Inspection/Repair**
d. **Lubrication**
e. **Reassembly**

INITIAL SETUP:

Tools
(ARMY) Small Arms Repairman Tool Kit (item 3, app B)
Bolt Carrier Key Tool (item 11, figure C-17)

References
TM 9-1005-319-10 (operator's manual)

Equipment Conditions
2-34 Bolt carrier assembly removed

General Safety Instructions
Bolt cam pin must be installed or rifle will blow up while firing the first round. If the bolt cam pin is not installed, injury to, or death of, personnel may result.

Do not interchange bolt assemblies from one rifle to another. Doing so may result in injury to, or death of, personnel.

a. DISASSEMBLY

CAUTION

Do not spread or close legs of firing pin retaining pin (1).

1. Remove firing pin retaining pin (1). Tilt key and bolt carrier assembly (2) and catch firing pin (3) as it drops out.
2. Rotate bolt cam pin (4) one quarter turn and lift straight up to remove.
3. Remove bolt assembly (5) from key and bolt carrier assembly (2).

NOTE
For disassembly of bolt assembly (5), see page 2-38.

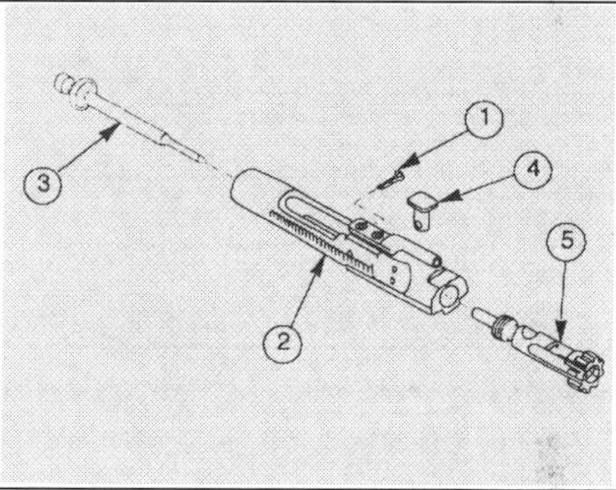

12-13. BOLT CARRIER ASSEMBLY (CONT).

b. CLEANING

Clean all items (operator's manual). Remove carbon deposits.

c. INSPECTION/REPAIR

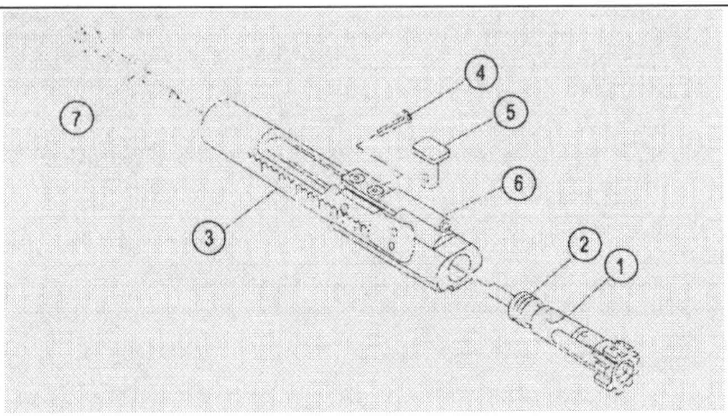

1. Inspect bolt assembly (1) as follows:

   a. Check to see that the bolt ring (2) gaps are staggered.

   b. Install bolt assembly (1) into key and bolt carrier assembly (3). Check for worn rings by holding the key and bolt carrier assembly (3) with the bolt assembly facing down. If bolt assembly falls out after firing pin retaining pin (4) and bolt cam pin (5) are removed, the rings are worn. Evacuate to support maintenance.

2. Inspect carrier key (6) for dents, distortion, or looseness. If dented or loose, evacuate to support maintenance.

3. Inspect firing pin retaining pin (4) and bolt cam pin (5) for cracks, damage, or excessive wear. Replace if unserviceable.

4. Inspect firing pin (7) for damage or If tip is chipped. Evacuate rifle to support maintenance if unserviceable

5. Inspect key and bolt carrier assembly (3) for damage or wear. If unserviceable, evacuate to support maintenance.

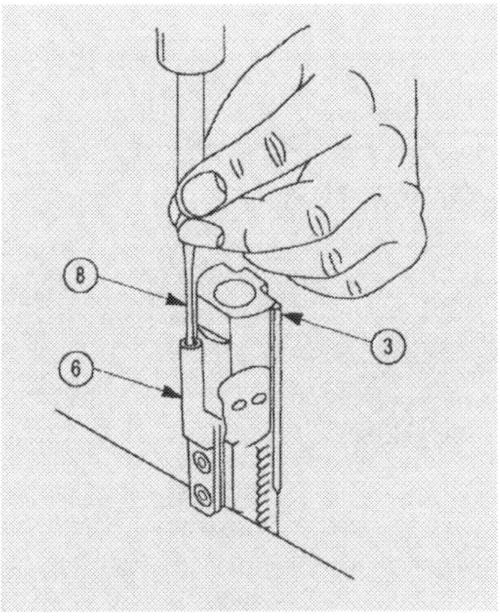

CAUTION

Extreme care must be exercised during the following procedure to assure that the striking force is not directed to the attaching screws and that the tube portion is not enlarged or flared beyond original requirement. Such enlargement would permit loss of gas pressure when the key and gas tube come together during functioning.

6. Repair small dents and/or distortions in carrier key (6) using bolt carrier key tool as follows:

a. Place the key and bolt carrier assembly (3) in a vertical position, supported In a manner that contact is made with the rear surface of the carrier key (6).

b. Insert the small end of the key tool (8) into the tube portion of the carrier key (6).

c. Strike the large end of the key tool (8) lightly with a 3 ounce, soft-brass hammer.

d. Repeat striking (gently) until carrier key (6) is reformed to original configuration.

e. If carrier key (6) cannot be reformed to original configuration, evacuate the weapon to support maintenance.

d. LUBRICATION

Lubricate all Items (p 2 33) (operator's manual)

e. REASSEMBLY

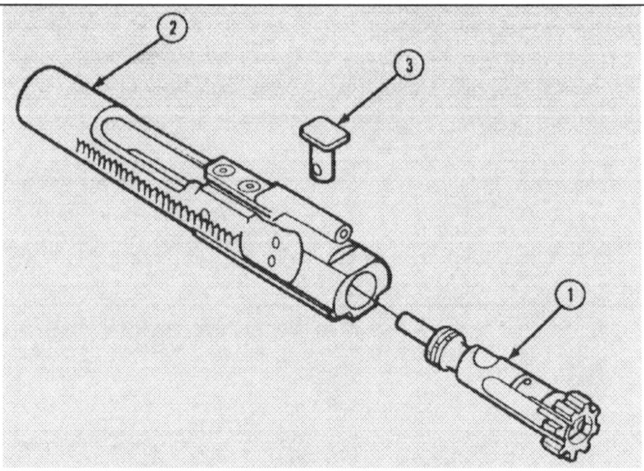

WARNING

Do not Interchange bolt assemblies from one rifle to another Doing so may result In Injury, or death of, personnel.

Bolt cam pin must be Installed or rifle will blow up while firing the first round. If the bolt cam pin is not Installed, Injury to, or death of, personnel may result

NOTE

Before installing bolt assembly, check to see that the bolt ring gaps are staggered to prevent loss of gas pressure

1. Install bolt assembly (1) In key and bolt carrier assembly (2)

2. Install bolt cam pin (3) and rotate one quarter turn

2-13. BOLT CARRIER ASSEMBLY (CONT).

e. REASSEMBLY (CONT)

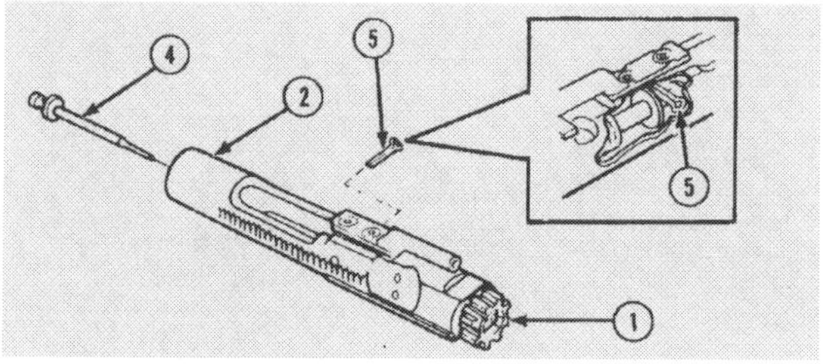

3. Hold key and bolt carrier assembly (2) with bolt assembly (1) down and drop in firing pin (4).

4. Install firing pin retaining pin (5) from left side only to ensure proper installation Check for proper installation by holding key and bolt carrier assembly (2) with bolt assembly (1) up, then attempt to shake out firing pin.

5. Reassemble rifle, refer to page 2-68.

2-14. BOLT ASSEMBLY.

This task covers:

a. **Removal**
b. **Cleaning**
c. **Inspection**
d. **Lubrication**
e. **Reassembly**

INITIAL SETUP:

Tools
(ARMY) Small Arms Repairman Tool Kit (item 3, app B)

Equipment Conditions
2-35 Bolt assembly removed

General Safety Instructions
Do not Interchange bolt assemblies from one rifle to another. Doing so may result in injury to, or death of, personnel.

To avoid injury to your eyes, use care when removing and installing spring-loaded parts.

a. DISASSEMBLY

NOTE

Do not separate cartridge extractor and extractor spring assembly unless replacement of either or both is required

Do not remove the rubber Insert from the extractor spring assembly

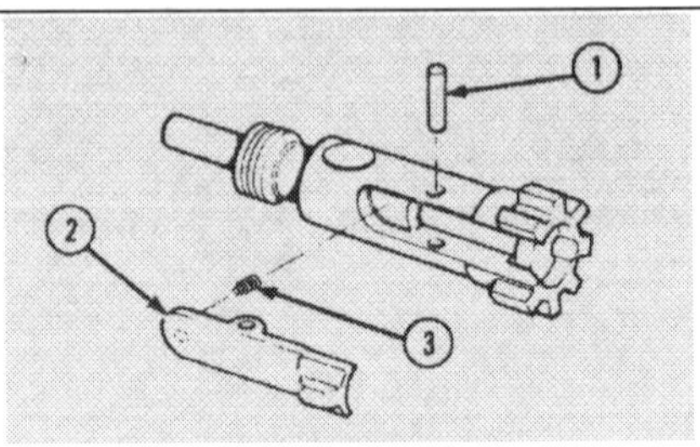

1. Push out extractor pin (1) and remove cartridge extractor (2) and extractor spring assembly (3) as a unit

2. If required, twist extractor spring assembly (3) counterclockwise to remove from cartridge extractor (2).

CAUTION

Be sure to use vise law protective caps

3. Hold bolt body In vise and remove spring pin (4) using 1 16 Inch punch and hammer

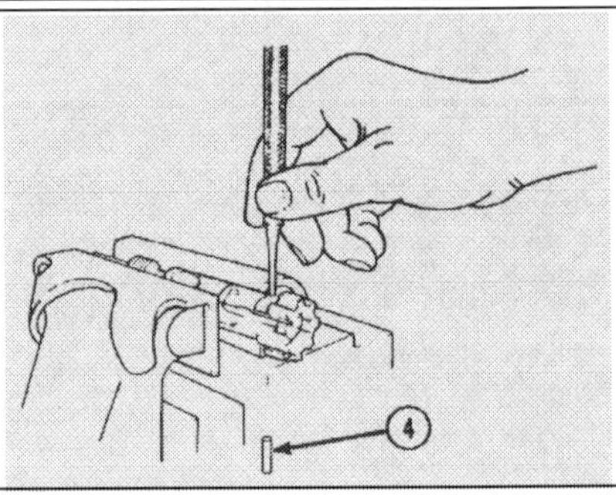

2-14. BOLT ASSEMBLY (CONT).

a. DISASSEMBLY (CONT)

WARNING

To avoid injury to your eyes, use care when removing and installing spring-loaded parts.

4. Remove punch, be sure to catch cartridge elector (5) and ejector spring (6) to prevent loss

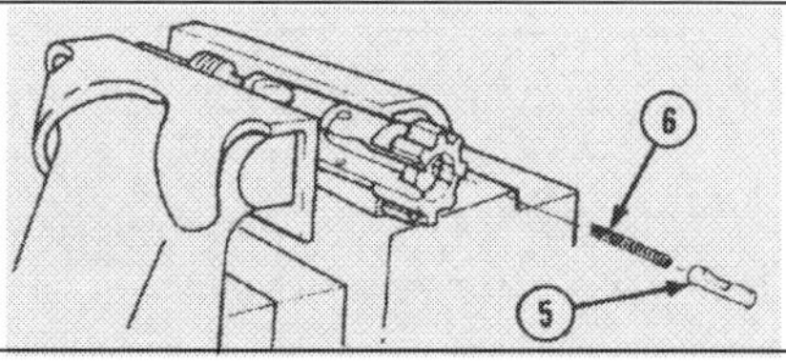

b. CLEANING

CAUTION

Do not distort extractor spring assembly during cleaning.

Clean all Items (operator's manual). Remove carbon deposits.

c. INSPECTION/REPAIR

1. Check bolt rings (1) for damage and proper stagger (gaps approximately 3 turn apart).
2. Inspect for cracks or damage, especially around bolt cam pin hole (2) and locking lugs (3). If cracked or damaged evacuate to support maintenance.
3. Inspect cartridge extractor (4), extractor spring assembly (5), and extractor pin (6) for cracks, breaks, chips, and other damage. Pay close attention to cartridge extractor lip (7). If damaged, replace.
4. Inspect cartridge ejector (8) and ejector spring (9) for cracks, breaks, and chips If damaged, replace.

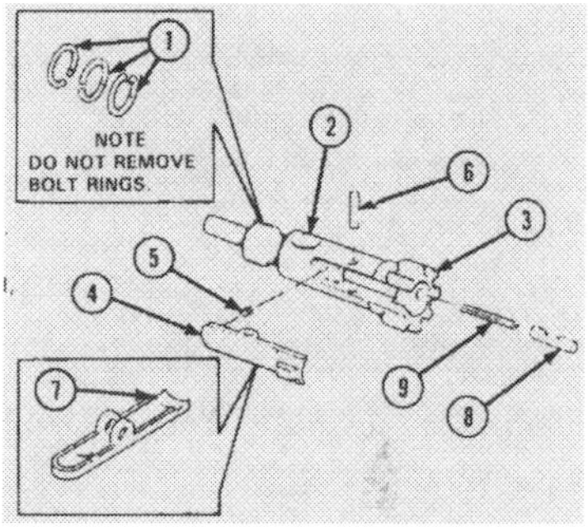

d. LUBRICATION

Lightly lubricate all items (p 2 33) (operator's manual)

e. REASSEMBLY

WARNING

To avoid Injury to your eyes, use care when removing and Installing spring loaded parts.

Do not Interchange bolt assemblies from one rifle to another Doing so may result In injury to, or death of, personnel.

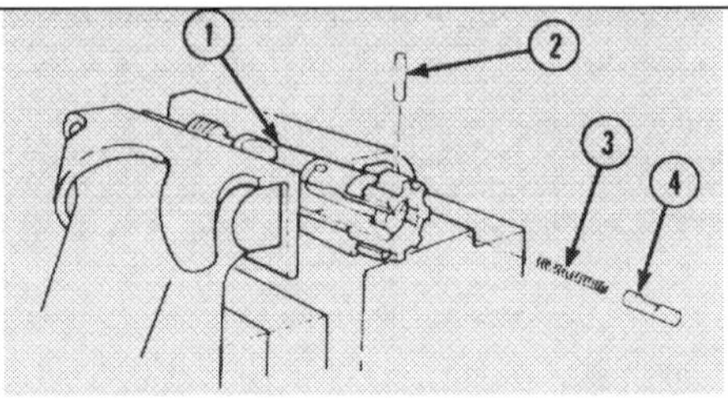

CAUTION

Be sure to use vise jaw protective caps

1. Place bolt body (1) in a vise and start spring pin (2) In hole.
2. Install ejector spring (3) and cartridge ejector (4). Align groove on cartridge ejector (4) so that spring pin 12) can be Installed
3. Compress and hold ejector spring and cartridge ejector In place with a 3 8 inch punch.
4. Using hammer and 1 16 Inch punch, complete Installation of spring pin (2) so that the ends are flush with the out side of bolt body (1).

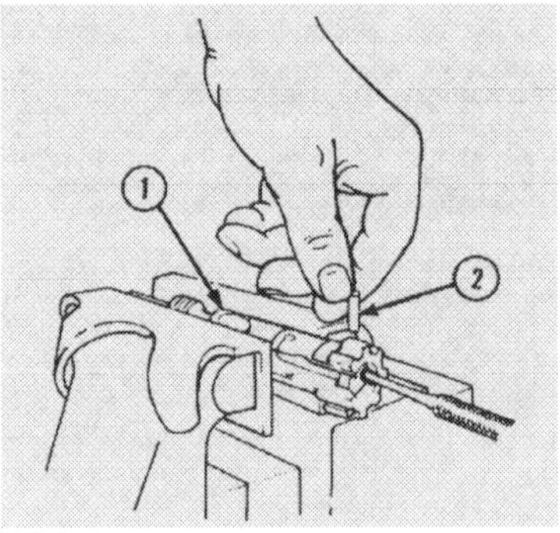

2-14. BOLT ASSEMBLY (CONT).

e. REASSEMBLY (CONT)

NOTE

Do not disassemble rubber Insert from extractor spring assembly.

5. If removed, Insert large end of extractor spring assembly (5) into cartridge extractor (6) and seat by pushing and turning clockwise.

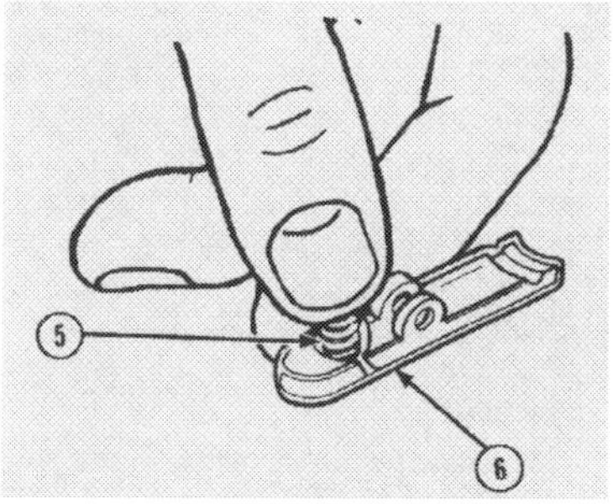

6 Position cartridge extractor (6) and ex tractor spring assembly (5) on bolt body (1).

7 Compress extractor spring assembly (5) and cartridge extractor (6) to align holes.

8. Install extractor pin (7) by hand.

9. Reassemble rifle, refer to page 2-68.

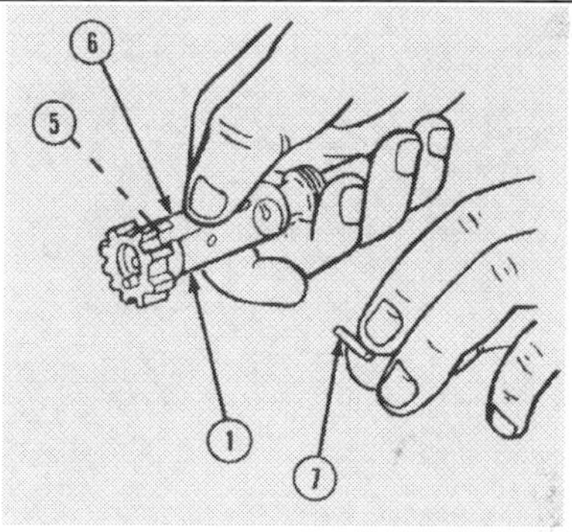

2-15. CHARGING HANDLE ASSEMBLY.

This task covers:

- a. **Disassembly**
- b. **Cleaning**
- c. **Inspection/Repair**
- d. **Lubrication**
- e. **Reassembly**

INITIAL SETUP:

Tools
(ARMY) Small Arms Repairman Tool Kit (item 3, app B)

Equipment Conditions
2-34 Charging handle assembly removed

References
TM 9-1005-319-10 (operator's manual)

a. DISASSEMBLY

WARNING

To avoid injury to eyes, use care when removing and installing spring-loaded parts.

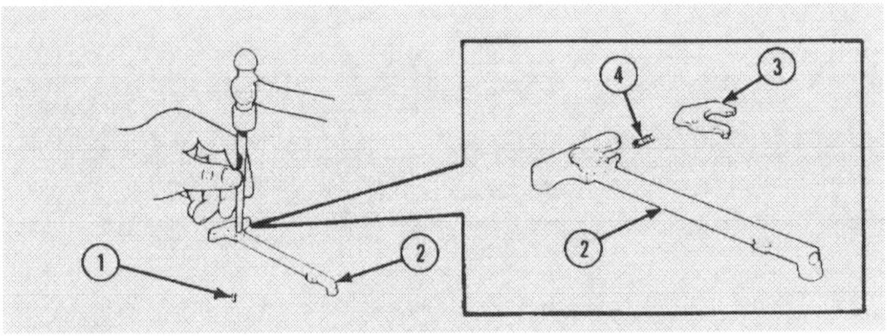

1. Remove spring pin (1) from charging handle (2) using a hammer and 1/16 inch punch.
2. As punch is withdrawn, catch charging handle latch (3) and helical spring (4) to prevent loss.

2-15. CHARGING HANDLE ASSEMBLY (CONT).

b. CLEANING

Clean all Items (operator's manual) Remove carbon deposits.

c. INSPECTION/REPAIR

Inspect all Items for breaks, cracks, or damage Replace all unserviceable Items

d. LUBRICATION

Lightly lubricate all Items (p 2-33) (operator's manual)

e. REASSEMBLY

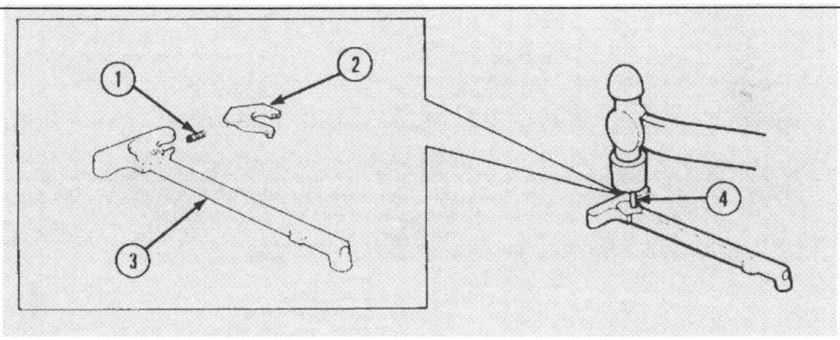

1. Position helical spring I1) and charging handle latch (2) in chdrgiin(J handle 13) Align holes and hold In position

2. Install spring pin (4) using a hammer Make sure spring pin is flush

3. Reassemble rifle, refer to page 2 68

2-16. UPPER RECEIVER AND BARREL ASSEMBLY, RIFLE BARREL ASSEMBLY, AND UPPER RECEIVER ASSEMBLY.

This task covers:

- **a.** Disassembly
- **b.** Cleaning
- **c.** Inspection
- **d.** Repair
- **e.** Lubrication
- **f.** Reassembly

INITIAL SETUP

Tools
(ARMY) Small Arms Repairman Tool Kit (item 3, app B)
Front sight post removal and installation tool (fig. E-2, app E)
Front sight detent depressor (fig. E-1, app E)

Materials/Parts
Lubricants (app D)

References
TM 9-1005-319-10 (operator's manual)

Equipment Conditions
2-34 Upper receiver and barrel assembly removed from lower receiver

General Safety Instructions
To avoid injury to your eyes, use care when removing and installing spring-loaded parts.

a. DISASSEMBLY

CAUTION

Do not use a screwdriver or any other tool when removing the handguard assemblies. Doing so may damage the handguard assemblies and/or slip ring. Do not remove heat shield for any reason. Doing so will damage the heat shield and the handguard assemblies will have to be replaced.

NOTE

Refer to operator's manual for "buddy system" procedure on removing handguard assemblies.

2-16. UPPER RECEIVER AND BARREL ASSEMBLY, RIFLE BARREL ASSEMBLY, AND UPPER RECEIVER ASSEMBLY (CONT).

a. DISASSEMBLY (CONT)

1. Push down on handguard slip ring (1) and lift upper handguard assembly (2) up and out.

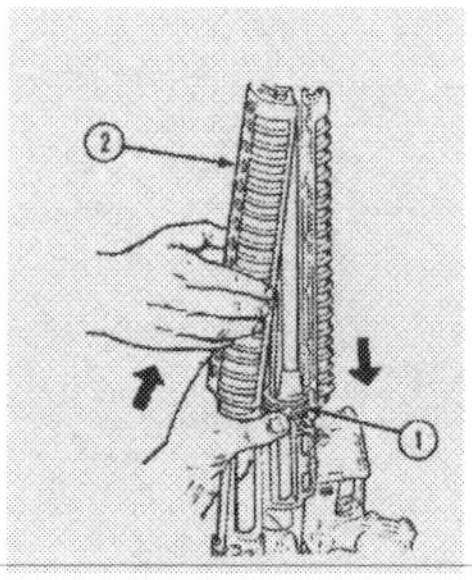

2. Push down on handguard slip ring (1) and lift the lower handguard assembly (3) up and out.

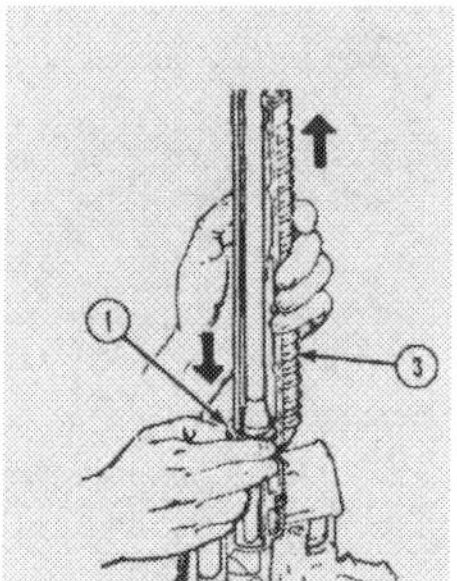

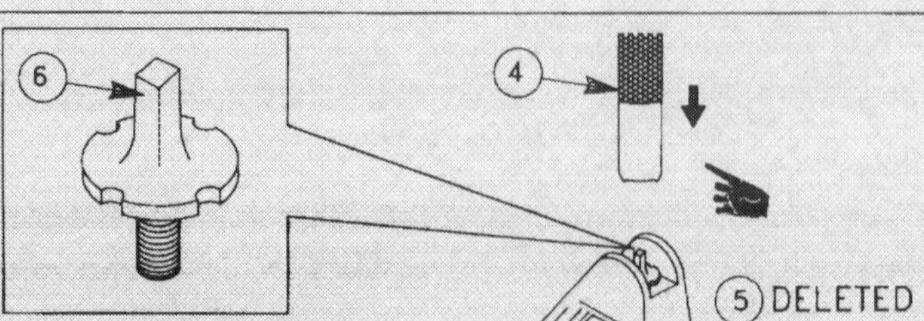

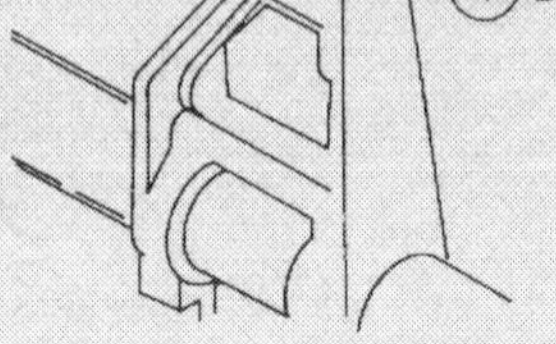

WARNING

To avoid injury to your eyes, use care when removing and installing spring-loaded parts.

NOTE

Remove front sight posts only if parts are damaged.

3. Using a dummy cartridge or a font sight post removal and installation tool (4).removed front sight post (6) by turning counter clockwise.

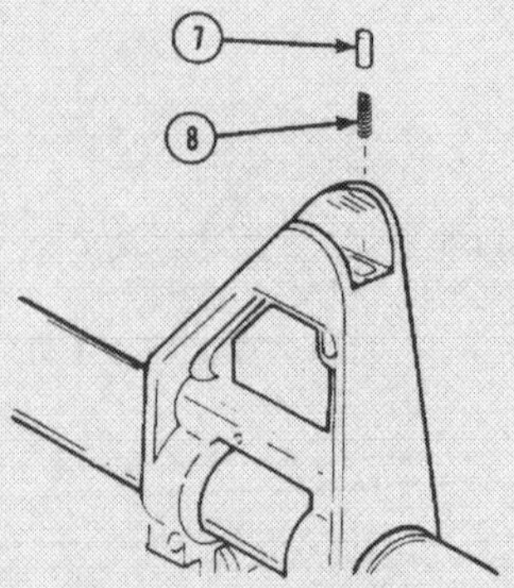

4. Catch front sight detent(7) and helical spring (8) to prevent loss.

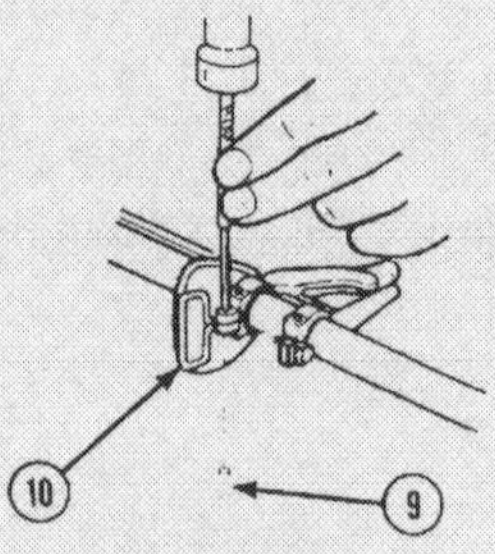

RIFLE ONLY

NOTE

Disassemble small sling swivel only if repair is necessary.

5. If necessary, knock out tubular rivet (9) with a hammer and punch and remove small sling swivel (10). Discard tubular rivet (9).

2-16. UPPER RECEIVE AND BARREL ASSEMBLY, RIFLE BARREL ASSEMBLY, AND UPPER RECEIVER ASSEMBLY (CONT).

a. DISASSEMBLY (CONT)

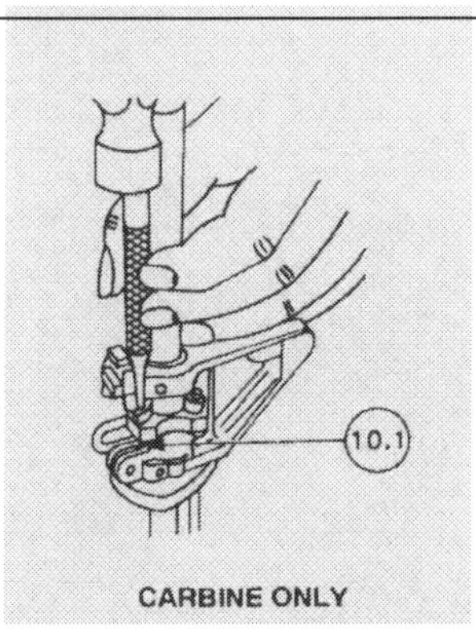

5A. Remove two spring pins (10 1)

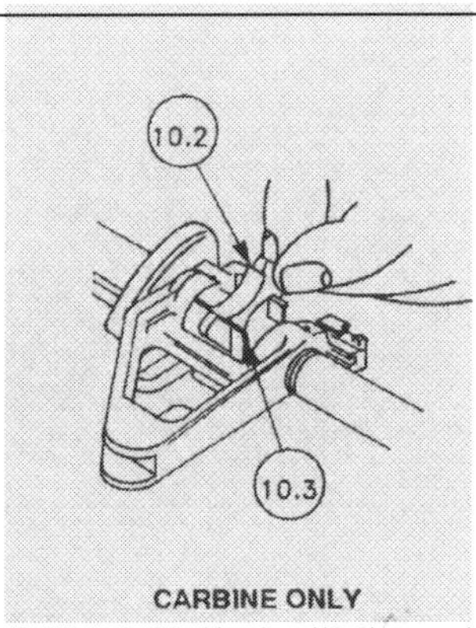

5B. Lift swivel locking bar (10 2) up and out of swivel mount (10 3)

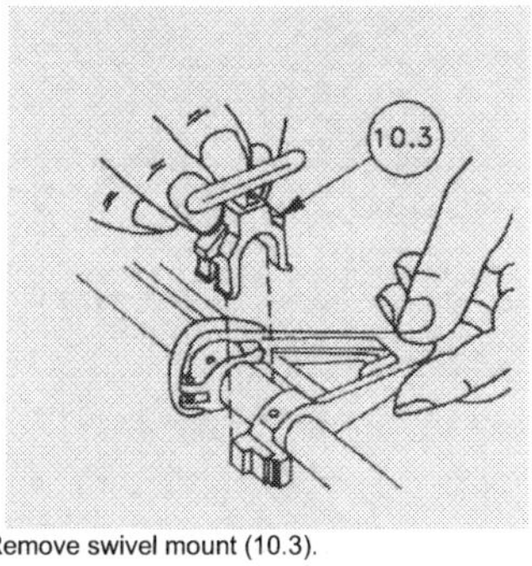

5C. Remove swivel mount (10.3).

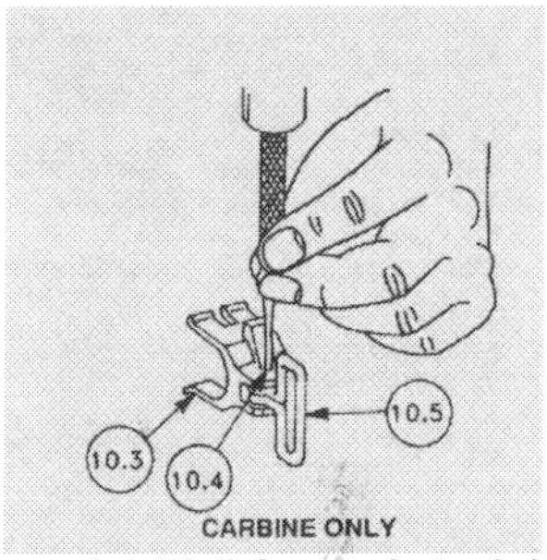

5D. Remove rivet (10 4) Separate front swivel (10 5) from swivel blank) mount (10 3)

2-16. UPPER RECEIVER AND BARREL ASSEMBLY, RIFLE BARREL ASSEMBLY, AND UPPER RECEIVER ASSEMBLY (CONT).

a. DISASSEMBLY (CONT)

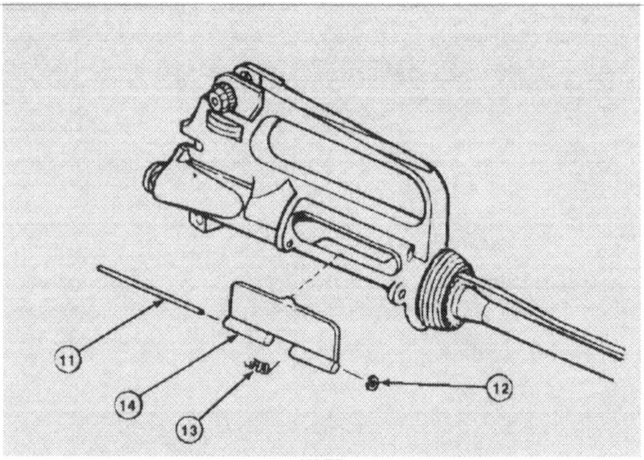

NOTE

Do not disassemble further unless repair is necessary. Headless grooved pin (11) may bind may bind against the forward assist housing and require some additional force to remove.

6. Remove retaining ring (12) and slide headles grooved pin (11) out to the rear.
7. Catch cover spring (13) and ejection port cover (14) to prevent loss as headless grooved pin is with-drawn.

M4/M4A1 ONLY

NOTE

Do not remove the carrying handle assembly nut (2 each) or clamp for normal maintenance. If carrying handle assembly nut or clamp are missing or the clamp is damaged to the point of being non-functional they may be replaced.

7.1. Disassemble as follows only if required to replace damaged or missing carrying handle assembly nut or clamp.

a. Observe the position of the clamp (15) in relation to the carrying handle (16).
b. Using care not to damage the surface finish, remove carrying handle assembly nut(s) (17) and clamp as required.
c. Inspect mating surfaces of the carrying handle (16) and the clamp (15) for damage. If damaged replace the damaged part.
d. Inspect for other damage, rust, etc. Repair or replace as required.
e. Apply a light coat of lubricant to all mating surfaces.
f. Install clamp (15) paying close attention to ensure that it is installed in its original position. The forward end of the clamp (15) must not extend past the forward end of the carrying handle (16).
g. Taking care not to cross thread, install handle assembly nut (17).

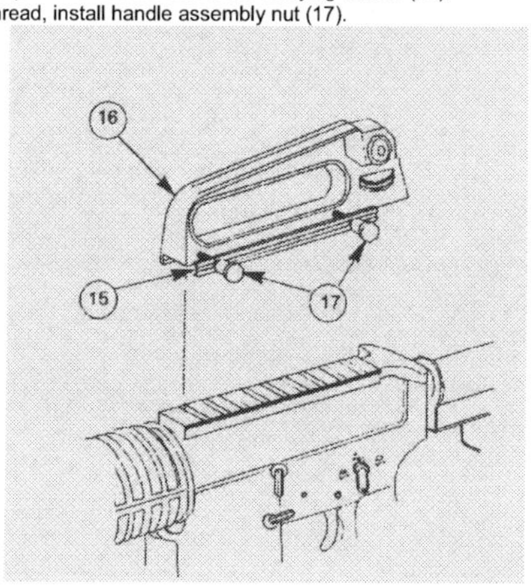

b. CLEANING

Clean all items (operator's manual).

c. **INSPECTION**

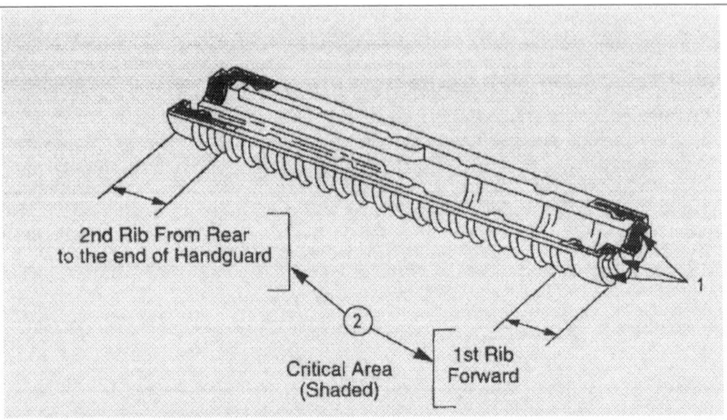

1. Inspect handguard assemblies for breaks separation, and cracks using the following guidelines:
   - **a.** Breaks and separations of material which prevent proper retention or interfere with functioning of the rifle will be cause for handguard assembly rejection and replacement.
   - **b.** M16A2 handguard assemblies may have up to two of the three front retaining tabs missing (1). M4/M4A1 may not have any front retaining tabs (1) missing. If all front retaining tabs for the M1 6A2 are missing, or any of the tabs for the M4/M4A1 are missing, handguard assemblies must be replaced.
   - **c.** Cracks up to one inch in length are acceptable provided they do not extend into the retaining flange (CRITICAL AREA) (2).
   - **d.** Replace severely cracked or damaged handguard assemblies. Handguard assemblies which have a heat shield which is loose enough to rattle when installed on the rifle must be replaced.

2. Inspect front sight assembly for chips, breaks, and cracks. Evacuate to support maintenance if broken, cracked, or bent.
3. Inspect front sight area for evidence of gas leakage around gas tube. Evacuate to support maintenance if short recoil results from gas leakage.
4. Inspect front sight post, front sight detent, and helical spring for damage. If damaged, replace.

2-16. UPPER RECEIVER AND BARREL ASSEMBLY, RIFLE BARREL ASSEMBLY, AND UPPER RECEIVER ASSEMBLY (CONT).

c. INSPECTION (CONT)

5. Inspect barrel for pits in bore, burrs, broken or worn locking lugs, and surface cracks and defects.
   a. Pits no wider than a land or groove and 3/8 inch (0.953 cm) or less in length are allowable in the bore.
   b. Uniformly fine pits in a densely pitted area of the bore are allowable.
   c. Lands that appear dark due to coating of gliding metal from projectiles are allowable.
   d. Striping of lands and grooves shall not be cause for rejection unless support maintenance determines by use of the barrel erosion gage.
   e. For pits other than mentioned above, broken or burred locking lugs, or surface cracks, evacuate to support maintenance.
6. Inspect bore for ringing. Definitely ringed bores or bores ringed sufficiently to bulge the outside surfaces of the barrel are causes for rejection. Evacuate to support maintenance.
7. Inspect chamber for pitting. Fine pits, or fine pits in a densely pitted area, are allowable. Pits 1/8 inch (0.318 cm) in length are cause for rejection. Evacuate to support maintenance.
8. Hand check compensator (3) for looseness on barrel. The third (middle) slot (4) must be straight up at top dead center (TDC). The alignment may vary as much as one half the width of the slot either direction. If loose or out of alignment, evacuate to support maintenance.
9. Inspect all items for serviceability in and tightness of latch assembly on ejection port cover. If items are damaged or nonfunctional, they are unserviceable.

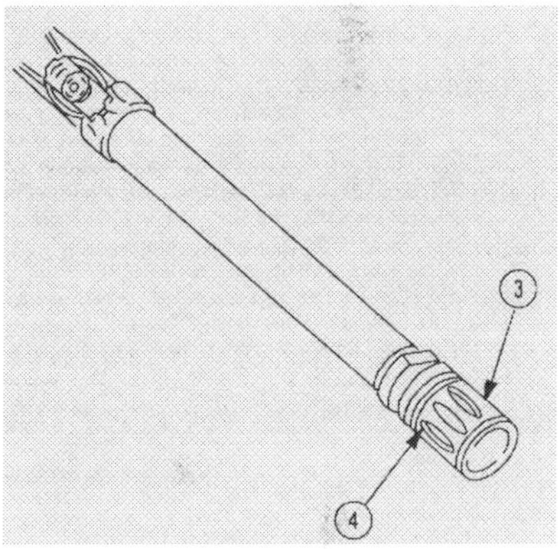

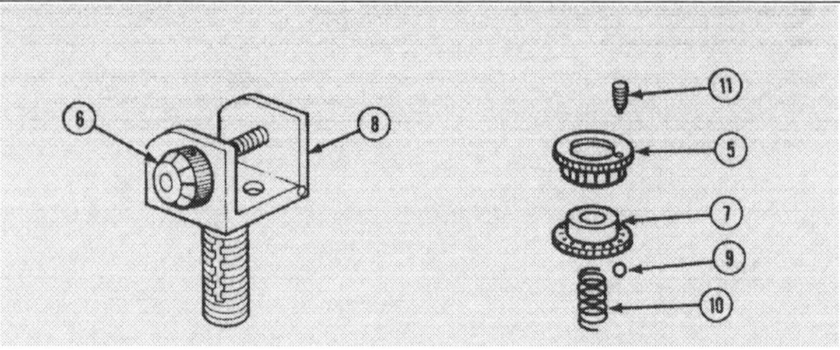

10. Rotate and test elevation index (5) and windage knob (6) for ease of functioning and legibility of markings.

11. Inspect elevation knob zero as follows:

a. Rotate elevation knob (7) counterclockwise until the rear sight assembly is all the way down. If a whole click is not felt as the rear sight assembly stops, the rear sight assembly has bottomed out and will not pivot freely.

b. Position elevation knob back slightly to Its last whole click so the rear sight assembly base (8) is under tension of the ball bearing (9) and helical spring (10). The 300 meter mark should align with the mark on the receiver.

c. If the 300 meter mark is not aligned with the mark on receiver, slip the range scale in the following manner:

(1) Position the 300 meter mark with the mark on the receiver.

(2) Insert a 1/16 inch allen wrench through the access hole of the rear sight assembly base and into the Index screw (11).

(3) Loosen the index screw three turns and leave the wrench in place.

(4) Rotate lower portion of elevation knob counterclockwise until it stops (range scale should not have moved). Elevation knob should be positioned on its last whole click.

(5) Tighten index screw and remove wrench.

(6) Check for proper setting.

2-16. UPPER RECEIVER AND BARREL ASSEMBLY, RIFLE BARREL ASSEMBLY, AND UPPER RECEIVER ASSEMBLY (CONT).

d. REPAIR

1. Replace all authorized unserviceable parts
2. Evacuate to support maintenance.

e. LUBRICATION

Lightly lubricate all Items (p 2 311 (oper ator's manual)

f. REASSEMBLY

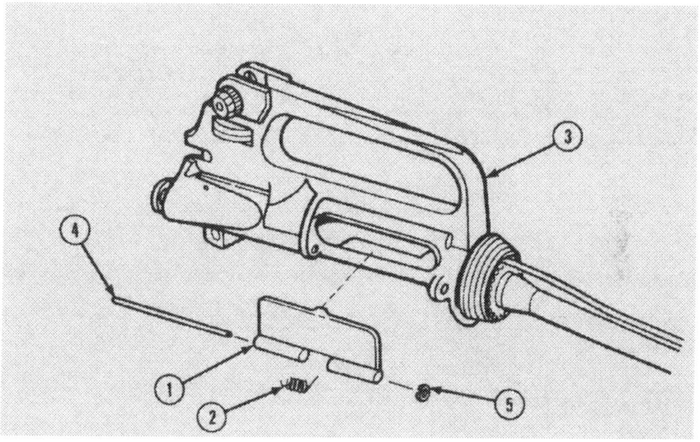

1. If previously disassembled, position election port cover 11) and helical spring (2) on Lip per receiver (3) with short leg of helical spring to the rear on Inside of ejection port cover

NOTE
Long legs of helical spring must be positioned and pretensioned before the headless grooved pin is installed

2. Hold helical spring (2) short leg In this position and turn long leg one half turn (180 degrees) with fingers of right hand
3. Position long leg of helical spring (21 against election port cover (1) Hold helical spring and election port cover in this position and Install headless grooved pin (4) Check for proper spring tension during installation of retaining ring (5)

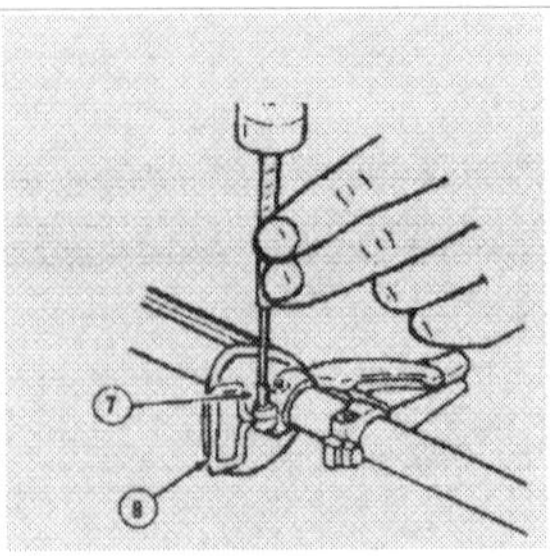

RIFLE ONLY

4. If previously disassembled, position small sling swivel (6) and install new tubular rivet (7) using center punch and hammer to spread and flare the hollow head of the tubular rivet.

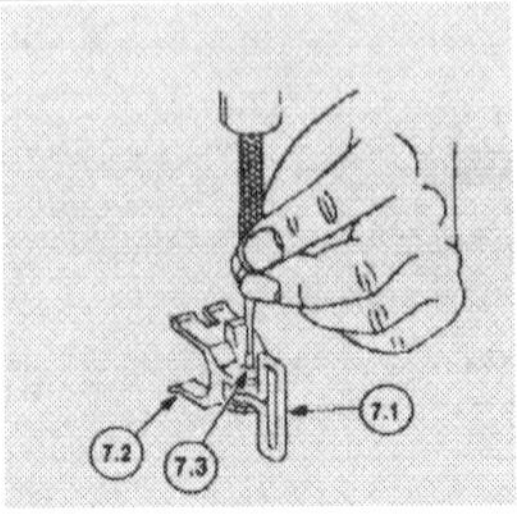

CARBINE ONLY

4A. Install front swivel (7.1) to swivel mount (7.2) with new rivet (7.3) using center punch and hammer to spread and flare the hollow head of the tubular rivet.

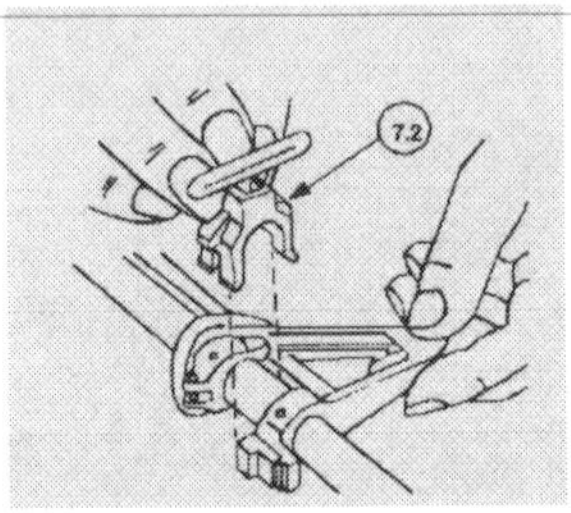

CARBINE ONLY

NOTE

It is recommended that the swivel mount be placed on the left side of the carbine for right handed operators and on the right side for left handed operators. This will keep the sling out of the way when the carbine is used.

4B. Install swivel mount (7.2).

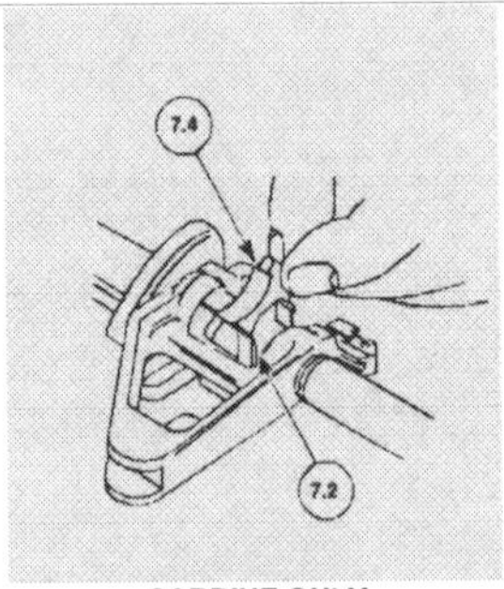

CARBINE ONLY

4C. Place swivel locking bar (7.4) in swivel mount (7.2).

2-16. UPPER RECEIVER AND BARREL ASSEMBLY, RIFLE BARREL ASSEMBLY, AND UPPER RECEIVER ASSEMBLY (CONT).

f. REASSEMBLY (CONT)

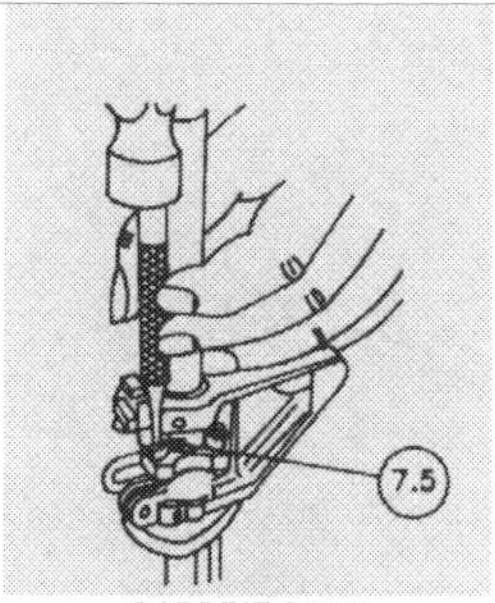

CARBINE ONLY

4D. Install two spring pins (7 5)

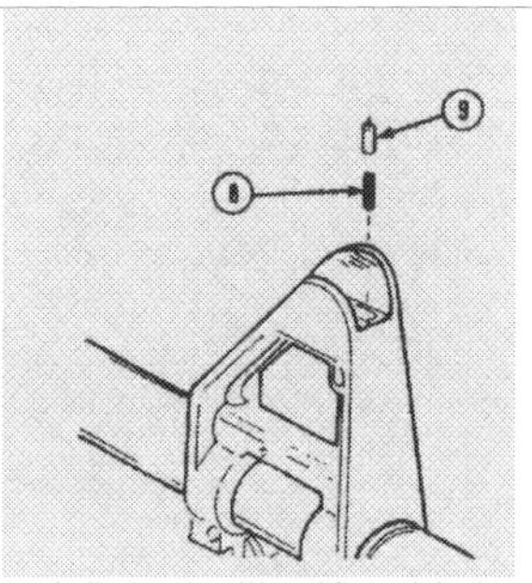

5. Position helical spring (8) and front sight detent (9).

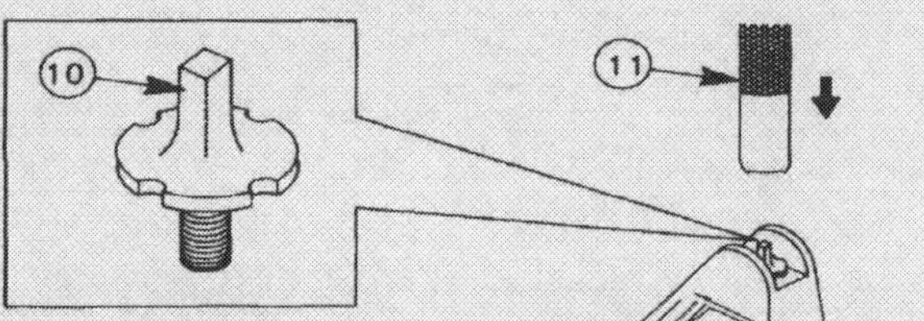

WARNING

To avoid injury to your eyes, use care when removing and installing spring-loaded parts.

6. Install front sight post (10) by turning front sight post removal and installation tool (11) clockwise until front sight post base is flush with or slightly below front sight frame.

2-16. UPPER RECEIVER AND BARREL ASSEMBLY, RIFLE BARREL ASSEMBLY, AND UPPER RECEIVER ASSEMBLY (CONT).

t. REASSEMBLY (CONT)

7. Mechanical Zero Procedures (A.F. Only).
   a. Mark a piece of plastic card stock or rigid paper with lines from 1 to 5mm in 1 mm increments Set the card on the front sight frame and check the height of the top of the front sight post.

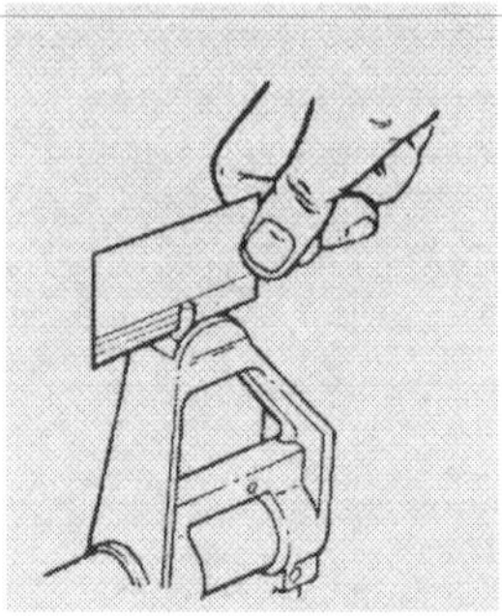

   b. Using a dummy round or front sight post tool adjust the front sight so the top of the front sight post is 5mm above the machined surfaces of the front sight frame.

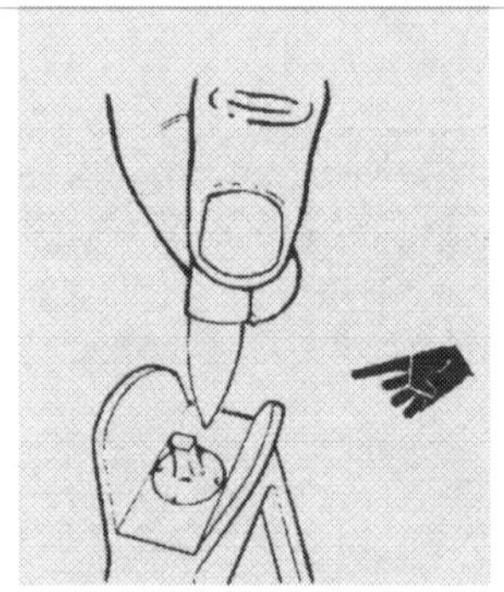

c. Visually check the front sight post top height by using the marked plastic or paper card. Card must set level on the machined surfaces of the front sight frame to obtain an accurate reading

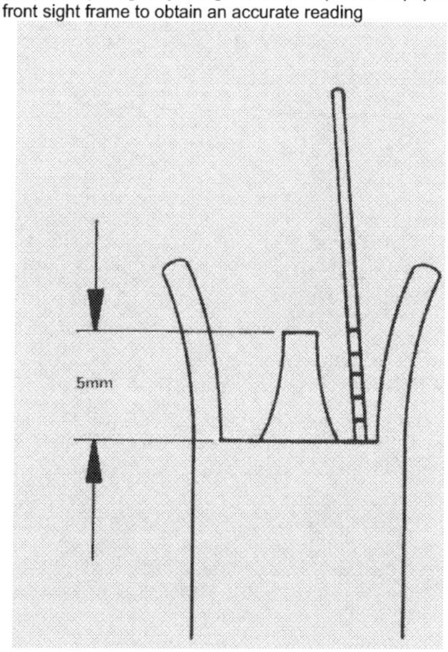

NOTE

This procedure, when used In conjunction with rear sight mechanical zero adjustment. will give an approximate battle sight zero to most M16A2 rifles The above steps can also be used before firing a new or newly assigned rifle Use the procedures to check rifles stored In preferred packaging during routine Inspections This will help ensure people armed with the rifles will stand a better chance of hitting an enemy If the rifles must be used before a live fire zero can be made Whenever possible, zeroing of the rifle should be accomplished using ball ammunition on a 25 meter zeroing target using the "L" aperture.

2-16. UPPER RECEIVER AND BARREL ASSEMBLY, RIFLE BARREL ASSEMBLY, AND UPPER RECEIVER ASSEMBLY (CONT).

f. REASSEMBLY (CONT)

NOTE

Refer to operator's manual for 'buddy system" procedure on installing hand guard assemblies

8. Install top of upper handguard assembly (12) In tube cap (13) while pushing down on handguard slip ring (14). Push bottom of upper handguard assembly (12) in place and release handguard slip ring (14) to lock handguard assembly In place

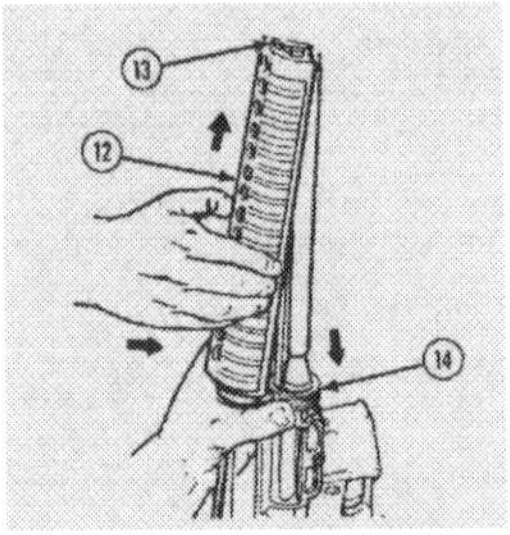

9. Install top of lower handguard assembly 1 5) In tube cap (13) while pushing down on handguard slip ring (14) Push bottom of lower handguard assembly (15) In place and release handguard slip ring {14) to lock both handguard assemblies In place

10. Reassemble rifle, refer to page 2-68

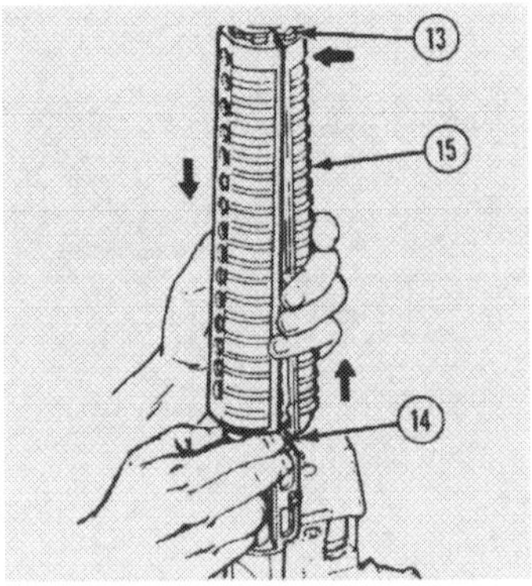

2-17. LOWER RECEIVER AND BUTTSTOCK ASSEMBLY.

This task covers:

a. Disassembly
b. Cleaning
c. Inspection
d. Repair
e. Lubrication
f. Reassembly

INITIAL SETUP

Tools
(ARMY) Small Arms Repairman Tol Kit (item 3, app B)
Pivot Pin Removal Tool (fig. E-3, app E)
Pivot Pin Installation Tool (fig. E-4, app E)

Materials/Parts
Lubricant, solid film (SFL) (item 21, app D)
Screw, self-locking (item 6, p C-1 1)

References
TM 9-1005-319-10 (operator's manual)

Equipment Conditions
2-34 Lower receiver and buttstock assembly removed

General Safety Instructions
To avoid injury to your eyes, use care when removing and installing spring-loaded parts.
When using solid film lubricant or dichloro-methane, be sure the area is well ventilated.

a. DISASSEMBLY

1. Remove screw (1) and lockwasher (2).

WARNING
To avoid injury to you eyes, use care when removing and installing spring-loaded parts.

2. Carefully remove pistol grip (3) and catch helical spring (4) and safety detent (5) to prevent loss.

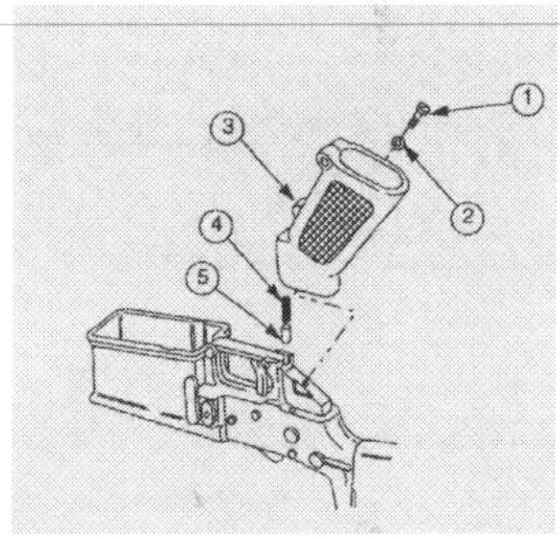

2-17. LOWER RECEIVER AND BUTTSTOCK ASSEMBLY (CONT).

a. DISASSEMBLY (CONT)

RIFLE ONLY
NOTE

If the self-locking screw is removed, it must be discarded and replaced with a new one

3. Remove self-locking screw (6).

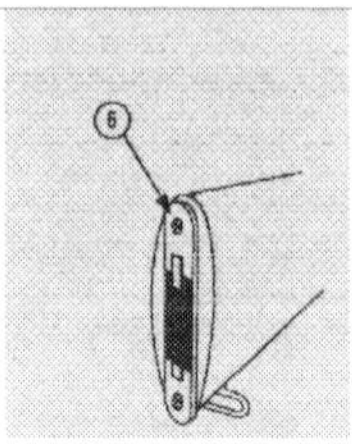

RIFLE ONLY
WARNING

To avoid injury to your eyes, use care when removing and installing spring-loaded parts.

4. Remove buttstock assembly (7) carefully and catch helical spring (8), takedown pin detent (9), takedown pin (10), and stepped spacer (11) to prevent loss.

NOTE

If detent (9) will not come out, use a wire to push it out.

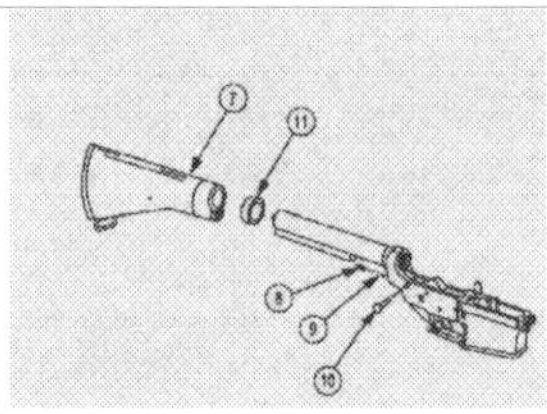

CARBINE ONLY

4A. Extend buttstock assembly (11.1).

4B Grasp the lock release lever (11.2) in the area of the retaining nut (11.3), pull down-ward, and slide buttstock to the rear to separate the buttstock assembly from the lower receiver extension

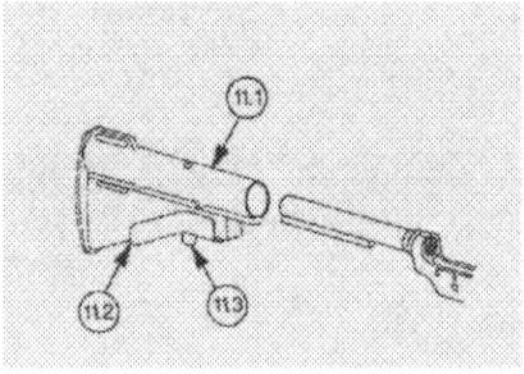

WARNING

To avoid injury to your eyes, use care when removing and installing spring-loaded parts

NOTE

Catch pivot pin detent and helical spring as pivot pin is removed (see step 6 on next page)

5. Depress pivot pin detent and spring and remove pin or Insert fabricated pivot pin removal tool (12) to compress pivot pin detent Turn pivot pin (13) a quarter turn Remove tool and pivot pin

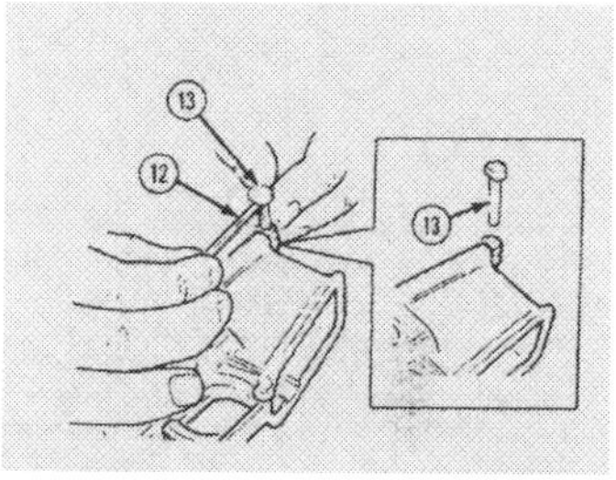

6. Be sure to hold cupped hand in front of pivot pin detent (14) and helical spring (15) to prevent loss of pivot pin detent and helical spring.

NOTE

If helical spring (15) will not come out, use a wire to pull it out.

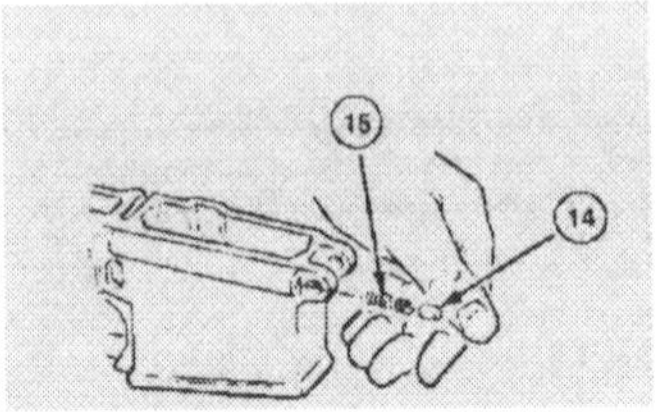

NOTE

Make sure the hammer is cocked and the selector lever is not set on BURST or AUTOMATIC before removing the buffer assembly.

7. Press buffer assembly (16) in about 1/4 inch (0.635 cm). Depress buffer retainer (17) and release buffer assembly (16) and action spring (18). Remove buffer assembly(16) and action spring (18) from receiver while depressing buffer retainer (17).

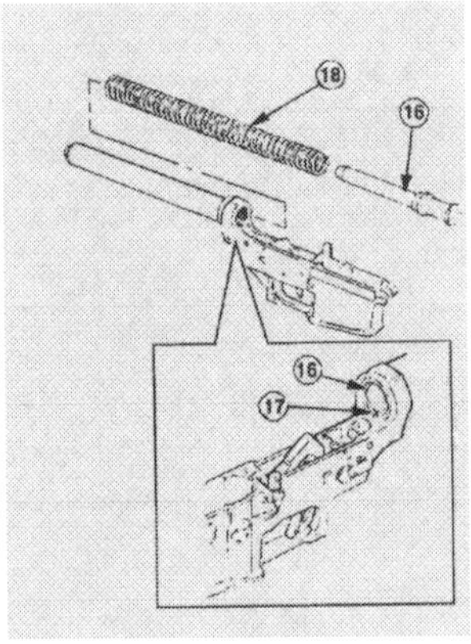

b. CLEANING

Clean all items (operator's manual). Remove carbon deposits

1 2-17. LOWER RECEIVER AND BUTTSTOCK ASSEMBLY (CONT).

c. INSPECTION I

1. Inspect buffer assembly (1, 2. 3, or 3 1) for cracks or damage

RIFLE ONLY

a. Some old buffers (1) have a hole with pin installed which protrudes equally on each side approximately 1/32 inch (O 08 cm)

b. Some buffers (2) have a hole In the housing but no pin

c. New buffers (3) do not have a hole In buffer body or a pin.

BOTH WEAPONS

2. If cracked or damaged, replace.

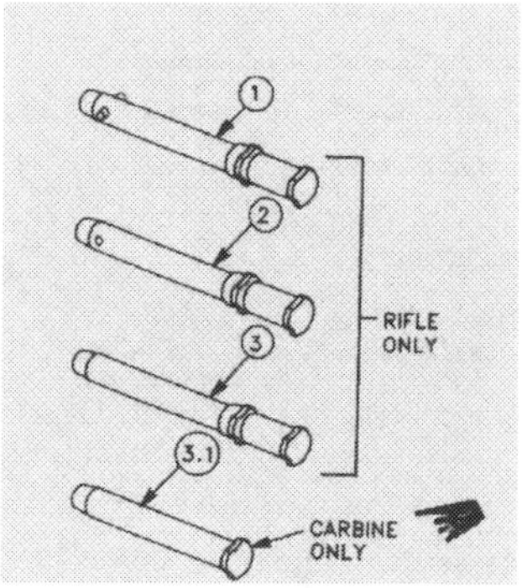

3. Check free length of action spring (4). The free length **FOR RIFLE ONLY** must be between 11 3/4 inches (29.85 cm) mini- mum and 13 1/2 inches (34 29 cm) maximum, **FOR CARBINE ONLY** must be between 10 1'/16 inches (25,56 cm) minimum and 11 114 Inches (28.58 cm) maximum, If not, replace Do not attempt to adjust the length by stretching the action spring

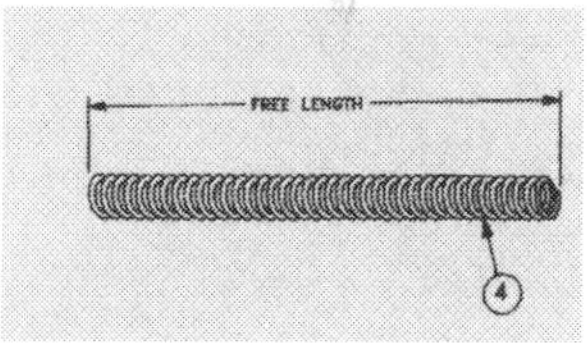

4. Inspect lower receiver (5) (without further disassembly) for legibility of serial number **ARMY ONLY**: If the serial number is hard to read, evacuate to direct support maintenance **AIR FORCE ONLY**: If the serial number is hard to read, evacuate to depot maintenance

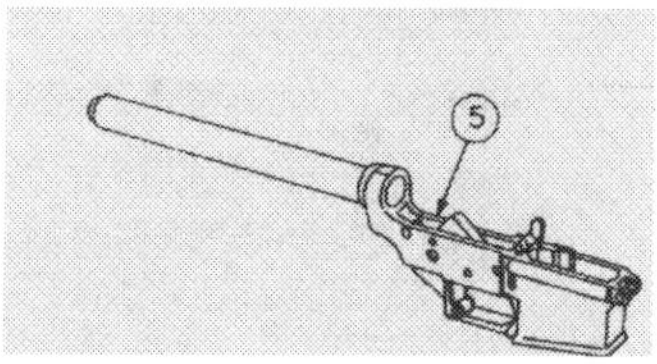

NOTE

AIR FORCE ONLY: Only depot maintenance is authorized to restamp the serial number.

5. Inspect for missing or damaged parts. Inspect finish of lower receiver for shiny spots. Touch up with solid film lubricant as required (p 2-34).

NOTE

If a M16A2 rifle or M4/M4A1 carbine lower receiver is missing one third or more of its exterior protective finish, resulting in an unprotected, light reflecting surface, it is candidate for overhaul. This missing finish will be considered a shortcoming. This shortcoming requires action to obtain a replacement weapon. Once a replacement has been received, evacuate the original weapon to depot maintenance for overhaul.

d. REPAIR

Replace all authorized unserviceable parts. If repair is not authorized at this level, evacuate to support maintenance.

e. LUBRICATION

Lightly lubricate all metal components (p 2-33) (operator's manual).

2-17. LOWER RECEIVER AND BUTTSTOCK ASSEMBLY (CONT).

f. REASSEMBLY

WARNING

To avoid Injury to your eyes, use care when removing and Installing spring loaded parts.

NOTE

Make sure the hammer Is cocked and the selector lever is not set on BURST before Installing the buffer assembly.

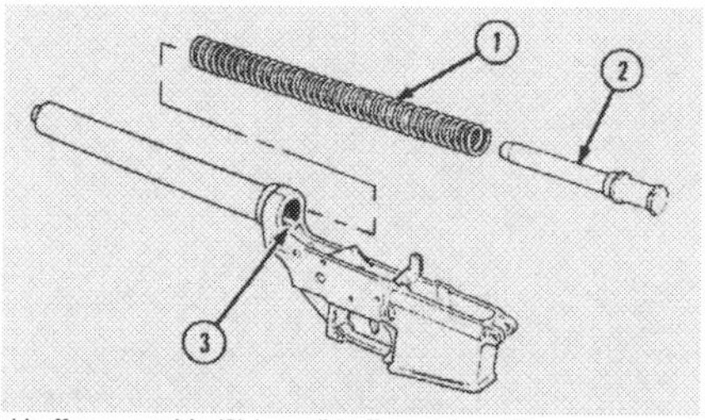

1. Press action spring (1) and buffer assembly (2) In until buffer retainer (3) snaps up and holds them In place.

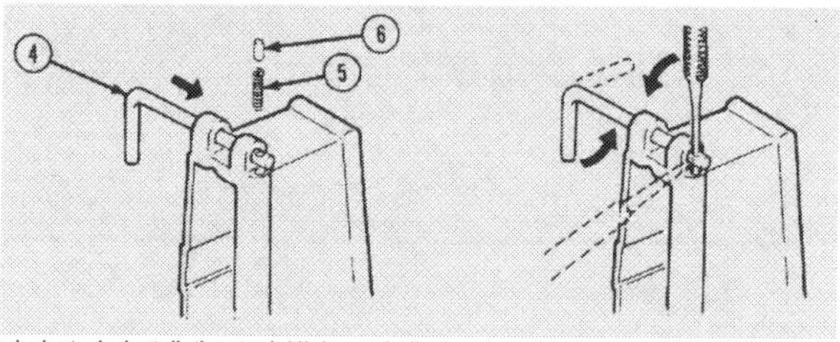

2. Install fabricated pivot pin installation tool (4) Insert helical spring 151 and pivot pin detent (6) Compress pivot pin detent in recess with 3 32 inch punch and rotate tool Remove 3'32 Inch punch

NOTE

Rounded end of pivot pin detent must be In the groove of the pivot pin (7) when assembly is complete.

3. Install pivot pin (7) while removing fabricated pivot pin Installation tool (4) Maintain pressure while sliding pivot pin (7) into hole Rotate pivot pin until pivot pin detent is inserted Into pivot pin groove

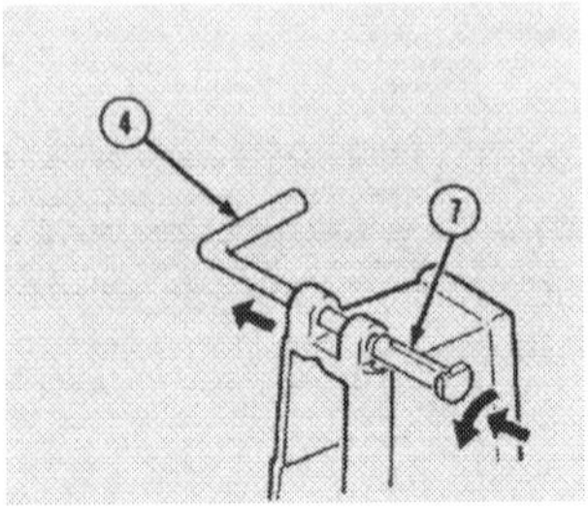

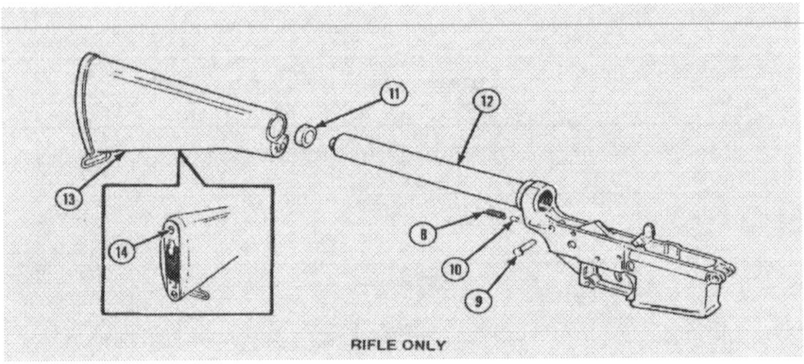

CAUTION

Do not kink helical spring (8) during assembly

4. Install takedown pin (9) with groove toward the rear Install takedown pin detent (10)and helical spring (8) from the rear
5. Install stepped spacer (11) on receiver extension (12) and carefully slide buttstock assembly (13) on to compress helical spring (8)

NOTE

Self-locking screw (14), If removed must be discarded and replaced with a new one 6. Install self-locking screw (14) to secure buttstock assembly (13)

6. Install self-locking screw (14) to secure buttstock assembly (13)

1 2-17. LOWER RECEIVER AND BUTTSTOCK ASSEMBLY (CONT).

f. REASSEMBLY (CONT)

CARBINE ONLY

6A. Grasp the lock release lever (14 1) In the area of the retaining nut (14 2) and pull to reinstall the buttstock assembly (14.3) onto the lower receiver extension (14 3) onto the lower receiver extension (14 4)

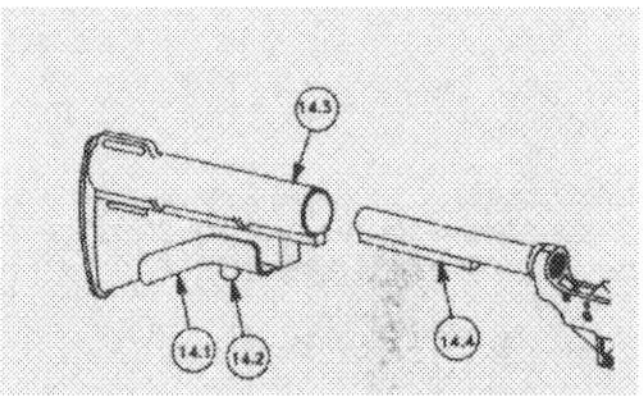

2-17. LOWER RECEIVER AND BUTTSTOCK ASSEMBLY (CONT).

f. REASSEMBLY (CONT)

WARNING

When utilizing the enhanced rifle grip (it has a bump between the second third finger for a better grip) PN 9349127, rifle grip screw. PN AN 501D416 18 (1-1/8 in.) or AN 501 D416 16 (1 in), is authorized to be(used with the enhanced grip Any screw longer than 1-1 8 In used with 9349127 could cause a hazardous situation Also, ensure the washer Is in 11place

CAUTION

Do not kink helical spring (15) during assembly

7. Install safety detent (16), pointed end first, and helical spring (1 5) into bottom of lower receiver (17)

NOTE

A portion of the helical spring will fit In a hole in the pistol grip

8. Carefully install pistol grip (18) to compress helical spring 115). Secure pistol grip 118) in place with lockwasher (19) and screw 120)

9. Reassemble rifle, refer to page 2 68

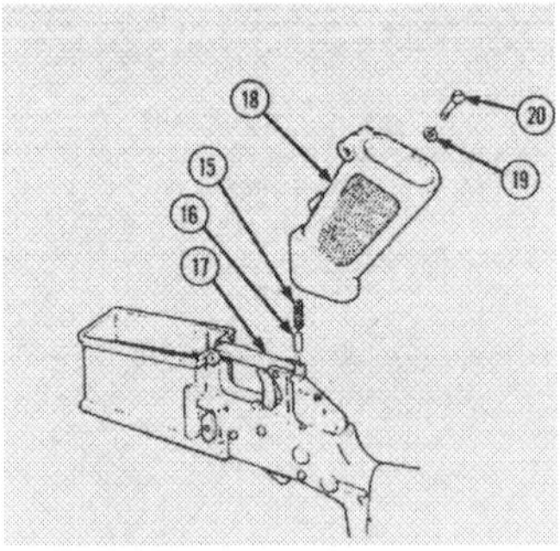

2-18. BUTTSTOCK ASSEMBLY.

This task covers:

a. Disassembly	d. Repair
b. Cleaning	e. Lubrication
c. Inspection	f. Reassembly

INITIAL SETUP

Tools	Equipment Conditions
(ARMY) Small Arms Repairman Tool Kit (item 3, app B)	2-57 Buttstock assembly removed from lower receiver and buttstock assembly

a. DISASSEMBLY

RIFLE ONLY

1. Remove self-locking screw (1), small sling swivel (2), and buttplate group (3) from buttstock (4)

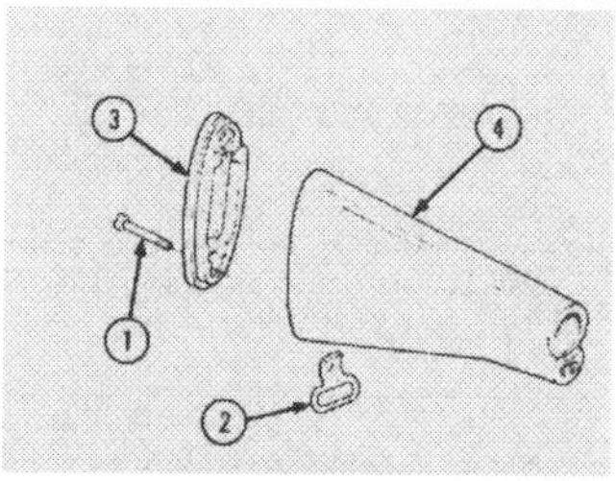

RIFLE ONLY

2. Push down on plunger (5) and lift door assembly (6) out of buttplate (7).

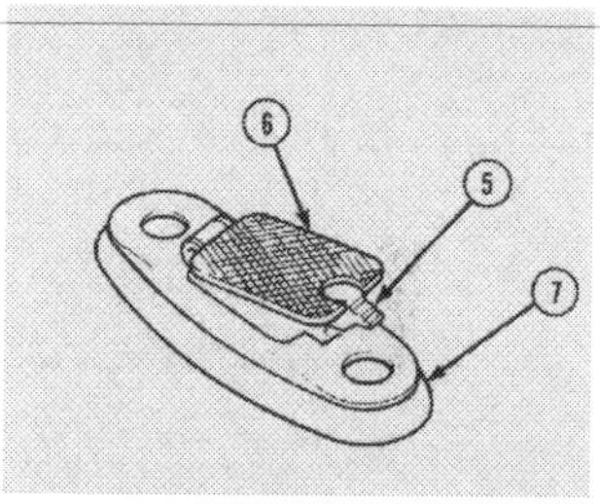

RIFLE ONLY

3. Remove straight pin 181 and separate hinge (9) and door assembly (61

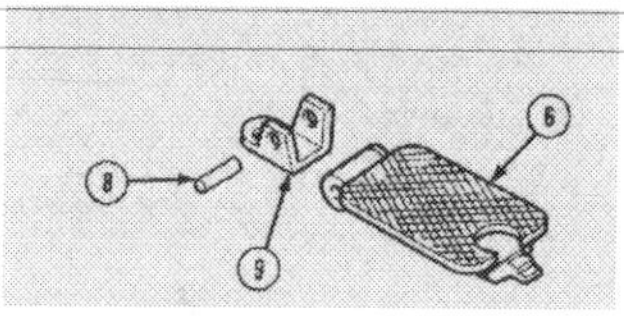

2-18. BUTTSTOCK ASSEMBLY (CONT).

a. DISASSEMBLY (CONT)

CARBINE ONLY

4. Disassemble the buttstock assembly by tapping out the spring pin (10) located in ,he oval slot of the retaining nut (1 1) This is done using a 1 16 Inch punch

5. Insert your Index finger into the forward end of the buttstock 112) and push down on the locking pin (13 Unscrew the retaining nut 11 and remove the release lever (14) locking pin (13) and spring (15)

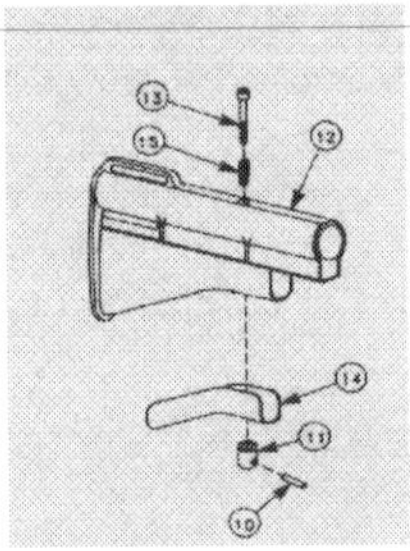

b. CLEANING (RIFLE ONLY)

Clean all parts with CLP (operator s manual) Use brush to clean knurled surface of door assembly

2-18. BUTTSTOCK ASSEMBLY (CONT).

c. INSPECTION

NOTE

M16A2 buttstocks, PN 9349121, with unauthorized markings may be used under the following conditions:

a. The only authorized markings are those which are temporary In nature, i e , paint, tape, etc

b. When marking a buttstock, only use temporary markings

c. Buttstocks with unauthorized markings that have been stamped into the surface of the buttstock will not be used

d. Unauthorized markings that have previously been scratched, etched, carved, etc may continue In use if the marks do not extend Into the fiber of the buttstock Cutting Into the fiber of the buttstock may weaken it

e. These marks may be at any location on the buttstock Unauthorized markings are not desirable. However, if previously applied, they will be allowed to continue In use due to the cost of the buttstock.

1. Inspect buttstock for cracks using the following guidelines:

   a. Under the following conditions, hairline cracks (no chipped away material allowed) originating from buttplate end of buttstock are acceptable.
      (1) One hairline crack, not to exceed 1 in. 12.54 cm) In length, per side of buttstock.
      (2) Two additional hairline cracks up to 0 25 in (0 64 cm) In length, per side of buttstock
      (3) A total of three cracks per side of buttstock, originating from buttplate end, are allowable.

   b. Cracks In the critical area at the front end of the buttstock are not acceptable and these buttstocks must be replaced.

2. While buttplate Is Installed on rifle, Inspect for cracks around the mounting holes Check for cracks In excess of 0 25 In. (O 64 cm) In length which extend through the buttplate Replace If cracked.

3. Inspect door assembly for cracks, corrosion, stuck plunger, separations on outer face, or other damage Replace If defective.

d. REPAIR

Replace all authorized unserviceable Items Unserviceable Items are those Items which are damaged.

e. LUBRICATION

Lubricate all metal components (p 2-33) (operator's manual)

f. REASSEMBLY

RIFLE ONLY

1. Position hinge (1) on door assembly (2) and Install straight pin (3)

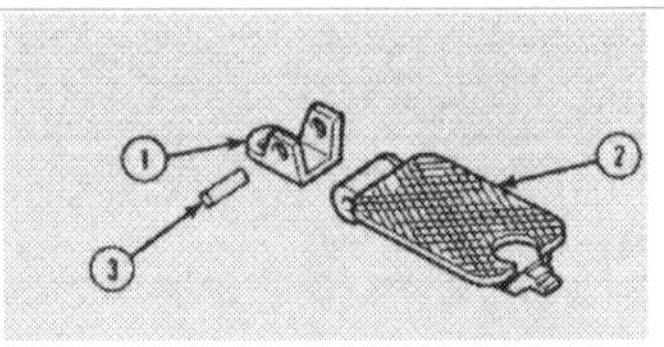

RIFLE ONLY

2. Install door assembly (2) into buttplate (4) and press plunger (5) to lock

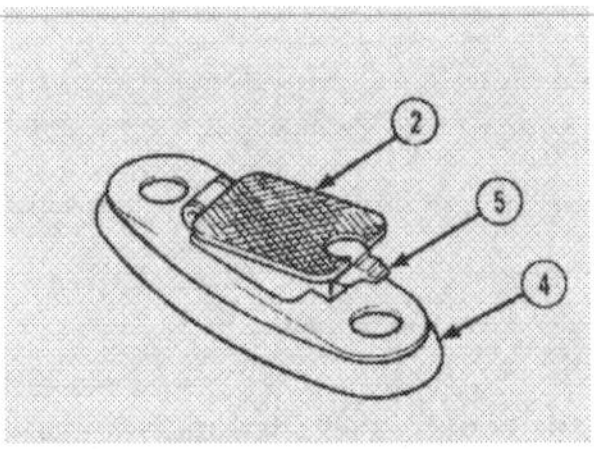

RIFLE ONLY

3. Position buttplate group (6) and small sling swivel (7) to the buttstock (8) and secure with self-locking screw (9).

NOTE

See page 2-62, reassembly, for reassembly of buttstock assembly to lower receiver

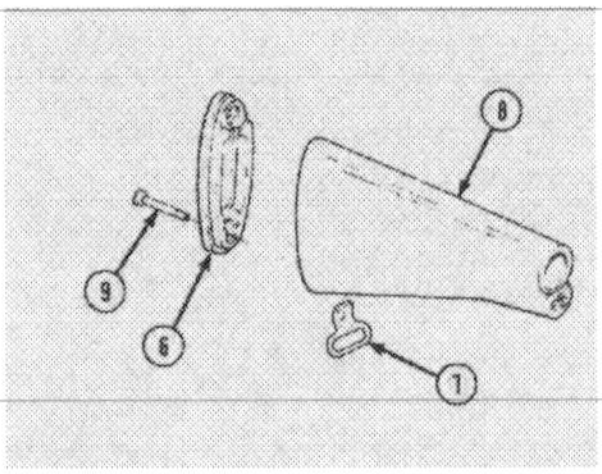

2-18. BUTTSTOCK ASSEMBLY (CONT).

f. REASSEMBLY (CONT)

CARBINE ONLY

4. Insert locking pin spring (10) onto locking pin (11)

5. Insert locking pin (11) and spring (10) Into the hole on top of the buttstock (12), threaded end first

6. Insert your index finger into the forward end of the buttstock (12) and push down on locking pin (11)

7. Install the release lever (13) onto the threaded portion of the locking pin (11) protruding through the bottom of the buttstock (12)

8. Screw on the locking nut (14) until flush with locking pin (11) Align the slot In locking nut (14) with the spring pin hole in the locking pin (11)

9. Lightly tap in spring pin (15) until flush on both sides of the retaining nut (14)

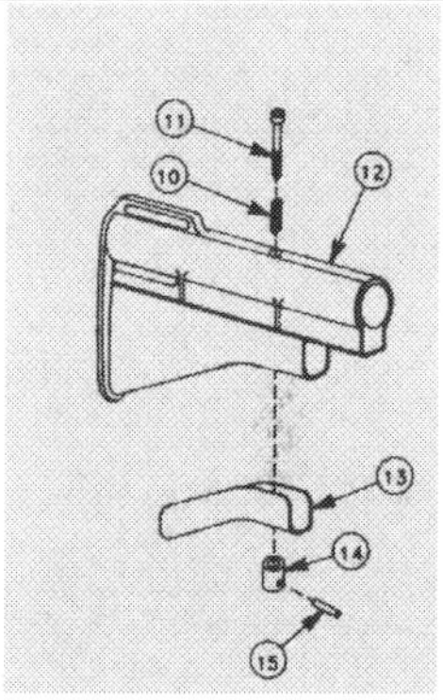

2-19. MAJOR COMPONENTS OF M16A2 RIFLE.

This task covers

a. Reassembly
b. Inspection
c. Stowage

INITIAL SETUP

References
TM 9 1005-319 10

Equipment Conditions
Rifle disassembled into major components

General Safety Instructions
To avoid injury to eyes, use care when removing and installing spring-loaded parts

Do not Interchange bolt assemblies from one rifle to another Doing so may result In Injury to, or death of. personnel

Do not keep live ammunition near the work area

a. REASSEMBLY

Refer to operator's manual and assemble lower receiver and buttstock assembly (1) upper receiver and barrel assembly (2). charging handle assembly (3), bolt carrier assembly (4), sling (5), and magazine(6).

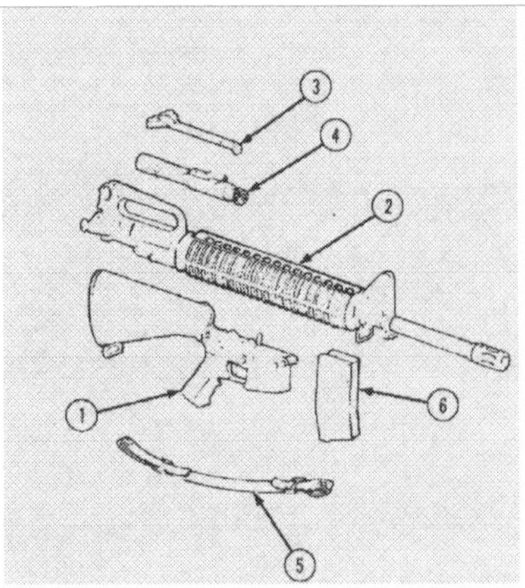

b. INSPECTION

Perform the following function checks on assembled weapon:

1. Remove magazine if installed. Pull charging handle assembly to rear. Check that chamber is clear. Let bolt and bolt carrier close. Do not pull trigger. Leave hammer in cocked position.

WARNING

If rifle fails any of the following tests, continued use of the rifle could result in injury to, or death of, personnel.

2. Place selector lever in SAFE position and pull trigger. Hammer should not fall.

3. Place selector lever in SEMI position. Pull trigger. Hammer should fall.

M16A2 and M4 ONLY
NOTE

For the purpose of the following check "SLOW" is defined as 1/4 to 1/2 the normal rate of trigger release.

4. Hold trigger to the rear, charge weapon, and release the trigger with a slow, smooth motion, without hesitations or stops, until the trigger is fully forward; an audible click should be heard. Hammer should not fall.

5. Repeat the SEMI position test five times. The weapon must not malfunction during any of these five tests. If the weapon malfunctions during any of these five tests, evacuate rifle to support maintenance for repair.

6. Place selector lever in BURST position. Charge weapon and pull trigger. Hammer should fall.

7. Hold trigger to the rear, pull charging handle assembly to rear and release three times. Release trigger. Hammer should not fall. The burst disconnector should have held the hammer to the rear while the trigger was in the pulled position.

8. Pull trigger. Hammer should fall. This should be the first round of a three round burst.

9. With hammer in forward position, attempt to place the selector lever in the SAFE position. If selector lever can be placed on SAFE, evacuate the weapon to support maintenance.

M4A1 ONLY

10. Place selector lever in AUTO position. Charge weapon and pull trigger. Hammer should fall.

11. Hold trigger to the rear, charge weapon and release trigger. Pull trigger. Hammer should not fall. AUTO sear should have released hammer while holding trigger in the squeezed position before releasing and resqueezing the trigger.

12. With hammer in forward position, attempt to place the selector lever in the SAFE position. If selector lever can be placed on SAFE, evacuate the weapon to support maintenance.

1 2-19. MAJOR COMPONENTS OF M16A2 RIFLE (CONT).

c. STOWAGE

Prior to stowing the weapon in arms room, perform the following procedures:

NOTE

Weapon passed into arms room issue window should be passed butt first with the bolt locked to the rear.

1. Clear. Refer to operator's manual.
2. Place selector lever in SEMI position.
3. Pull trigger. Hammer should fall.
4. Close ejection port (dust) cover.
5. Place weapon in rack.

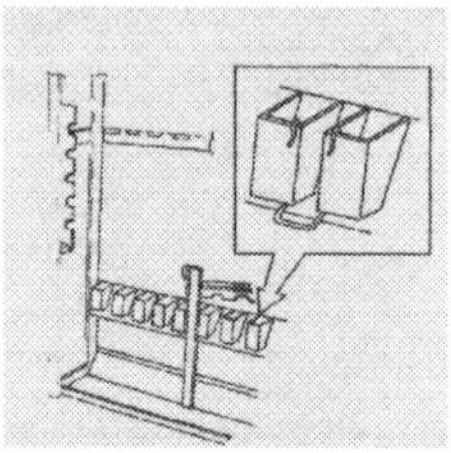

6. M4/M4A1 Carbine Only, use M12 Arms Rack.

a. The M12 arms rack is the correct arms rack in which to store the M4/M4A1 Carbine. The carbine must be stored with buttstock extended. When storing the M4/M4A1 Carbine in the M12 arms rack a mounting bracket, NSN 5340-01-230-3181, (app D, item 23.1) may be used for each M4/M4A1 Carbine being stored. This option is for the convenience of the person who opens and closes the arms rack to store the carbines.

b. To install the mounting bracket on the M12 arms rack, for use with the M4/M4A1 Carbine, install the bracket with the hooks of the bracket facing toward the carbine, so that the lower receiver extension will contact the bent end of the bracket. The bent end of the bracket will hold the carbine upright when the arms rack is opened. The bracket can be turned around when not in use for the carbine to allow storage of the M16A2 rifle.

c. When storing the M4/M4A1 Carbine in the M12 arms rack, an adapter bar (fig E-6, app E), MUST be used for security reasons. To install the adapter bar to the M12 arms rack:

(1) Remove all weapons from the rack and position the rack to gain access to the back.

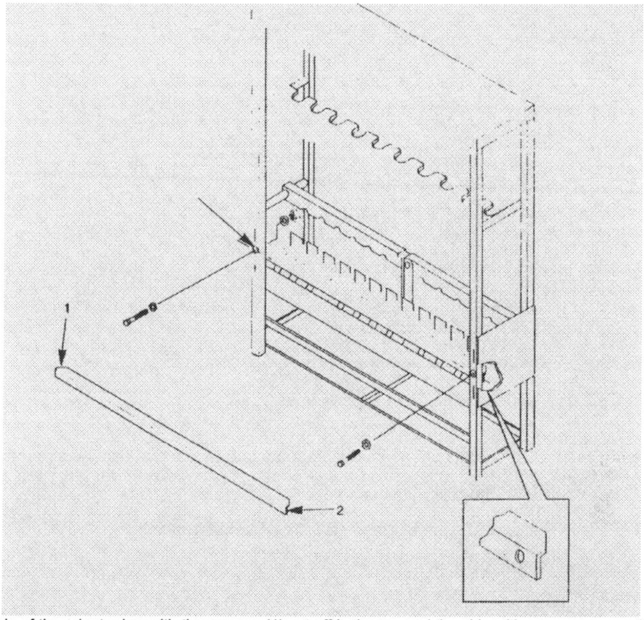

(2) The side of the adapter bar with the corners (1) cut off is the top and the side with square corners (2) is the back. The portion of the bar must be placed so the cut off corners (1) face the front of the rack.

(3) Holding the adapter bar at an angle, place one end into position inside the rear leg of the arms rack. Lower the other end of the bar into position. Allow the adapter bar to rest on the rack.

(4) Clamp both ends of the adapter bar into position. Mark the center line of the leg and adapter bar where they meet (see illustration) Using a center punch, mark the location of the holes to be drilled where the center lines cross. The holes must be centered on both the leg and the adapter bar. Drill a 1/8 inch pilot hole through both arms rack legs and the ends of the adapter bar. Drill a 3/8 inch hole through both arms rack legs and the ends of the adapter bar using the pilot hole as a guide. Remove the adapter bar. File the edges of all holes smooth Paint all bare metal surfaces with olive drab enamel paint, NSN 8010-01-350-5249, or equivalent.

2-19. MAJOR COMPONENTS OF M16A2 RIFLE (CONT).

c. STOWAGE (CONT)

(5) Reinsert the adapter bar into position on the arms rack. Using two 3/8 inch x 2 inch machine screws (MS35206-315, NSN 5305-00-984-5695) or equivalent, four washers (MS21306-1C, NSN 5310-01- 327-9713) or equivalent and two nuts (MS35649-2382, NSN 5310-00-056-3395) or equivalent assemble adapter bar to arms rack and tighten securely. The bolts can be inserted from either the back or the front to meet your requirements. If the rack is placed close to a wall or another rack It is recommended that the bolts be inserted from the back

(6) Tack weld, braze, or peen the threaded end of the bolt to the nut to prevent easy removal

(7) Place rack back into position and replace the weapons

NOTE

Adapter bar must be removed from arms rack prior to turning in the arms rack to the supply system.

d. To remove the adapter bar from the arms rack, remove the head of the bolt (or the nut), using a hand held grinder or file to avoid damage to the aluminum arms rack legs. A hammer and cold chisel may be used If no other way to remove the adapter exist

e. It is recommended that the rail protector NSN 1005-01-394-7677, (app D, item 26.1) be used during storage of the carbines when the carrying handle assembly or some other accessory is not installed on the upper receiver to prevent damage to the mounting surface on the upper receiver.

Section VI. PREPARATION FOR STORAGE OR SHIPMENT

2-20. PREPARATION FOR STORAGE OR SHIPMENT.

a. Packaging of the M16A2 Rifle and the M4/M4A1 Carbine shall be in accordance with the following

ARMY ONLY: Army users shall package the rifle and the carbine in accordance with each respective Packaging Data Sheet (PDS) for shipment or storage which may exceed 90 days The PDS is part of the Army Master Data File Retrieval Microform System (ARMS) Packaging
File

AIR FORCE ONLY: Air Force users shall package the rifle in accordance with each respective Special Packaging Instruction (SPI) 00-856-6885 for shipment or storage which may exceed 90 days The SPIs are part of the Army Master Data File Retrieval Microform System (ARMS) Packaging File

b. Packaging, if required, for shipping/storage which will not exceed 90 days shall be as follows

(1) Clean in accordance with operator's manual

(2) Wrap with MIL-B-121 waterproof material.

(3) Place in barrier bag MIL-B-117, Type I, Class C, or wrap with MIL-B-121, Type I, Grade A, and seal with tape, PPP-T-76.

(4) Place one or more of item in minimum size container. Block and brace in accordance with MIL-STD-1186. Cushion the M16 and similar weight items with PPP-C-843, and use PPP-F-320 as filler, to create a tight pack.

(a) Fiber board containers shall be in accordance with PPP-B-636 and may be Class Domestic. Gross weight and size of material shall determine grade of fiberboard container. PPP-B-640 may also be used.

(b) Wood containers shall be in accordance with PPP-B-601 or PPP-B-621.

(5) Equivalent materials may be used.

c. NSNs are not assigned to all the specified material. If it is necessary to specify an NSN in the TMs, the packing materials will have to be spared and part numbers and NSNs assigned.

d. The specifications used are:

(1)	MIL-B-121	Barrier Material, Greaseproofed, Waterproofed, Flexible (NSN 8135-00-753-4661)
(2)	MIL-B-117	Bag, Sleeve and Tubing -- Interior Packaging (NSN 8135-00-543-6574)
(3)	PPP-B-636	Boxes, Shipping, Fiberboard
(4)	PPP-B-601	Boxes, Wood, Cleated Plywood
(5)	MIL-STD-129	Marking for Shipment and Storage
(6)	PPP-T-76	Tape, Packaging, Paper
(7)	MIL-STD-1186	Cushioning, Anchoring, Bracing, Blocking and Waterproofing; with Appropriated Test Methods
(8)	PPP-C-843	Cushioning Material, Cellulosic
(9)	PPP-F-320	Fiberboard, Corrugated and Solid Sheet Stock (Container Grade), and Cut Shapes
(10)	PPP-B-640	Boxes, Fiberboard, Corrugated, Triple-Wall
(11)	PPP-B-621	Boxes, Wood, Nailed and Locked -- Corner

CHAPTER 3

DIRECT SUPPORT MAINTENANCE INSTRUCTIONS

CHAPTER OVERVIEW

This chapter provides information and instructions to keep the weapon in good repair and contains the following sections:

a. Repair Parts and Special Tools
b. Direct Support Troubleshooting
c. Direct Support Maintenance Procedures for the M1 6A2 Rifle and M4/M4A1 Carbine
d. Preparation for Storage or Shipment
e. Preembarkation Inspection of Material in Units Slated for Overseas Movement

Section I. REPAIR PARTS, SPECIAL TOOLS, TMDE, AND SUPPORT EQUIPMENT

3-1. COMMON TOOLS AND EQUIPMENT. For authorized common tools and equipment, refer to the Modified Table of Organization and Equipment (MTOE) applicable to your unit. Air Force users must maintain the following common tools:

Flat tip screwdriver
Socket wrench handle and socket head screw socket wrench
Vise jaw caps
Machinist's vise
Solid center punch
Hammer
Combination wrench
Torque wrench
Retaining ring pliers
(Two) 8-inch adjustable wrenches
Flat file
Ball-peen hammer
Trigger pull test fixture rod and weights
5/64-inch drive pin punch
1/8-inch drive pin punch
1/16-inch drive pin punch
3/32-inch drive pin punch

3-2. SPECIAL TOOLS, TMDE, AD SUPPORT EQUIPMENT. Special tools required for direct support maintenance are listed in appendixes B and C. Fabricated tools are listed and illustrated in appendix E.

3-3. REPAIR PARTS. Repair parts are listed and illustrated in appendix C of this manual.

NOTE

Bolt assemblies, and/or barrel assemblies may be interchanged , at the Direct Support Maintenance level, from one rifle to another, under the provisions of the note on page C-3: If these parts are interchanged the rifle must be checked/inspected as depicted on pages 3-17, 3-21, and 3-33. While performing these checks and inspections, pay special attention to the headspace requirements on page 3-47.

Section II. TROUBLESHOOTING

3-4. GENERAL.

a. This section contains direct support level troubleshooting information for locating and correcting most of the operating troubles which may develop in the M16A2 rifle and M4/M4A1 carbine. Each malfunction for the individual part or assembly is followed by a list of tests or inspections which will help you to determine the corrective action to take. You should perform the tests/inspections and corrective actions in the order listed.

b. This manual cannot list all malfunctions that may occur, nor all tests or inspections and corrective actions. If a malfunction is not listed or is not corrected by listed corrective actions, see individual repair sections in the maintenance procedures on each major assembly.

3-5. TROUBLESHOOTING PROCEDURES. Refer to troubleshooting table for malfunctions, tests, and corrective actions. The symptom index is provided for a quick reference of the malfunctions covered in the table.

SYMPTOM INDEX

	Troubleshooting Procedures Page
Failure of magazine to lock in rifle	3-3
Failure to feed	3-3
Failure to chamber	3-3
Failure to lock	3-4
Failure to fire	3-5
Failure to unlock	3-6
Failure to extract	3-6
Failure to eject	3-6
Failure to cock	3-7
Short recoil	3-9
Rifle cannot be zeroed	3-10
Failure to cycle with selector lever set o n BURST	3-11
Fires two rounds with one pull of trigger with selector lever set on SEMI (double firing)	3-13
Fires with selector lever on SAFE on when trigger is released with selector lever on SEMI	3-13
Hammer pin "walks"	3-14
Bolt assembly fails to lock to rear after firing last round	3-14

TROUBLESHOOTING

MALFUNCTION
TEST OR INSPECTION
CORRECTIVE ACTION

1. FAILURE OF MAGAZINE TO LOCK IN RIFLE.
Step 1. Dirty or corroded magazine catch (1).
Disassemble and clean.
Step 2. Defective magazine catch spring (2).
Replace magazine catch spring (2) (p 3-62).
Step 3. Worn or broken magazine catch (1).
Replace magazine catch (1) (p 3-62).

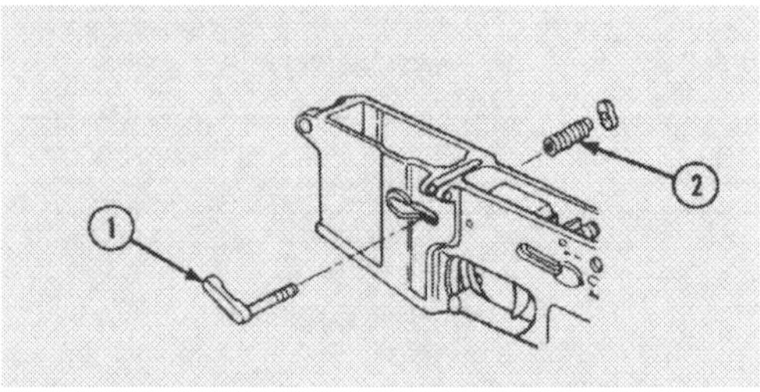

2 FAILURE TO FEED
Step 1 Magazine catch spring (1) weak or broken.
Replace magazine catch spring l1) (p 3-62)

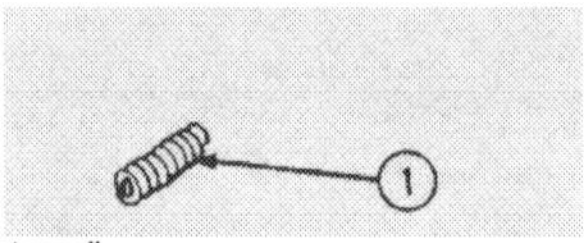

Step 2 Short recoil
Refer to page 3-9

3 FAILURE TO CHAMBER
Short recoil.
Refer to page 3-9.

3-5. TROUBLESHOOTING PROCEDURES (CONT).

TROUBLESHOOTING (CONT)

MALFUNCTION
TEST OR INSPECTION
CORRECTIVE ACTION

4. FAILURE TO LOCK.

Step 1. Damaged bolt carrier key (1).
Repair or replace bolt carrier key (1) (p 3-25)

Step 2 Loose screws (2) on bolt carrier key (1).
a. Disassemble and repair (p 3-25)
b. Reassemble using new screws.

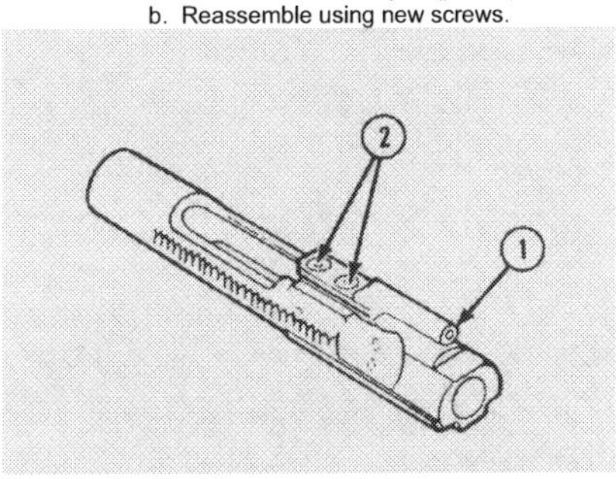

Step 3 Bent gas tube (3).
a Adjust by bending gas tube 13) In area of handguards
b Replace gas tube (3) and check alignment (p 3-29)

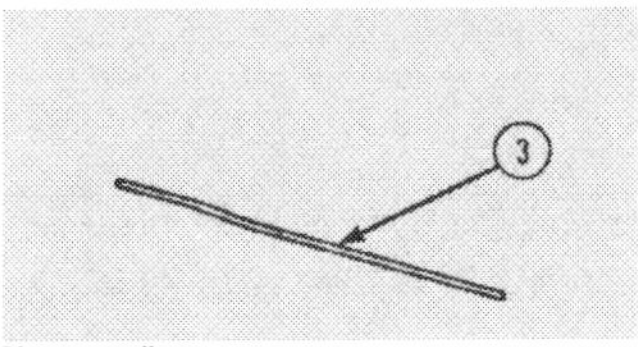

Step 4. Short recoil
Refer to page 3 9

TROUBLESHOOTING (CONT)

MALFUNCTION
TEST OR INSPECTION
CORRECTIVE ACTION

5. FAILURE TO FIRE.

Step 1. Broken hammer (1).
Replace hammer (1) (p 3-73).

Step 2. Weak or broken hammer spring (2).
Replace hammer spring (2) (p 3-73).

Step 3. Hammer spring (2) improperly assembled.
Assemble properly.

Step 4. **(M16A2 and M4 ONLY)** Burst cam (3) and/or cam spring (4) frozen or improperly assembled
Disassemble, clean, lubricate, and reassemble correctly (p 3-73).

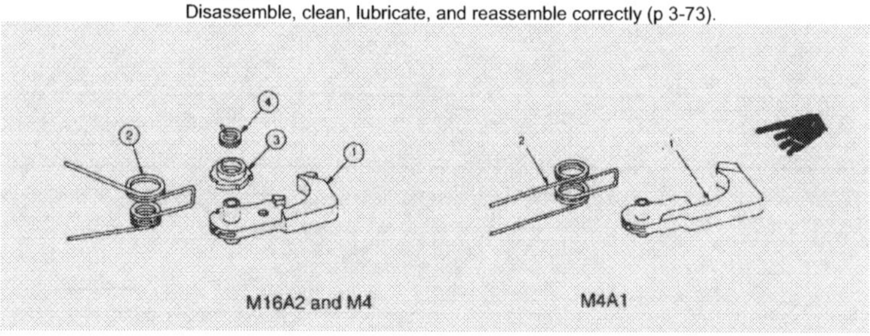

Step 5. Selector lever (5) frozen on SAFE position.
Disassemble and clean (p 3-62)

Step 6. Broken firing pin (6) or firing pin does not meet gage protrusion requirement.
Replace firing pin (6) (p 3-16).

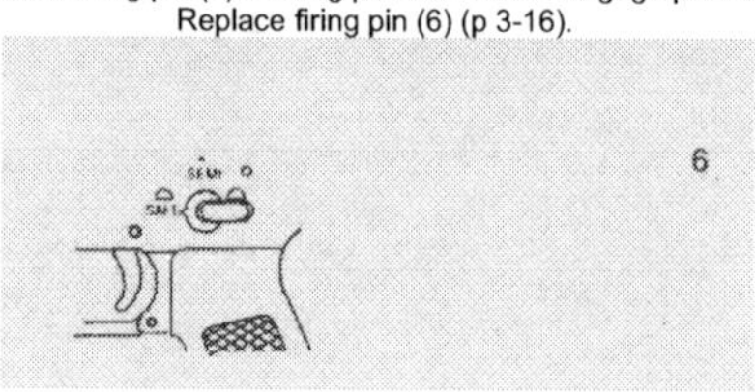

3-5. TROUBLESHOOTING PROCEDURES (CONT).

TROUBLESHOOTING (CONT)

MALFUNCTION
TEST OR INSPECTION
CORRECTIVE ACTION

6. FAILURE TO UNLOCK.

Step 1 Burred locking lugs (1) on bolt assembly.
Remove burrs.

Step 2 Burred lugs (2) on barrel extension.
Remove burrs.

Step 3. Short recoil.
Refer to page 3-9.

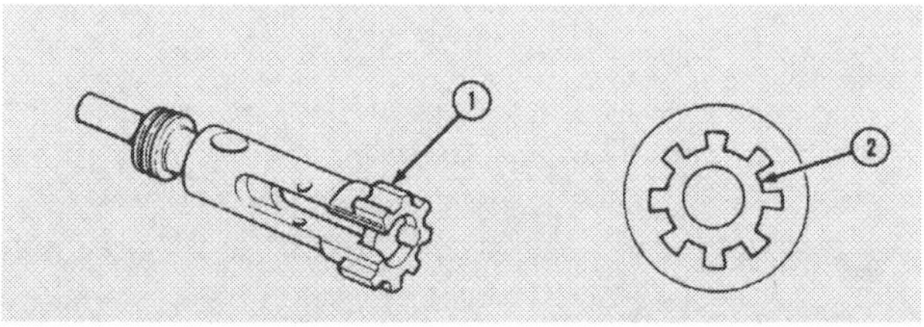

7. FAILURE TO EXTRACT

Step 1. Inspect cartridge extractor and extractor spring assembly.
Replace if cracked or broken (p 2-35).

Step 2. Inspect badly pitted chamber with reflector tool (Item 2, fig. C-1 7).
Replace barrel assembly If chamber Is badly pitted (p 3-29).

8 FAILURE TO EJECT.

Short recoil.
Refer to page 3-9.

TROUBLESHOOTING (CONT)

MALFUNCTION
TEST OR INSPECTION
CORRECTIVE ACTION

9. FAILURE TO COCK.

Step 1. Worn or broken trigger nose (1) or trigger spring (2).
Replace trigger (3) or defective trigger spring (2) (p 3-62).

Step 2. Worn or broken hammer trigger notch (4).
Replace hammer (5) (p 3-62).

Step 3. Worn or broken hammer disconnector hook (6).
Replace hammer (5) (p 3-62).

Step 4. Worn or broken hammer automatic sear hook (7).
Replace hammer (5) (p 3-62).

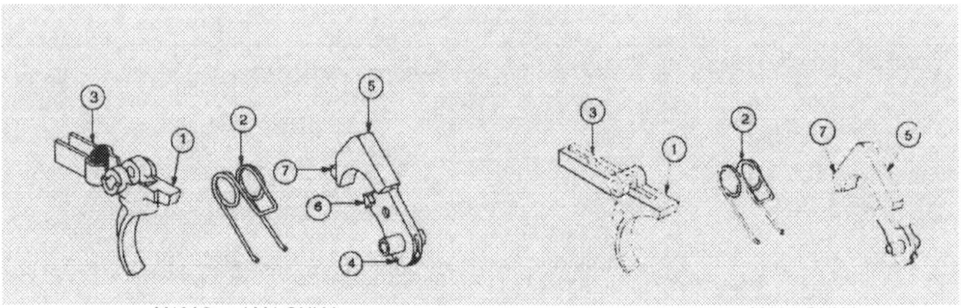

M16A2 and M4 ONLY M4A1 ONLY

3-5. TROUBLESHOOTING PROCEDURES (CONT).

TROUBLESHOOTING (CONT)

MALFUNCTION
TEST OR INSPECTION
CORRECTIVE ACTION

Step 5 **(M4A1 has only 1 disconnector)** Worn or broken disconnector hooks (8).
Replace defective disconnectors (9) (p3-62).

Step 6 **(M4A1 has only 1 spring)** Weak, broken, or missing disconnector springs (10).
Replace disconnector springs (10) (p 3-62).

Step 7. Worn, broken, or missing automatic sear (11).
Replace automatic sear (11) (p 3-62).

Step 8 Weak or broken automatic sear spring (12).
Replace automatic sear (11) (p 3-62).

Step 9. Long leg (13) of automatic sear spring incorrectly assembled in receiver.
Remove automatic sear assembly (11) and install correctly (p 3-62).

Step 10. **(M16A2 and M4 ONLY)** Burst cam (14) or clutch spring (15) frozen or improperly assembled.
Disassemble, inspect, clean, lubricate, or replace as required (p 3-73).

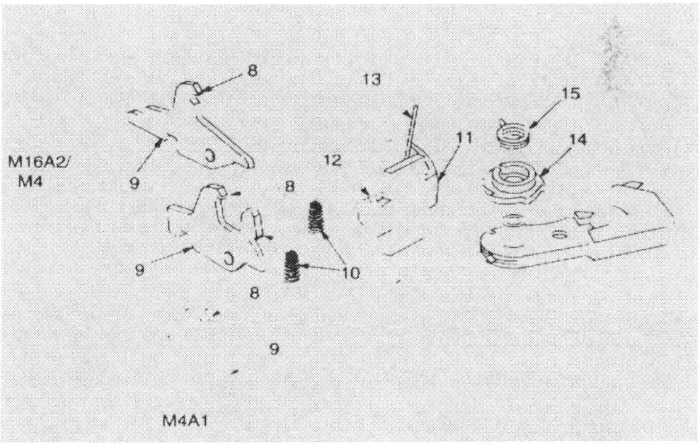

TROUBLESHOOTING (CONT)

MALFUNCTION
TEST OR INSPECTION
CORRECTIVE ACTION

10. SHORT RECOIL.

Step 1. Improper gap space or worn, missing, or broken bolt rings (1).

a. Stagger bolt ring gaps (p 3-21).

b. Replace bolt rings (1) and stagger gaps (p 3-21).

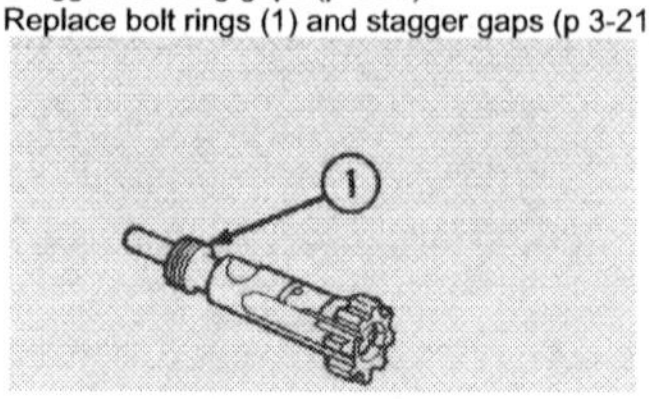

Step 2. Broken or bent gas tube (2).

Adjust by bending in area of handguards or replace gas tube (2) (p 3-29).

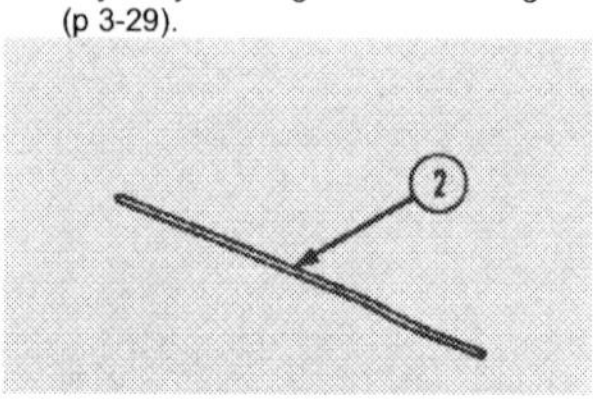

Step 3. Gas tube spring pin (3) missing from front sight (4).

Replace gas tube spring pin (3) (p 3-29).

Step 4. Partially plugged gas system because of carbon build-up in the gas tube (2).

Replace gas tube (2) (p 3-29).

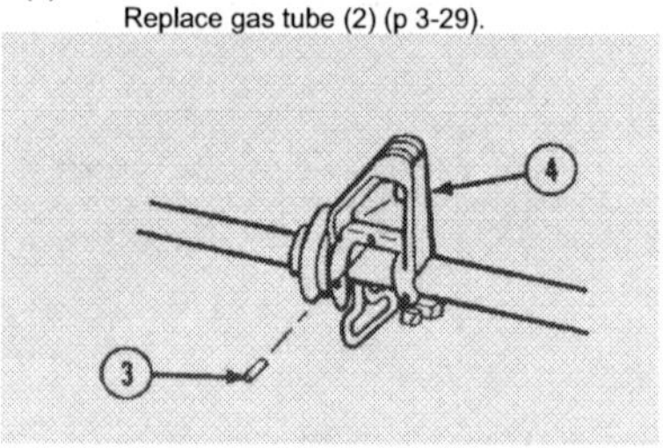

3-5. TROUBLESHOOTING PROCEDURES (CONT).

TROUBLESHOOTING (CONT)

MALFUNCTION
TEST OR INSPECTION
CORRECTIVE ACTION

10 SHORT RECOIL (CONT).

WARNING

When using carbon removing compound (Item 8, app D), avoid skin contact. If it comes in contact with the skin, wash off thoroughly with running water. The use of a good lanolin base cream after exposure to compound Is helpful. The use of gloves and protective equipment Is required. Step 5. Carbon build-up in barrel gas port (5). Remove carbon build-up by soaking barrel In carbon removing compound (item 8, app D). Use rubber gloves (Item 18, app D) with carbon removing compound. Use a bore small arms cleaning brush (Item 4, app D).

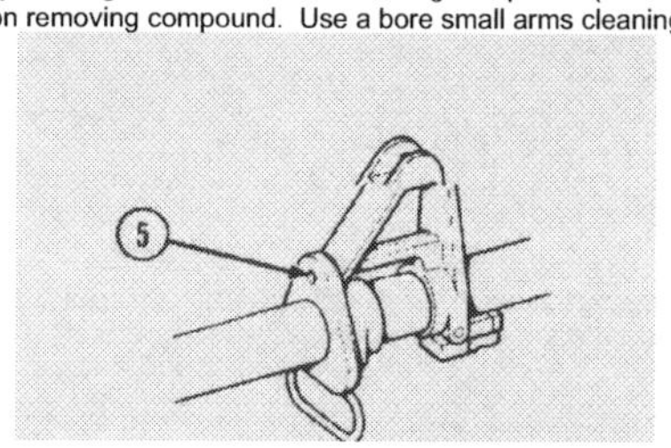

11. RIFLE CANNOT BE ZEROED.

Step 1 Inspect for defective or bent barrel assembly (1) (p 3-29).
Replace barrel assembly (1) (p 3-29).

Step 2 (For windage) barrel assembly (1) out of alignment with rear sight on up per receiver.
Align barrel assembly (1) and upper receiver (p 3-291

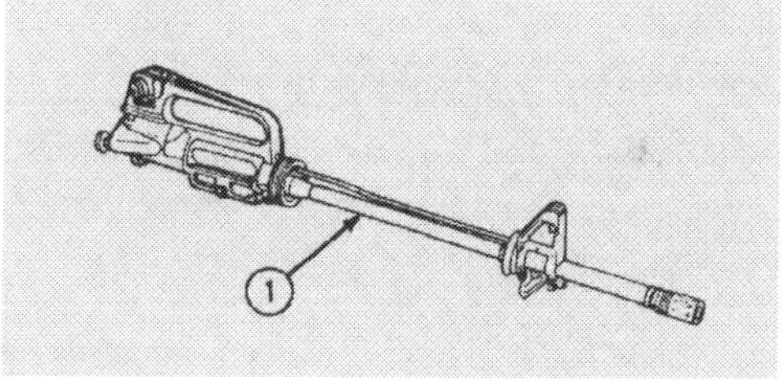

TROUBLESHOOTING (CONT)

MALFUNCTION
TEST OR INSPECTION
CORRECTIVE ACTION

Step 3. (For elevation) defective front sight (2) or rear sight (3).
Repair as required (p 3-48).

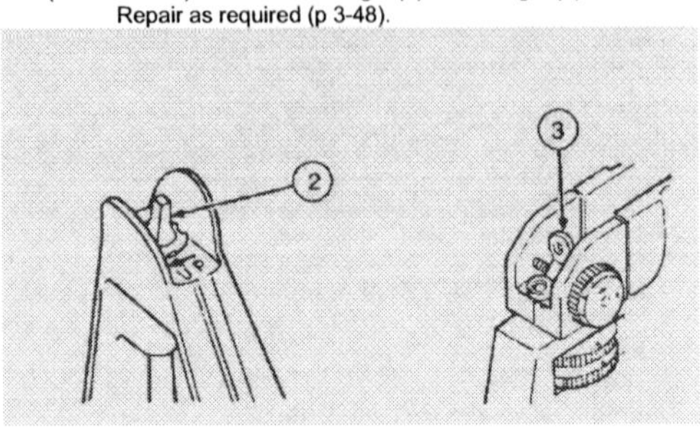

12. FAILURE TO CYCLE WITH SELECTOR LEVER SET ON BURST (M16A2 and M4 ONLY).

Step 1. Broken automatic sear (1) or spring (2).
Replace automatic sear assembly (1) (p 3-62).

Step 2. Faulty selector lever (3).
Replace selector lever (3) (p 3-62).

Step 3. Broken tooth on burst cam (4).
Replace burst cam (4) (p 3-73).

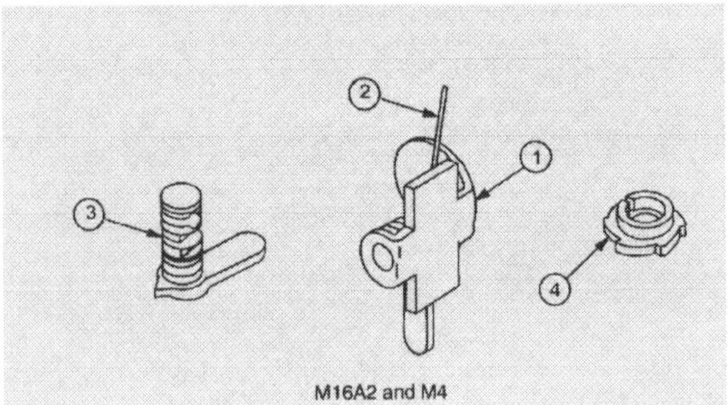

M16A2 and M4

3-5. TROUBLESHOOTING PROCEDURES (CONT).

TROUBLESHOOTING (CONT)

MALFUNCTION
TEST OR INSPECTION
CORRECTIVE ACTION

12 FAILURE TO CYCLE WITH SELECTOR LEVER SET ON BURST (M16A2 and M4 ONLY) (CONT).

Step 4 Broken cam clutch spring (5) Cam clutch spring should be bent and properly formed without any sharp edges or corners
Inspect and replace If required.

Step 5 The bend In the cam clutch spring (5) installed backwards (toward outside).
Install cam clutch spring (5) properly with the bend to the inside (p 3-73).

NOTE

When hammer Is rotated back to cocked position, cam should rotate to allow the burst disconnector to latch in the next notch.

Step 6. Cam clutch spring (5) fails to "clutch" and burst cam (4) fails to rotate back with hammer (6).
Replace cam clutch spring (5) (p 3-73). If problem continues, replace hammer (6) and cam (4) (p 3-73).

Step 7. Broken front hook (7) on burst disconnector (8).
Replace burst disconnector (8) (p 3-62).

Step 8. Short recoil.
Refer to page 3-9

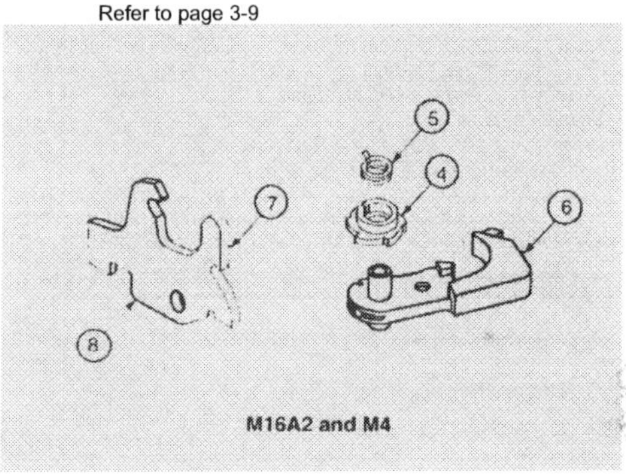

TROUBLESHOOTING (CONT)

MALFUNCTION
TEST OR INSPECTION
CORRECTIVE ACTION

12.1. FAILURE TO CYCLE WITH SELECTOR LEVER SET ON AUTO (M4A1 ONLY)

Step 1. Broken automatic sear (1) or spring (2).
Replace automatic sear assembly (1).

Step 2. Faulty selector lever (3).
Replace selector lever (3).

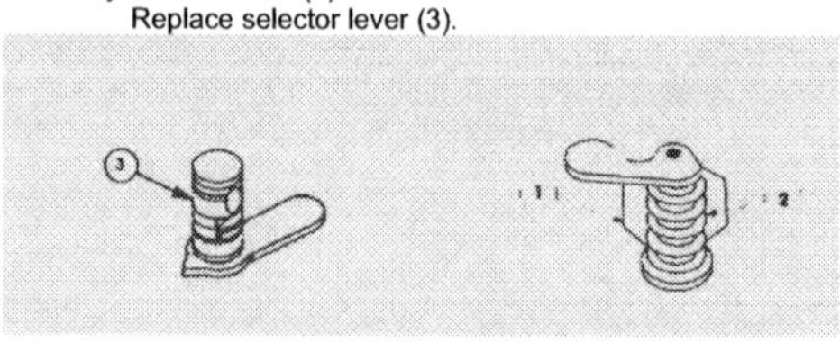

Step 3. Short recoil.
Refer to page 3-9.

13. FIRES TWO ROUNDS WITH ONE PULL OF TRIGGER WITH SELECTOR LEVER SET ON SEMI (DOUBLE FIRING).

Step 1. Defective semiautomatic disconnector (1). Replace semiautomatic disconnector (1) (p 3-62).

Step 2 Worn or broken trigger notch (2) of hammer (3) (searing portion). Replace hammer (3) (p 3-62).

Step 3. Worn or broken disconnector notch (4) of hammer (3). Replace hammer (3) (p 3-62).

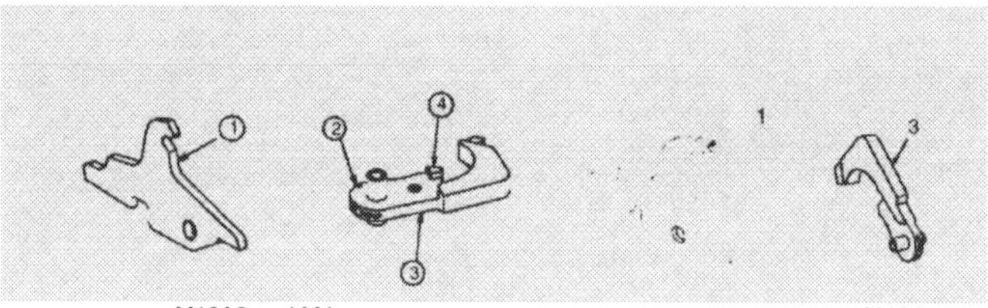

M16A2 and M4 M4A1

3-5. TROUBLESHOOTING PROCEDURES (CONT).

TROUBLESHOOTING (CONT)

MALFUNCTION
TEST OR INSPECTION
CORRECTIVE ACTION

13 FIRES TWO ROUNDS WITH ONE PULL OF TRIGGER WITH SELECTOR LEVER SET ON SEMI (DOUBLE FIRING) (CONT).

Step 4 Worn or broken trigger (5) (searing portion).
Replace trigger (5) (p 3-62).

Step 5 Worn trigger or hammer pin hole (6).
Gage trigger pin hole (7) and hammer pin hole (6) (p 3-62). If test falls, replace weapon.

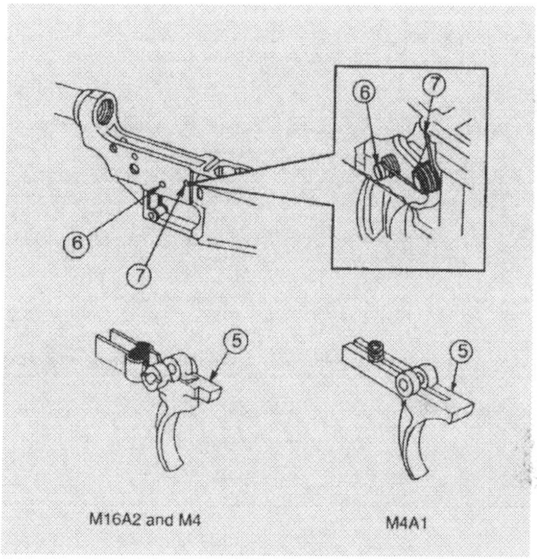

TROUBLESHOOTING (CONT.)

MALFUNCTION

TEST OR INSPECTION

CORRECTIVE ACTION

14. FIRES WITH SELECTOR LEVER ON SAFE OR WHEN TRIGGER IS RELEASED WITH SELECTOR LEVER ON SEMI.

Step 1. Defective selector lever (1).

Replace selector lever (1) (p 3-62).

Step 2. Worn or broken trigger (rear portion) (2).

Replace trigger (3) (p 3-62).

3-5. TROUBLESHOOTING PROCEDURES (CONT.).

TROUBLESHOOTING (CONT.)

MALFUNCTION

TEST OR INSPECTION

CORRECTIVE ACTION

15. HAMMER PIN "WALKS".

Hammer pin (1) "walks" or works loose during firing or hammer pin Is very easy to push out of receiver when hammer Is Installed.

Replace hammer assembly (p 3-62).

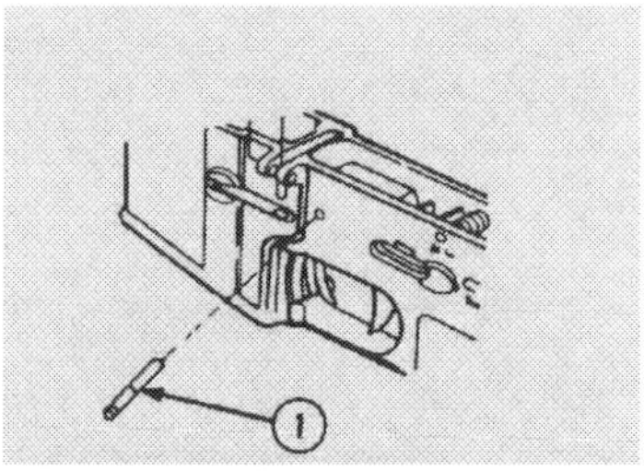

16. BOLT ASSEMBLY FAILS TO LOCK TO REAR AFTER FIRING LAST ROUND.

Step 1. Broken bolt catch (1).

Replace bolt catch (1) (p 3 62,.

Step 2. Weak or broken bolt catch spring (2).

Replace bolt catch spring (2) (p 3-62)

Step 3 Restricted movement of bolt catch (1).

Disassemble and clean.

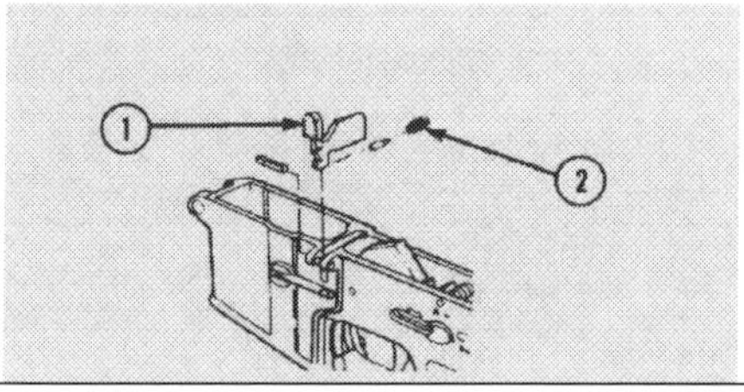

Section III. MAINTENANCE PROCEDURES FOR THE M16A2 RIFLE AND M4/M4A1 CARBINES

3-6. MAJOR COMPONENTS OF M16A2 RIFLE AND M4/M4A1 CARBINES.

This task covers disassembly.

INITIAL SETUP

Tools
(ARMY) Small Arms Repairman Tool Kit
(item 3, app B)

References
TM 9-1005-319-10 (operator's manual)

Equipment Conditions
Weapon assembled

General Safety Instructions
Before starting an inspection, be sure to clear the weapon Do not pull the trigger until the weapon has been cleared.
Inspect the chamber and receiver to ensure no ammunition is present Do not keep live ammunition near work area.

DISASSEMBLY

a. Refer to operator's manual.

b. Remove magazine (1), sling (2), bolt carrier assembly (3), charging handle assembly (4), and upper receiver and barrel assembly (5), carrying handle assembly (5.1), from lower receiver and buttstock assembly (6).

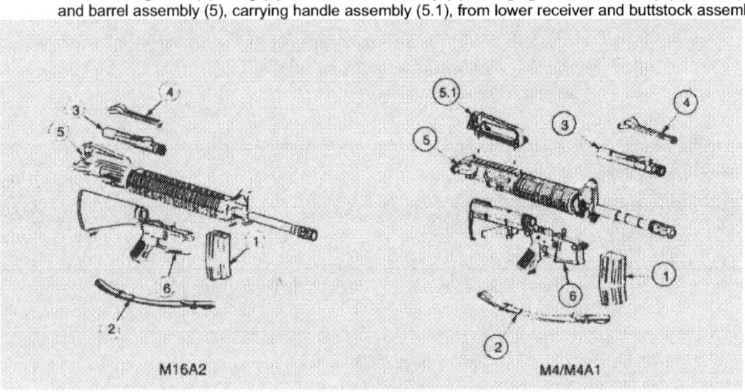

NOTE

Solid Film Lubricant (SFL) is the only authorized touchup for the M16A2 rifle and M4/M4A1 carbines and may be used on up to one third of the exterior finish of the weapon. **FOR ARMY CONUS USE ONLY AND AIR FORCE TRAINING WEAPONS ONLY**: SFL may be used as touchup without limitation on the upper receiver and barrel assembly. This is to say that units which DO NOT fall under the category of Divisional Combat Units or rapid deployment type units may have up to 100 percent of the exterior surface of the upper receiver and barrel assembly protected with SFL if necessary.

3-7. BOLT CARRIER ASSEMBLY.

This task covers:

a	Disassembly	**d**	Test
b	Cleaning	**e**	Repair
c	Inspection	**f**	Reassembly

INITIAL SETUP

Test Equipment
- Tool and Gage Set (item 2, app B)

Tools
- (ARMY) Small Arms Repairman Tool Kit (item 3, app B)

Materials/Parts
- Cleaner, lubricant, and preservative (CLP) (item 9, app D)
- Cleaning compound, rifle bore (RBC) (item 12, app D)
- Pipe cleaner (item 11, app D)

Equipment Conditions
- 3-1 5 Bolt carrier assembly removed

a. DISASSEMBLY

1. Remove firing pin retaining pin (1).
2. Tip key and bolt carrier assembly (2) allowing firing pin (3) to drop out. Catch the firing pin.
3. Rotate bolt cam pin (4) one quarter turn and lift straight up to remove.
4. Remove bolt assembly (5) from key and bolt carrier assembly (2)

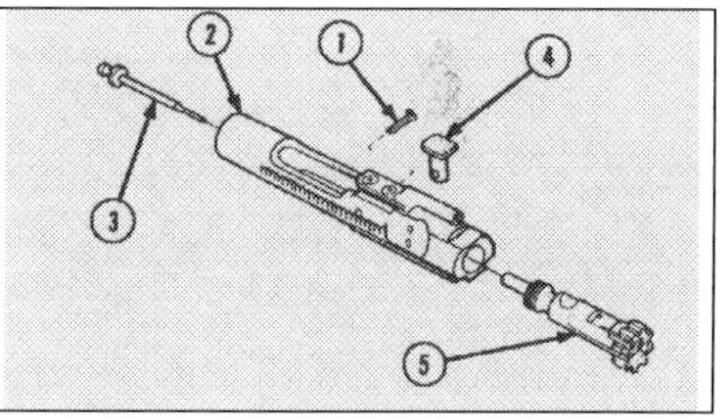

b. CLEANING

Clean carrier key and gas relief ports using CLP or RBC and pipe cleaner

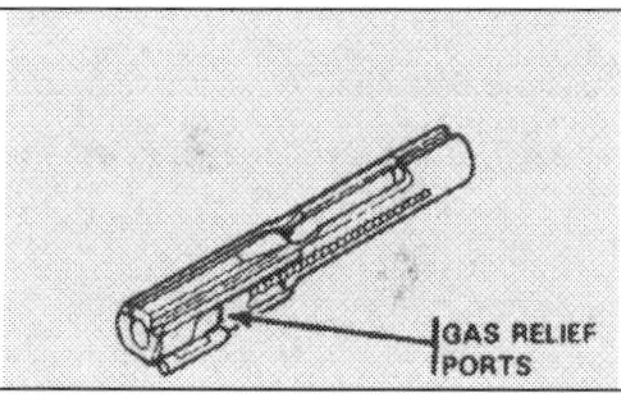

c. INSPECTION

1. Inspect bolt carrier assembly using the following guidelines.

(a) Inspect bolt carrier assembly (1) for burrs, cracks, wear, and evidence of gas loss.

(b) Visually inspect the carrier and key screws (2) for looseness and proper staking as shown below.

NOTE

Do not attempt to retorque if there is no loosening of the screws indicated by the staking marks.

Surface "A" must not Indicate distortion or damage which Impairs parallelism.

NOTE

A maximum of 0.025 in. (0.064 cm) protrusion In an upward direction is permissible.

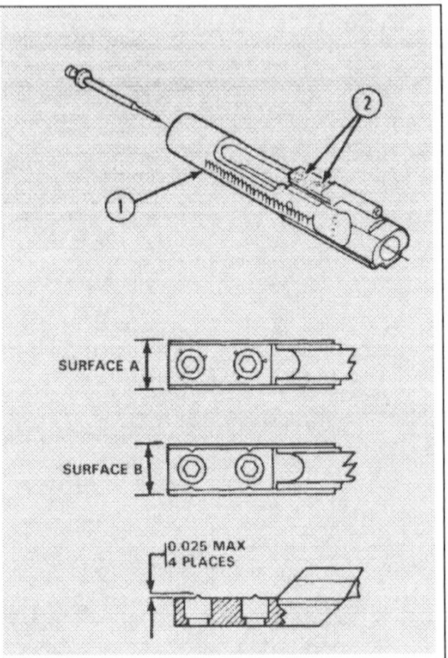

NOTE

There are bolts and bolt carriers on fielded rifles, some with chrome-plated exterior surface finishes and some with phosphate coating Both finishes are acceptable under certain operational requirements and or restrictions Phosphate-coated bolt carriers are required for divisional combat units Chrome plated bolt carriers are acceptable for divisional noncombat units and training center units. Chrome-plated and phosphate-coated bolt assemblies, bolt carrier assemblies, and repair parts for these assemblies may be intermixed In any combination, with the following exception:

Phosphate-coated bolt carriers are required for all deployable and deploying units Chrome-plated bolt carriers are acceptable for nondeployable and training center units.

3-7. BOLT CARRIER ASSEMBLY (CONT.).

c. INSPECTION (CONT.)

2. Inspect firing pin 131 tip for proper con tour Inspect for pitting, wear, and burrs Pits or wear in area (4) Is permissible Replace firing pin If defective

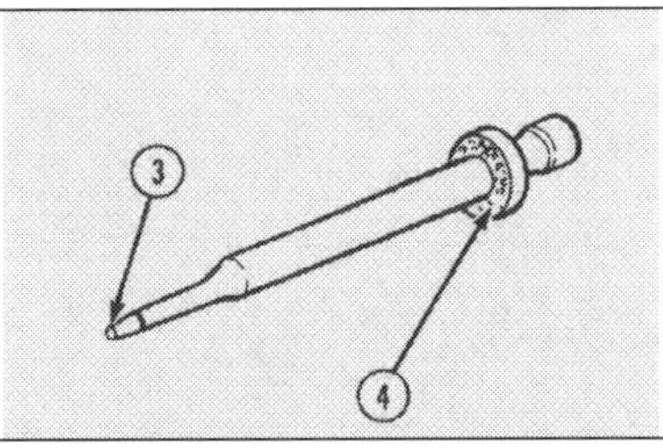

3. Prior to reassembly, insert bolt assembly (5) Into key and bolt carrier assembly (6) (do not insert bolt cam pin) and exercise bolt assembly in and out of key and bolt carrier assembly Check for binding

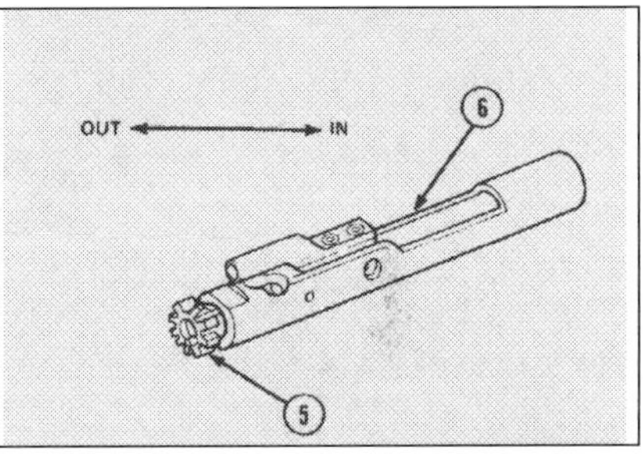

4. Check bolt assembly (5) for proper fit with bolt cam pin removed Turn key and bolt carrier assembly (61 and suspend so the bolt assembly is pointed down

NOTE

The bolt assembly must not drop out If weight of bolt assembly allows It to drop out of key and bolt carrier assembly, replace bolt rings (p 3-21)

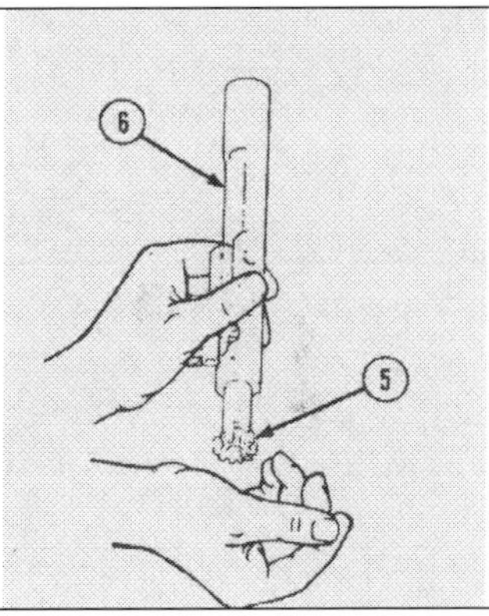

This page is blank. Procedures and figure were moved to page 2-36.

3-7. BOLT CARRIER ASSEMBLY (CONT.).

d. TEST

1. Insert firing pin (1) through bolt (2).

2. Position firing pin protrusion gage (3) PN 7799735 to check for proper firing pin (1) protrusion (minimum 0.028 in. (0 07 cm)-maximum 0 036 In. (0.09 cm))

NOTE

Firing pin should touch the gage on minimum but should not touch on maximum.

3. Replace a defective firing pin.

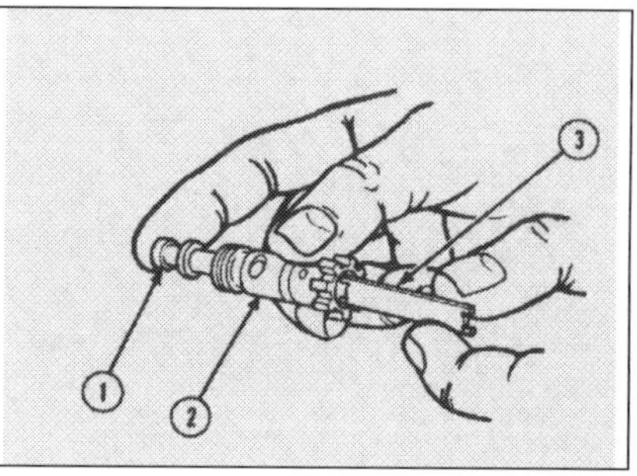

e. REPAIR

Replace all authorized unserviceable items. Retest all replaced parts.

f. REASSEMBLY

1. Install bolt assembly (1) into key and bolt carrier assembly (2).

2 Install bolt cam pin (3) and rotate one quarter turn to secure bolt assembly (1)

3. Hold key and bolt carrier assembly (2) with bolt assembly (1) down and drop In firing(pin (4)

4. Install firing pin retaining pin (5) from left side only to ensure proper Installation Check Installation by attempting to shake out firing pin

5. Reassemble rifle, refer to page 3-77

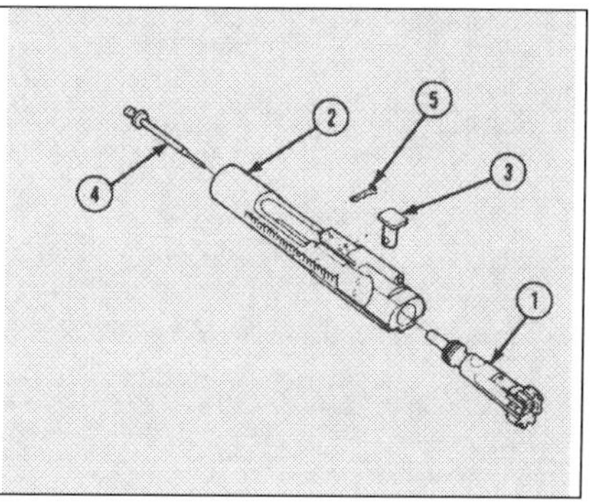

3-8. BOLT ASSEMBLY.

This task covers:

a. Disassembly
b. Inspection/Repair
c. Test
d. Reassembly

INITIAL SETUP

Test Equipment
Tool and Gage Set (Item 2, app B)

Tools
(ARMY) Small Arms Repairman Tool Kit (Item 3, app B)

Materials/Parts
Penetrant kit (item 25, app D)
Rag, wiping (item 26, app D)

Equipment Conditions
3-16 Bolt assembly removed

a. DISASSEMBLY

NOTE
Do not remove bolt rings unless they require replacement and three new replacement bolt rings are on hand.

Using small flat tip jeweler's screwdriver, remove the three bolt rings (1) from the bolt (2).

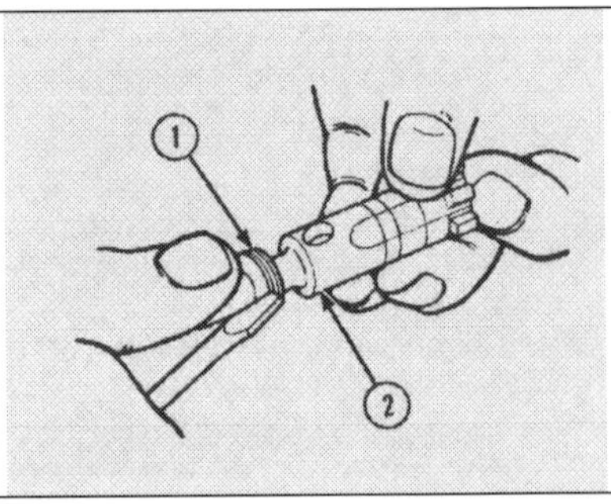

b. INSPECTION/REPAIR

1. Visually Inspect bolt rings for cracks, kinks, and bends. Replace all three bolt rings if one or more bolt rings is damaged. See page 3-16 for bolt ring wear check.

3-8. BOLT ASSEMBLY (CONT.).

b. INSPECTION/REPAIR (CONT.)

2. Inspect bolt for pits, burrs, and wear as follows.

(a) Bolt faces with a cluster of pits which are touching or tightly grouped, covering an area measuring approximately 1,8 Inch across, will be rejected and replaced.

(b) Bolts which contain individual pits or a scattered pattern will not be cause for rejection.

(c) Bolts that contain pits extending Into the firing pin hole will not be rejected unless firing pin hole gaging check determines excess wear.

(d) Rings on the bolt face (machine tool marks), grooves, or ridges less than approximately 0.010 inch will not be cause for rejection.

3. Inspect bolt for cracks In the locking lugs and the bolt cam pin hole area. Use black light If available; otherwise, use a glass of no more than 3X magnification or use a penetrant kit (item 25, app D). Pay close attention to the area where the locking lugs meet the body. Replace bolt assembly if bolt is defective.

WARNING

Dry cleaning solvent is flammable and toxic and should be used in a well-ventilated area. The use of rubber gloves is necessary to protect the skin when washing rifle parts.

4. Use penetrant kit (Item 25, app D) to check for cracks in bolt as follows:

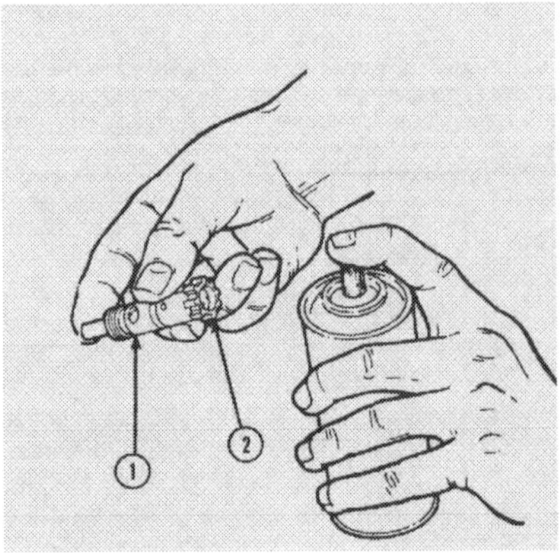

(a) The area to be inspected must be clean, free of oil, etc. Spray a small amount of remover on the area to be inspected, let dry and wipe off with a wiping rag.

(b) Spray penetrant (only enough to wet the area) on the area of the bolt (1) to be inspected.

(c) Spray developer over the penetrant and let the developer work. Cracks will be indicated by a change in color where there is a crack. If there are cracks, the component 1' unserviceable.

(d) Pay close attention to the area where the locking lugs (2) meet the body.

(e) If there are no cracks, spray remover on the area; let dry and wipe off with a wiping rag. Oil the area to prevent corrosion.

(f) Replace bolt assembly If bolt (1) is defective.

NOTE

Replacement of the bolt assembly will require that the headspace be tested (p 3-45, TEST).

3-8. BOLT ASSEMBLY (CONT.).

c. TEST

Test bolt (1) for elongated or oversized firing pin hole using special no-go plug gage (2) PN 12620101.

NOTE

Bolts with firing pin holes which permit the special no-go plug gage to fully penetrate at any position on the circumference will be rejected and replaced.

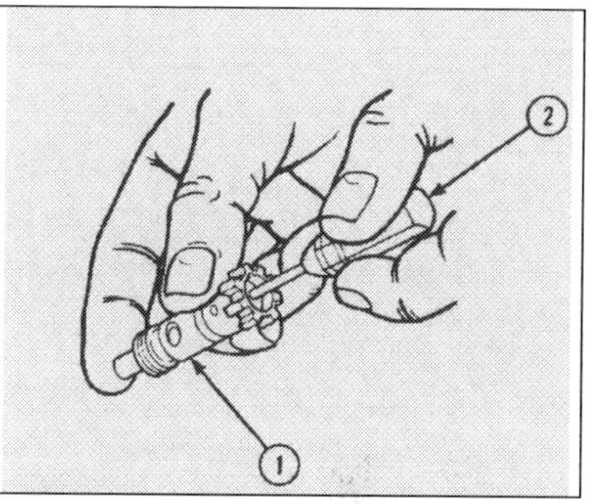

d. REASSEMBLY

NOTE

To install a bolt ring, carefully place one end in the bolt ring groove and hold In place with the thumb of one hand With the index finger of the other hand, gently guide and push the rest of the bolt ring into the groove a little bit at a time until the entire bolt ring is in place.

1. Install the three bolt rings (1) one at a time onto the bolt (2) using care not to bend or "spring" new bolt rings Stagger the bolt ring gaps (approximately ' turn apart).

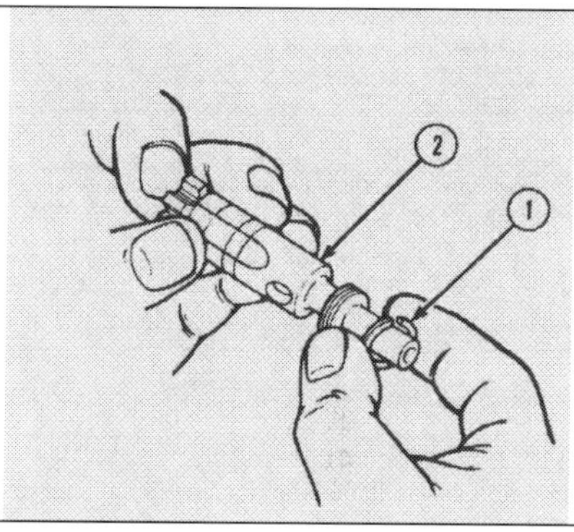

NOTE

Make certain bolt ring gaps are staggered to prevent loss of gas pressure. New bolt rings will make installing the bolt assembly difficult.

Lubricate inside key and bolt carrier assembly and use gentle pressure when installing.

2. Reassemble rifle, refer to page 3-77.

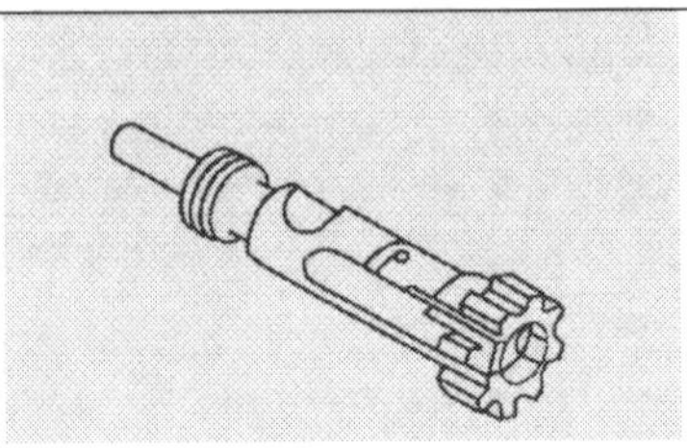

3-9. KEY AND BOLT CARRIER ASSEMBLY.

This task covers:

a Disassembly
b Repair
c Reassembly

INITIAL SETUP

Tools
- Field Maintenance Basic Less Power Small Arms Shop Set (item 1, app B)
- (ARMY) Small Arms Repairman Tool Kit (item 3, app B)

Materials/Parts
- Carrier and key screws (2) (8448508)

Equipment Conditions
- 3-16 Key and bolt carrier assembly removed

3-9. KEY AND BOLT CARRIER ASSEMBLY (CONT.).

a. DISASSEMBLY

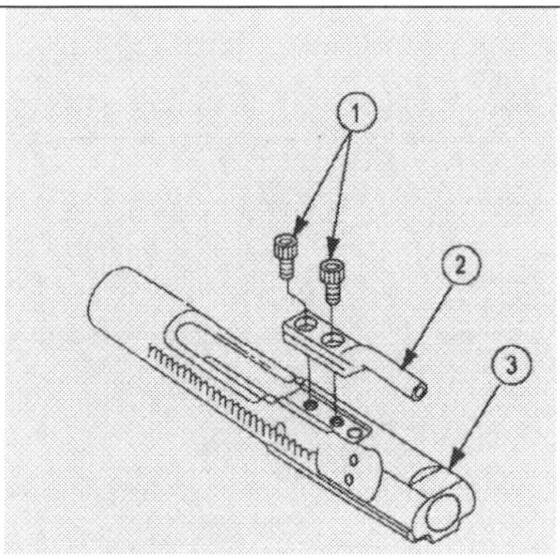

NOTE

Do not disassemble the key and bolt carrier assembly unless the bolt carrier key is defective as determined by inspection procedures described in section III, preceding.

1. Using socket wrench handle and tight fitting 1/8 inch socket head screw socket wrench attachment, remove two carrier and key screws (1).

NOTE

The heads and part of the bolt carrier key may be ground off in order to remove bolt carrier key from bolt carrier if carrier and key screws cannot otherwise be removed.

2. Remove bolt carrier key (2) from bolt carrier (3).

b. REPAIR

NOTE

Do not retorque carrier and key screws if staking marks do not indicate loosening screws.

Repair by replacing, torquing, and restaking carrier and key screws. Refer to the following reassembly procedures.

c. **REASSEMBLY**

NOTE

Do not reuse old carrier and key screws New carrier and key screws must be used at assembly.

1. If disassembled, place bolt carrier (1) In vise using vise law caps Install and position bolt carrier key (2) on bolt carrier (1).

2. Install two carrier and key screws (3). Always use new screws.

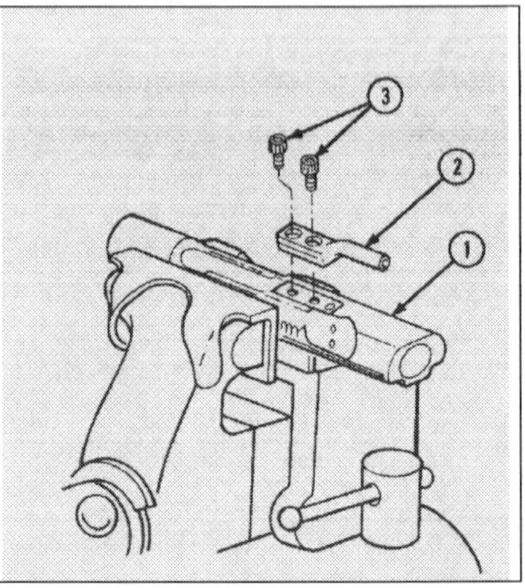

3. Use a tight-fitting 1 8 inch socket head screw wrench attachment and an Inch-pound torque wrench to torque the carrier and key screws (3) to 35 to 40 Inch-pounds (3 95 to 4 52 N m)

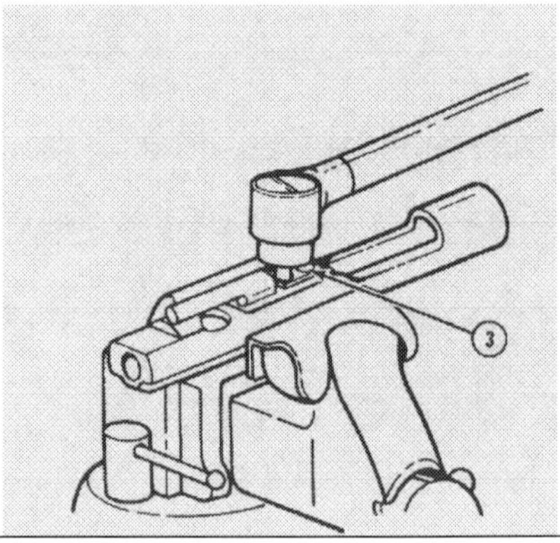

3-9. KEY AND BOLT CARRIER ASSEMBLY (CONT.).

c. REASSEMBLY (CONT.)

NOTE
Field staking method will be used by field units.

4. Use solid center punch and hand hammer to stake the two carrier and key screws (3) In three places.

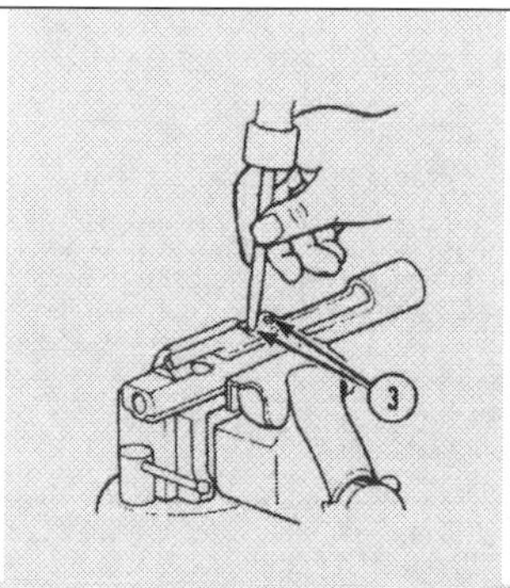

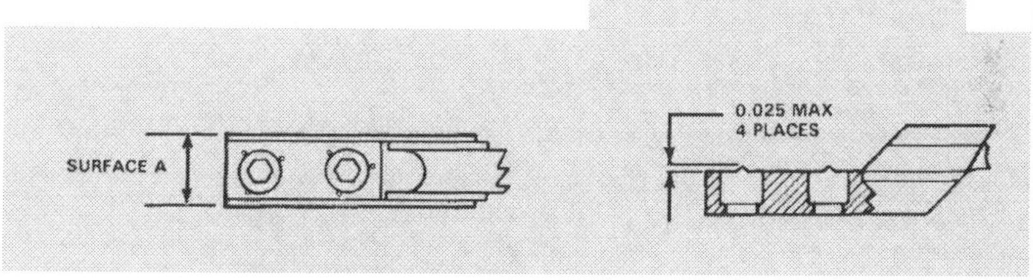

FIELD REPLACEMENT STAKING

5. Reassemble rifle, refer to page 3-77

CAUTION
If blanks are used, blank firing attachment (BFA) must be attached

NOTE
If the bolt carrier key is replaced, three to eight rounds of blank or ball ammunition must be fired to ensure a seal is created. Manual operation of the rifle may be required. If blank ammunition is utilized, M16A2 Blank Firing Attachment must be adapted

3-10. UPPER RECEIVER AND BARREL ASSEMBLY.

This task covers:

a Disassembly
b Inspection/Cleaning
c Repair
d Reassembly
e Test

INITIAL SETUP

Test Equipment
- Tool and Gage Set (item 2, app B)

Tools
- (ARMY) Small Arms Repairman Tool Kit (item 3, app B)
- Field Maintenance Basic Less Power Small Arms Shop Set (item 1, app B)
- Modified Needle Nose Pliers (fig E-7, app E)

Materials/Parts
- Brush, cleaning, small (item 3, app D)
- Carbon removing compound (item 8, app D)
- Cloth, abrasive (item 13, app D)
- Dichloromethane, technical (item 15, app D)
- Dry cleaning solvent (item 16, app D)
- Gloves, chemical and oil protective (item 18, app D)
- Grease, molybdenum disulfide (item 19, app D)
- Lubricant, solid film (item 21, app D)
- Pan, wash (item 24, app D)
- Polyethylene (item 32, app D)
- Sealing compound (item 28, app D)
- Target (item 31, app D)

References
FM 23-9
TM 9-1005-319-10

Equipment Conditions
- 3-15 Upper receiver and barrel assembly removed
- 2-43 Handguard assemblies removed

General Safety Instructions
- To avoid injury to your eyes, use care when removing and installing spring-loaded parts.
- When using solid film lubricant or dichloromethane, be sure the area is well ventilated.
- When using carbon removing compound, avoid skin contact If carbon removing compound comes in contact with the skin, wash thoroughly with running water Using a good lanolin base cream after exposure to the compound is helpful Using gloves and protective equipment is required.

3-10. UPPER RECEIVER AND BARREL ASSEMBLY (CONT.).

a. DISASSEMBLY

1. Using ball-peen hammer and 5/64 inch diameter drive pin punch, drive spring pin (1) (which retains the gas tube) out of front sight assembly (2).

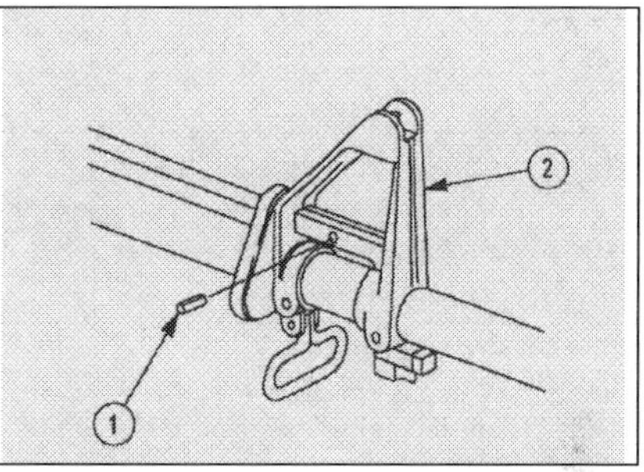

2. Slide gas tube (3) back into upper receiver assembly (4) to clear front sight assembly (2). Then lift slightly, pull forward, and remove gas tube (3).

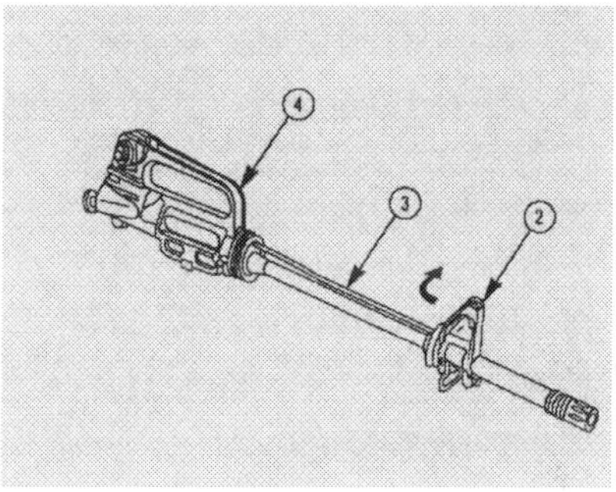

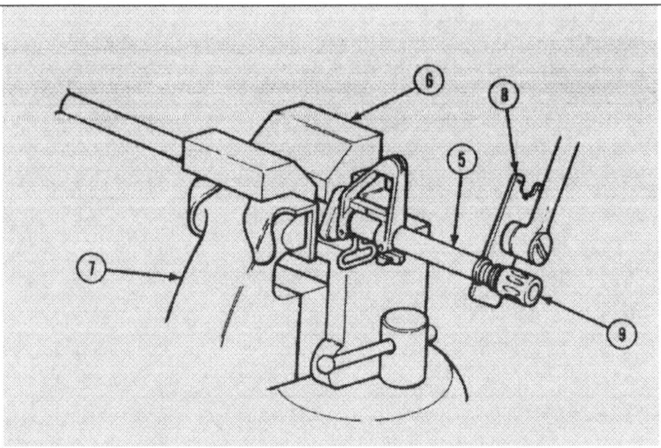

3. Position upper receiver and barrel assembly (5) in barrel removal fixture (6) and secure both In machinist's vise (7).

4. Using combination wrench (8) and 12 inch drive handle, remove compensator (9).

5. Remove peel washer (10) being careful not to lose or bend thin sections.

6. Remove upper receiver and barrel assembly (5) from barrel removal fixture and machinist's vise.

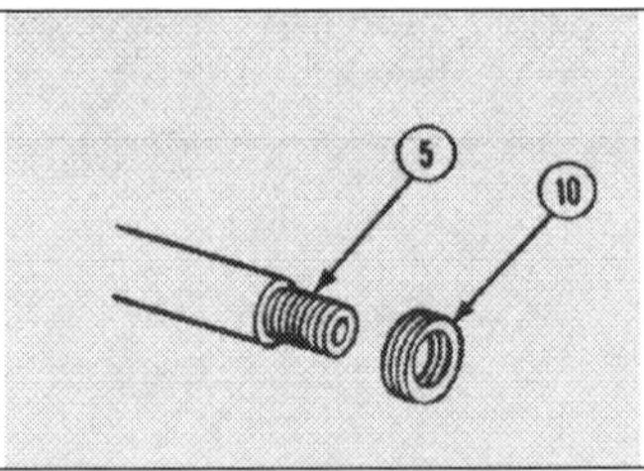

3-10. UPPER RECEIVER AND BARREL ASSEMBLY (CONT.).

a. DISASSEMBLY (CONT.)

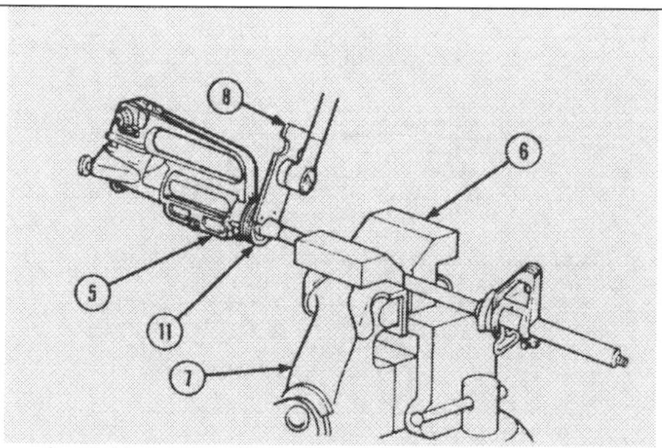

7. Place upper receiver and barrel assembly (5) Into barrel removal fixture (6) and clamp Into machinist's vise (7).

NOTE

Be sure all three drive pins on combination wrench are fully engaged with barrel nut assembly. Wrench must be pushed toward upper receiver assembly to compress the slip ring spring In barrel nut assembly. Do not use a torque wrench to loosen the barrel nut assembly.

8. Using '2 Inch drive handle and combination wrench (8), loosen barrel nut assembly (11)

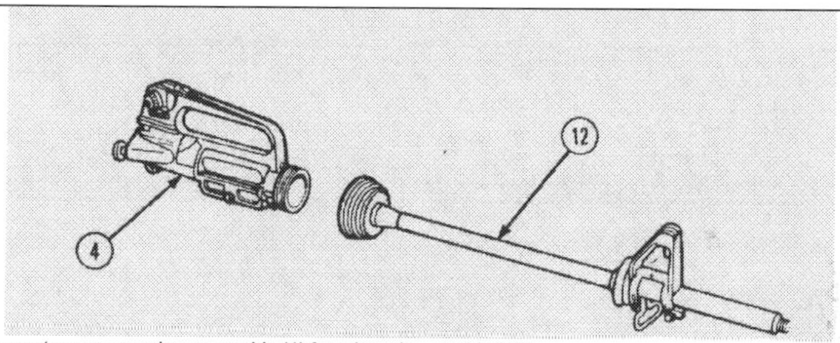

9. Separate upper receiver assembly (4) from barrel assembly (12)

10. Remove barrel assembly (12) from machinist's vise and barrel removal fixture

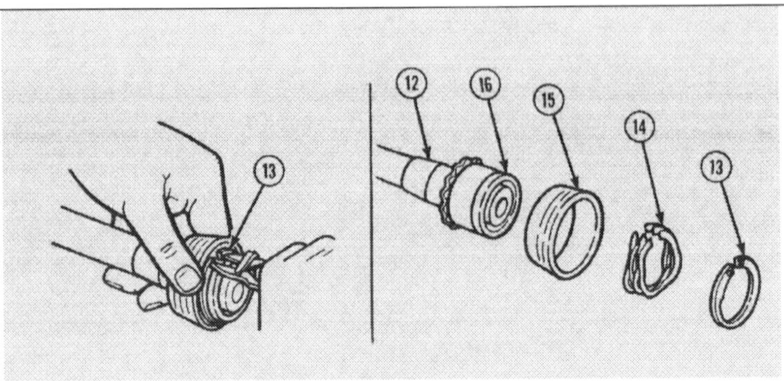

11. Remove retaining ring (13) using retaining ring pliers.

12. Remove slip ring spring (14) and handguard slip ring (15).

13. Do not remove barrel nut (16) from barrel assembly (12).

b. INSPECTION/CLEANING

WARNING

When using carbon removing compound, avoid skin contact. If carbon removing compound comes In contact with the skin, wash thoroughly with running water. Using a good lanolin base cream after exposure to compound is help ful. Using gloves and protective equipment is required.

1. Inspect gas tube for cracks. Replace if defective

2. Use carbon removing compound to remove carbon deposits from interior and exterior of gas tube. If a large amount of carbon is found and cannot be removed, replace the gas tube

NOTE

A small arms cleaning brush (bore) (item 4, app D) may be used to clean Interior of front sight assembly where gas tube is secured.

3-10. UPPER RECEIVER AND BARREL ASSEMBLY (CONT.).

b. INSPECTION/REPAIR (CONT.)

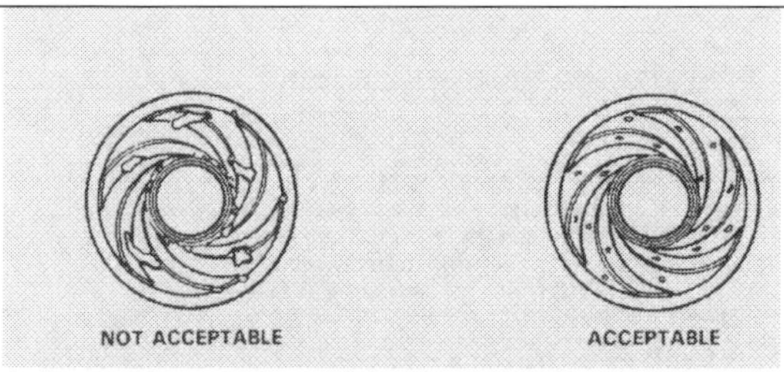

3. Inspect bore for burrs, cracks, rust, bulges, and pits using the following guidelines

(a) Pits no wider than a land or groove and no longer than 3'8 inch (0.95 cm) are al lowed in the bore.

(b) Lands that appear dark blue due to coating of gliding metal from projectiles are allowable.

(c) Definitely ringed bores or bores ringed sufficiently to bulge the outside surface of the barrel are cause for rejection. Replace barrel assembly if defective

4. If the upper receiver is separated from the barrel assembly, Inspect chamber for pits utilizing a flashlight. Pits 1'8 inch 10 32 cm) In length are cause for rejection Replace rifle barrel assembly If defective

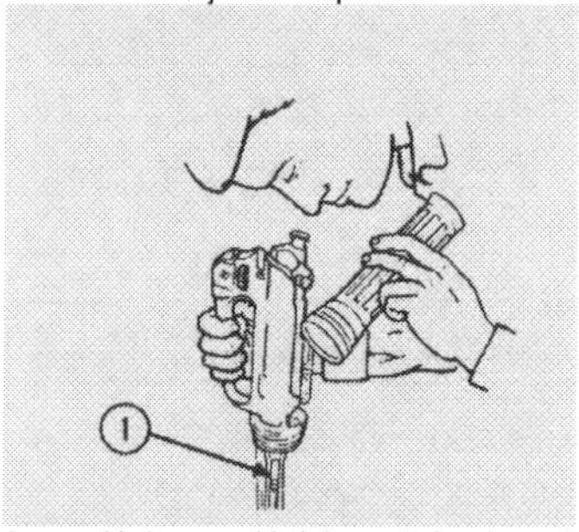

5. If upper receiver and barrel assembly is assembled, Inspect chamber using reflector tool (1) and flashlight. Pits 1 '8 inch (O 32 cm) In length are cause for rejection. Replace rifle barrel assembly If defective. If rifle barrel assembly Is replaced, Inspect headspace (p 3-45).

6. Inspect upper receiver assembly for cracks, corrosion, wear, or damage

- **(a)** Small dents or gouges that do not affect functioning will not be cause for rejection.
- **(b)** If upper receiver assembly contains cracks or holes, the upper receiver assembly will be replaced

7. Inspect all parts for damage and wear. Replace all defective parts.

NOTE

Damaged or missing teeth of the barrel nut is not cause for rejection provided the proper torque value can be obtained during installation using the tools depicted. If removal of the barrel is not possible with the combination tool, a pipe wrench or other such tool may be used during removal.

8. Inspect front sight guards for bends, if bent see page 3-37 for repair procedures.

c. REPAIR

1. Repair corroded upper receiver assembly surfaces as follows:

- **(a)** Sand corroded area with abrasive cloth and make sure all corrosion has been removed.

WARNING

When using solid film lubricant or dichloromethane, be sure the area is well ventilated

- **(b)** Wash area with technical dichloromethane (methylenechloride) to re move all dirt, grease, and foreign material
- **(c)** Apply sealing compound, mixed In accordance with manufacturer's directions, to areas to be filled
- **(d)** Spread sealing compound as smoothly as possible Into defective area using a putty knife or similar tool

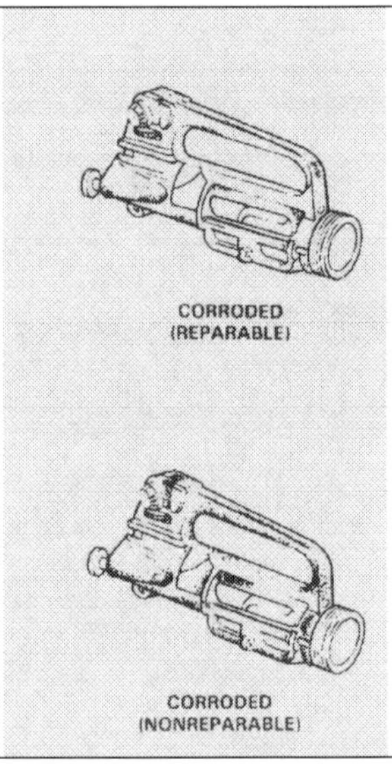

3-10. UPPER RECEIVER AND BARREL ASSEMBLY (CONT.).

c. REPAIR (CONT.)

NOTE
Do not feather edges

(e) Place a sheet of polyethylene (Item 32, app D), cut to size, over filled area Rub by hand to smooth.

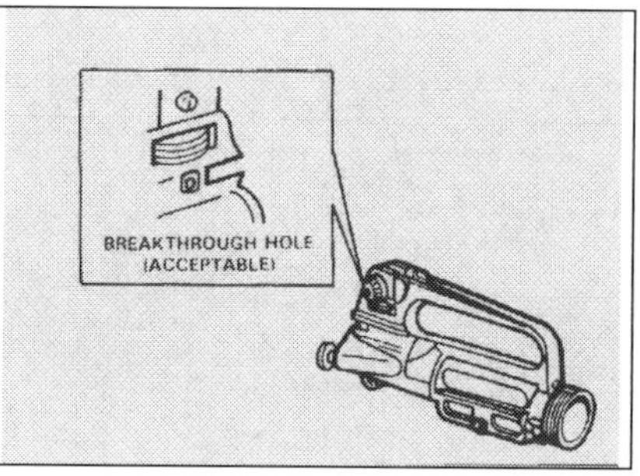

2. After curing, remove polyethylene sheet In accordance with Instructions by the manufacturer.

WARNING
When using solid film lubricant or dichloromethane, be sure the area is well ventilated.

CAUTION
Solid film lubricant is to be used only as an exterior surface protective finish and touchup If solid film lubricant comes in contact with recoiling parts or functional surfaces of the rifle, remove immediately by washing with technical dichloromethane.

3. Wash area with technical dichloromethane (methylenechloride) ; to remove dirt and grease, and foreign material.

4. Roughen area to be refinished with abrasive cloth and clean surface again Do not touch the area with fingers.

5. Repair shiny surfaces by spraying a coat of solid film lubricant in accordance with Instructions supplied by the manufacturer Dry 24 hours before handling.

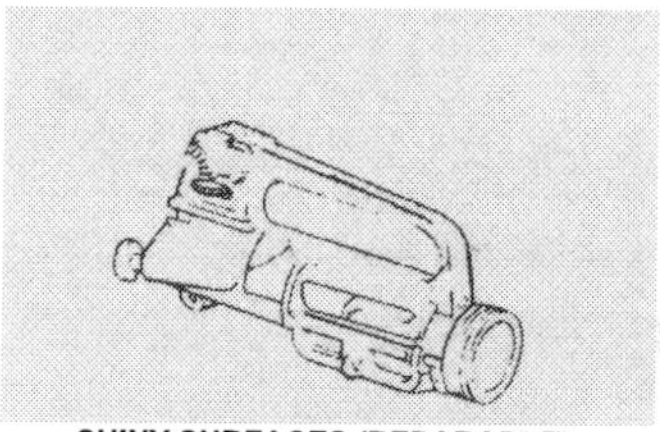

SHINY SURFACES (REPARABLE).

6. Straighten bent front sight guards (1) as follows:

NOTE

Remove spring before heating. (Heat will damage spring.) The sight post and plunger may be reused unless damaged.

(a) Remove front sight post, detent, and helical spring (see p 2-43).

NOTE

Use copper or brass caps (jaw Inserts on bench vise to prevent damage to front sight base (2) during clamping.

(b) Place front sight base (2) in a bench vise.

(c) Heat front sight guards (1) and bend with pliers. The front sight guards (1) should be put back as nearly as possible to the original position. Allow front sight housing to air cool.

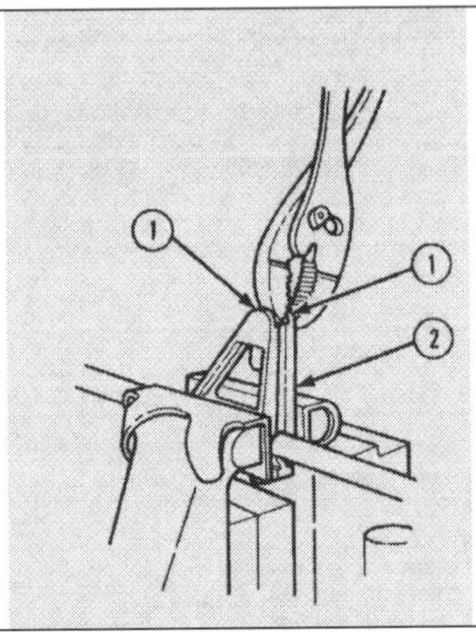

WARNING

Dry cleaning solvent is flammable and toxic and should be used in a well ventilated area. The use of rubber gloves is necessary to protect skin when washing rifle parts.

(d) Roughen any damaged surface of front sight guards with abrasive cloth and clean with dry cleaning solvent. Always wear rubber gloves when using dry cleaning solvent.

3-10. UPPER RECEIVER AND BARREL ASSEMBLY (CONT.).

c. REPAIR (CONT.)

CAUTION

Do not allow solid film lubricant to flow into front sight post threaded well.

(e) Apply solid film lubricant to cover the damaged finish.

(f) If front sight guards cannot be straightened utilizing the above procedures, reduce the rifle barrel assembly.

7. Slightly bent barrels may be straightened as follows'.

(a) Check straightness using straightness gage 8448202 (p 3 45) If the barrel falls the straightness test, and the gage remains in the barrel In the area of the front sight assembly, perform step (b) to determine If it may be straightened.

(b) With the gage remaining in the bore, hold the rifle in a vertical position with the end of the barrel into which the gage was inserted pointing up Ensure that if when the gage passes through the barrel it will not be damaged Using hand pressure' ONLY, flex the portion of the barrel between the front sight assembly and the compensator in all four directions (left, right, forward, and back) If the barrel is only slightly bent, the gage will drop through when the barrel is flexed in one of these directions Note the direction which allowed the gage to drop through the' barrel.

CAUTION

Remove the gage from the barrel before continuing

NOTE

If the gage does not pass through the barrel when it is flexed, replace the rifle barrel assembly.

(c) Place the barrel in a vise using appropriate protective laws . Clamp the barrel between the front sight assembly and the compensator approximately 1 inch (2.54 cm) from the front sight assembly The rifle barrel assembly should be in a horizontal position with the side noted in step (b) toward you.

CAUTION
Do not apply pressure to the receiver.

(d) Grasp the BARREL near the receiver so that when force is applied the barrel will flex In the same direction as noted In step (b).

(e) Give the barrel a sharp jerk of approximately 20 to 40 pounds of force (f) Remove the barrel from the vise and recheck straightness (step (al).

(g) If gage still will not pass through the barrel, perform step (b) to determine direction of bend If the barrel is still bent in the same direction as before, perform steps (c) through (f) using slightly more force. If the barrel Is now bent In the opposite direction, replace the rifle barrel assembly.

(h) If the gage passes freely through the barrel, the barrel shall be considered straight and continue In service.

(i) If the barrel has been straightened, the rifle must be targeted (p 3-45).

d. REASSEMBLY

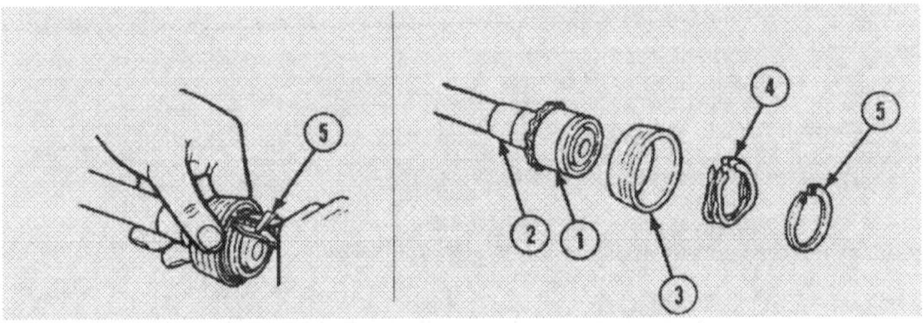

1. Position barrel nut (1) by sliding it to the rear of rifle barrel assembly (2) as far as possible

2. Slide handguard slip ring (3) over barrel nut (1).

3. Press slip ring spring (4) from both sides and insert It Into handguard slip ring (3)

4. Install retaining ring (5) against slip ring spring (4) using retaining ring pliers Snap retaining ring (5) to barrel nut (1).

NOTE
After cleaning, apply molybdenum disulfide grease to threads of barrel nut assembly before Installation.

3-10. UPPER RECEIVER AND BARREL ASSEMBLY (CONT).

d. REASSEMBLY (CONT)

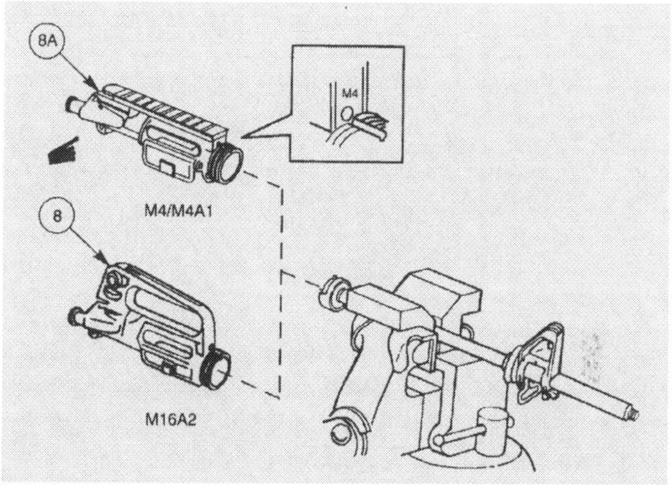

NOTE
The alignment pin (6) must not show any signs of looseness.

5. Position rifle barrel assembly (2) with alignment pin (6) up. Using barrel remover fixture (7), clamp rifle barrel assembly (2) in vise.

NOTE
The slot should fit the alignment pin perfectly with very little or no rotational play present. Note the play in a new rifle barrel assembly and new upper receiver assembly and use this as a guide.

6. Align upper receiver assembly (8) or (8A) using alignment pin (6) and the slot in upper receiver assembly (8) or (8A). Install over end of rifle barrel assembly (2).

7. Wipe upper receiver threads clean and ensure there are no burrs. Apply molybdenum disulfide grease to the threads before installation.

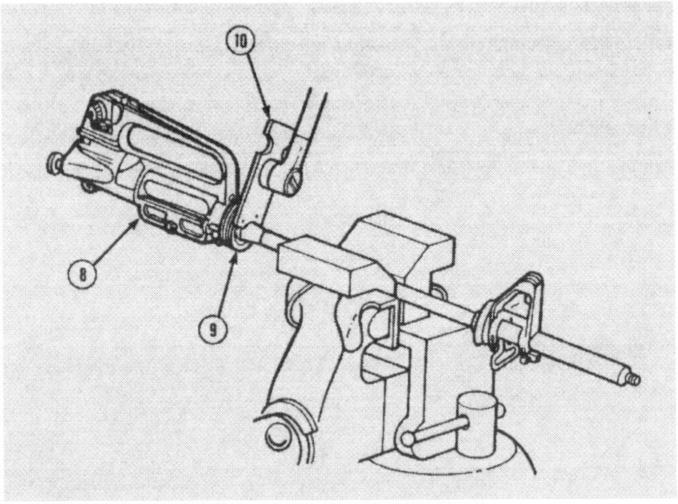

8. Engage threads of barrel nut assembly (9) with upper receiver assembly (8).

9. Using combination wrench (10) and torque wrench, torque barrel nut assembly (9) to 30 ft-lb (40.5 N-m). Torque is measured when both wrenches are used together.

NOTE

Three times torquing procedure provides for a better thread fit and prevents barrel nuts from becoming loose. Do not use the torque wrench for loosening.

10. Make certain all three drive pins on combination wrench are engaged with barrel nut assembly (91 Loosen and repeat torque operation. Then loosen the barrel nut again.

3-10. UPPER RECEIVER AND BARREL ASSEMBLY (CONT).

d. REASSEMBLY (CONT)

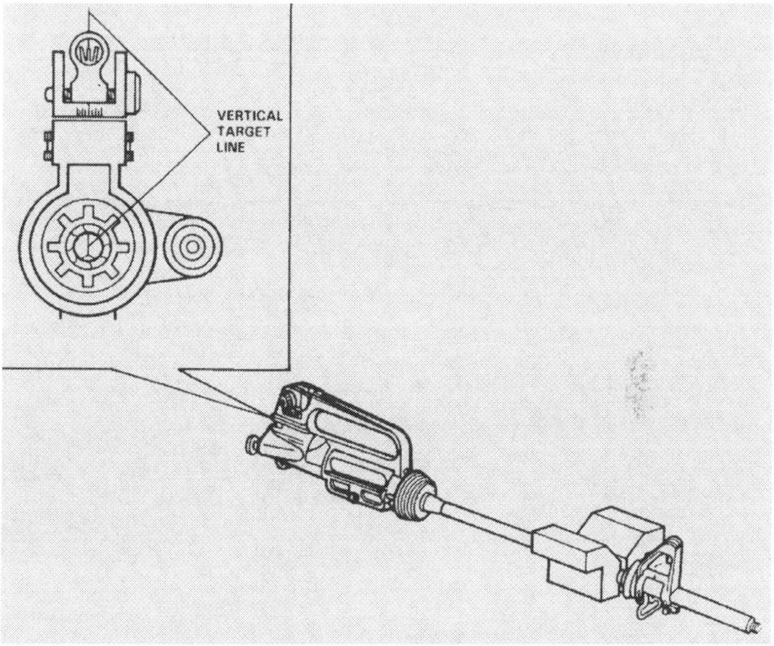

11. Front sight post must be Installed. Loosen the vise and align the bore on a distant vertical target Center the target In the bore from 12 o'clock through 6 o'clock The, front sight post should be on line and vertical with the target. Tighten vise Adjust the rear sight windage until a proper sight picture is obtained on the vertical target The rear sight aperture will be approximately In the center of the rear sight base If the rifle barrel assembly is properly aligned in the upper receiver assembly

NOTE

If rifle barrel assembly (usually new) is not properly aligned In the upper receiver assembly (usually an old part), excessive windage will be present and the upper receiver assembly will require replacement to obtain the proper fit between the alignment pin and slot.

CAUTION

Do not torque over 80 ft-lb (108 N-m) while tightening the barrel nut assembly to the next hole, to allow for proper alignment of gas tube.

NOTE

Do not attempt to hold the upper receiver assembly with a pry bar, however, If the rifle barrel assembly turns In the holding fixture, a pry bar may be used through the front sight assembly base to help prevent the rifle barrel assembly from turning In the holding fixture. Use care not to distort or bend front sight assembly or retaining pins. Use "buddy system" to hold pry bar.

12. Torque the barrel nut assembly again to 30 ft-lb (40.5 N-m) while maintaining sight alignment. The barrel nut assembly may be tightened beyond 30 ft-lb (40.5 N-m) to align the barrel nut assembly serrations for proper gas tube clearance Never loosen the barrel nut assembly to align for gas tube clearance.

13. Check alignment of barrel nut assembly (9) with upper receiver assembly (8). The front 8 inches (20.32 cm) of a gas tube may be used as an alignment tool (see Illustration). This Is Inserted Into the bolt carrier key and then Inserted into the rear of the receiver. If the parts of the barrel nut assembly are properly aligned, the tool will pass freely and lay top dead center along the top of the barrel. A number 15 twist drill (0.180 inch) may also be needed as an alignment tool. If necessary, tighten barrel nut assembly to next hole to allow proper alignment.

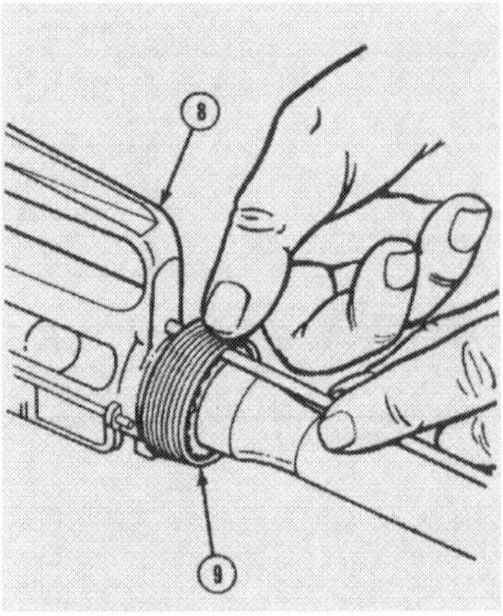

3-10. UPPER RECEIVER AND BARREL ASSEMBLY (CONT).

d. REASSEMBLY (CONT)

NOTE

The peel washer may be heated to remove thin sections. Always place thin section to rear.

14. Install peel washer (11) and compensator (12) on rifle barrel assembly (2).

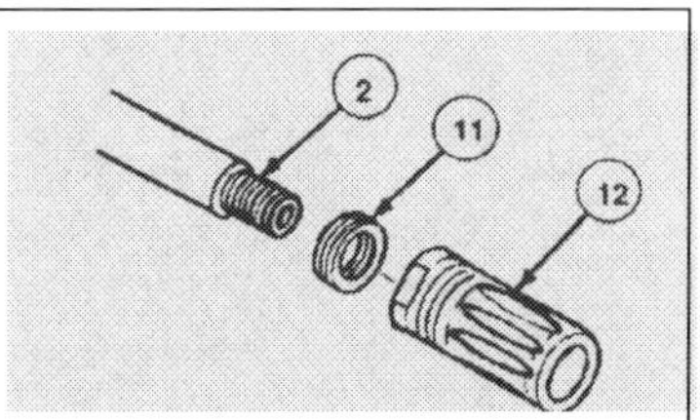

15. Torque compensator (12) to 15-20 ft-lb (20-27 N-m) using combination wrench (10) and torque wrench. Torque is measured when both wrenches are used together. Some portion of the openIng of the third or middle slot must be straight up (align with the front sight post) (12.1) at proper torque level. Thin sections of the peel washer may be removed or added as required. Save unused sections.

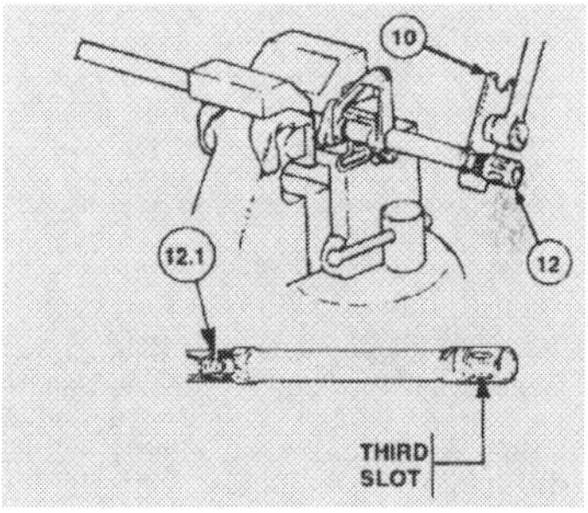

16. Slide gas tube (13) through the barrel nut assembly (9) and then slide forward, Inserting gas tube into hole In the front sight assembly base (14).

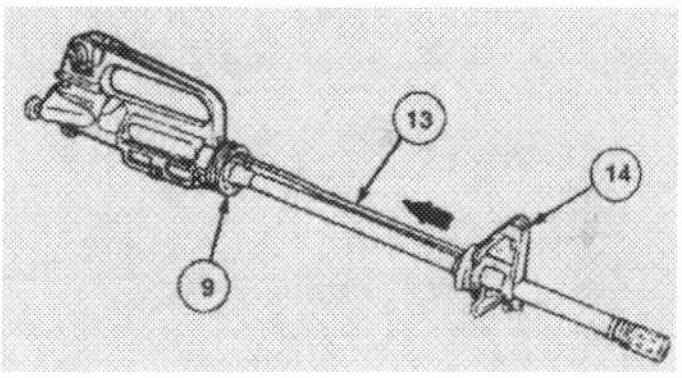

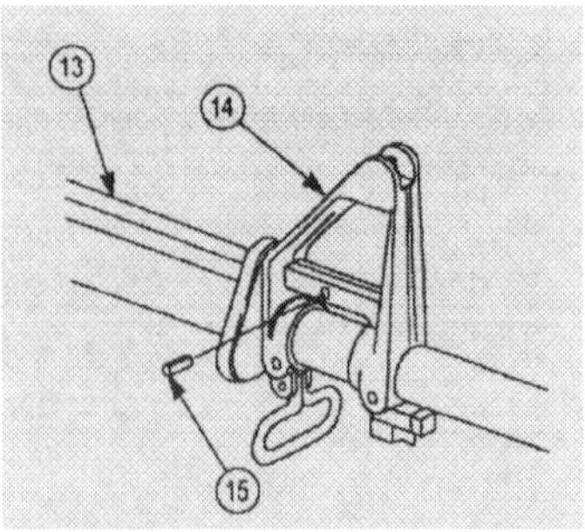

17. Align the holes in gas tube (13) and front sight assembly base (14).

NOTE

To assist in installing spring pin (1) modified nose pliers may be used.

18. Using ball-peen hammer and 5/64-inch diameter drive pin punch, drive spring pin (15) into front sight assembly base (14) to secure gas tube (13).

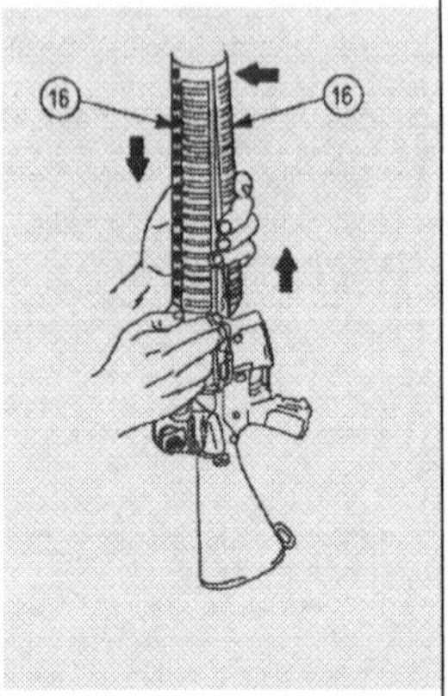

19. Install upper and lower handguard assemblies (16).

e. TEST

1. The following information pertains to the use of breech, bore, and other gages:

(a) All M1 6A2 barrels and chambers are chromed.

(b) Barrel erosion gage, PN 8448496 (normally used on M16A1 fully-chromed barrels), can be used to gage M16A2 barrels.

(c) The bore straightness gage, PN 8448202, is required for use on all barrels. The gage must pass through the barrel without being forced.

3-10. UPPER RECEIVER AND BARREL ASSEMBLY (CONT).

e. TEST (CONT)

2. Use barrel erosion gage (1) PN 8448496. Install key and bolt carrier assembly with bolt assembly and firing pin removed. Hold rifle vertical with receiver up. Insert gage into rear of key and bolt carrier assembly. The reject line must be read at the rear edge of the key and bolt carrier assembly.

3. If the reject mark passes beyond the rear surface of the key and bolt carrier assembly, the barrel is unserviceable and shall be replaced.

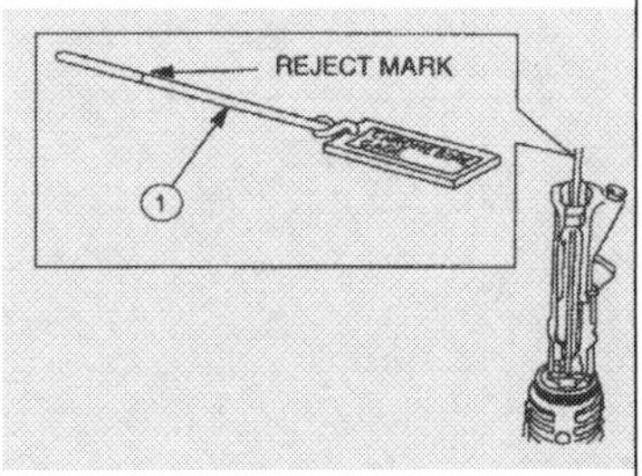

NOTE
Ensure barrel is clean prior to performing the following test.

4. Check straightness of bore using straightness gage (2) PN 8448202. Put gage in barrel. Tilt barrel and allow gage to fall through. Catch gage.

5. Gage must pass freely through barrel. If gage does not pass through barrel, recheck as follows. Hold upper receiver and barrel assembly (3) in vertical position with muzzle pointed down; insert gage into chamber end of barrel. Release gage and catch it as it exits muzzle end. If gage passes freely through the barrel, barrel is acceptable. If it does not, the barrel must be straightened or replaced. (See page 3-38 for straightening instructions/ procedures.)

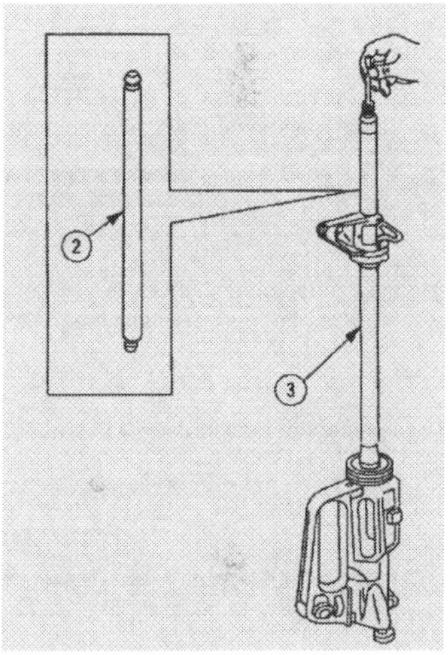

6. Assemble charging handle assembly (4), bolt assembly, and key and bolt carrier assembly (5) into upper receiver assembly (6).

7. Insert headspace gage (7) PN 7799734 in chamber.

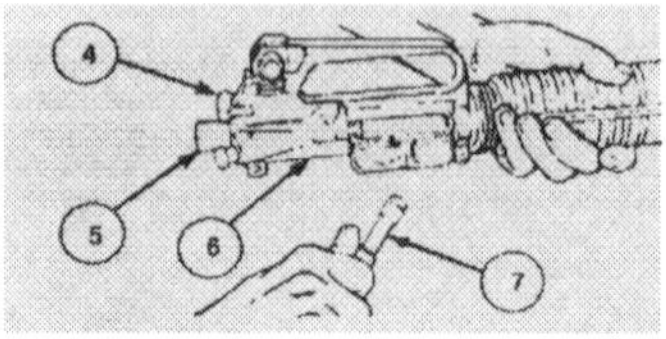

NOTE

For the purpose of this test "light finger pressure" is defined as 8 1/2 to 8 3/4 pounds.

8. Check headspace by pressing key and bolt carrier assembly (5) and charging handle assembly (4) forward using light finger pressure.

9. Bolt should not rotate to locked position. Key and bolt carrier assembly (5) must protrude from rear of upper receiver assembly (6) for proper headspace. If excessive headspace, first replace old bolt assembly with an unused bolt assembly and then recheck. If headspace is not corrected, replace rifle barrel assembly; then recheck with the original bolt assembly to determine if the bolt assembly is still good or if the bolt assembly should be replaced also.

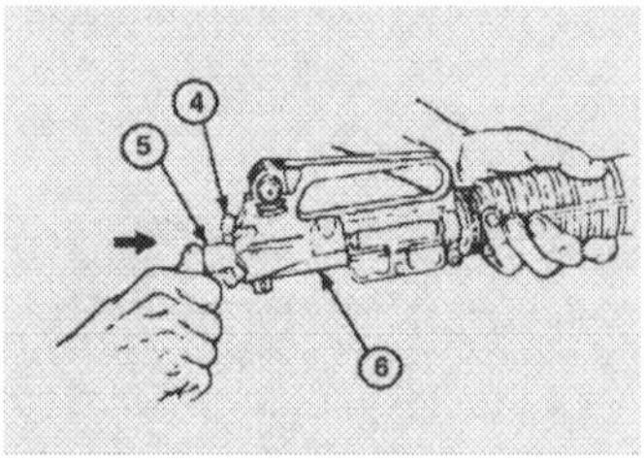

10. Remove key and bolt carrier assembly, bolt assembly, charging handle assembly, and headspace gage.

11. Reassemble rifle, refer to page 3-77.

NOTE

Rifles which have been rebarreled must be function-fired with seven rounds of 5.56mm ball ammunition. After rebarreling, the rifle must be targeted with three rounds of 5.56mm ball ammunition at 25 meter range using target. Refer to TM 9-1005-319-10 and FM 23-9

3-10. UPPER RECEIVER AND BARREL ASSEMBLY (CONT).

e. TEST (CONT)

12. (a) If the test fails using finger pressure, remove the gage and perform the test again as follows: With the muzzle down, stack 8 1/2 to 8 3/4 pounds of trigger weights (8) on a locally fabricated spacer/weight (9) on the bolt carrier assembly (10) Insert headspace gage (11) and test per above instructions.

(b) Remove trigger weights (8), spacer/weight (9), bolt carrier assembly (10), charging handle (12), and headspace gage (11).

(c) If excessive headspace, first replace bolt assembly and then recheck If headspace is not corrected, replace barrel assembly; then recheck with the original bolt to determine if the bolt is still good or If the bolt should be replaced also.

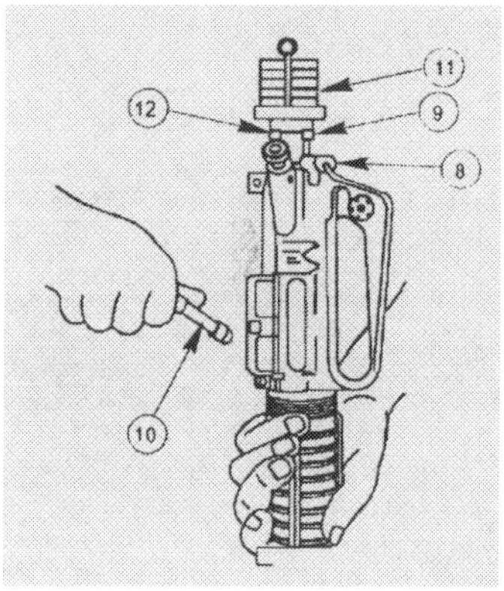

3-11. UPPER RECEIVER ASSEMBLY AND REAR SIGHT ASSEMBLY.

This task covers:

a. Disassembly
b. Inspection
c. Repair
d. Lubrication
e. Reassembly
f. Mechanical Zero Procedures (A.F. Only)

INITIAL SETUP:

Tools
(ARMY) Small Arms Repairman Tool Kit (item 3, app B)
Field Maintenance Basic Less Power Small Arms Shop Set (item 1, app B)

Materials/Parts
Cleaner, lubricant, and preservative (CLP) (item 9, app D)
Lubricant, solid film (item 21, app D)
Index screw (9349065)

Equipment Conditions
3-29 Upper receiver assembly removed.

General Safety Instructions
To avoid injury to your eyes, use care when removing and installing spring-loaded parts.
When using solid film lubricant or dichloromethane, be sure the area is well ventilated.

a. DISASSEMBLY

WARNING
To avoid injury to your eyes, use care when removing and installing springloaded parts.

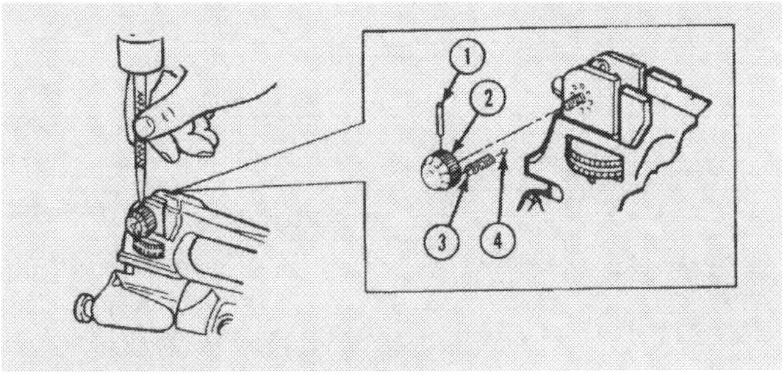

CAUTION

Be sure to catch small parts (1, 2, 3, and 4). AIR FORCE ONLY: Use magnet to keep from losing small parts (1, 3, and 4).

1. Drive out spring pin (1) using a hammer and 1/16 inch punch.
2. Catch rear sight assembly windage knob (2), helical spring (3), and ball bearing (4).

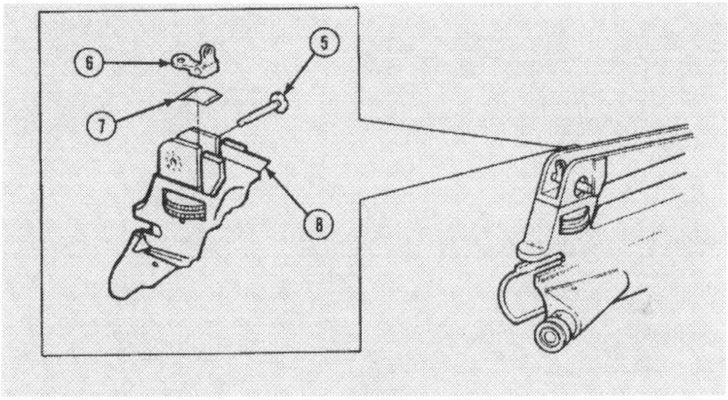

3. Using a flat-bladed screwdriver, remove rear sight assembly windage screw (5)
4. Remove sight aperture (6) and flat spring (7) from upper receiver (8).

3-11. UPPER RECEIVER ASSEMBLY AND REAR SIGHT ASSEMBLY (CONT).

a. DISASSEMBLY (CONT)

WARNING

To avoid injury to your eyes, use care when removing and installing spring-loaded parts.

5. Drive out spring pin (9) using a 3/32 Inch punch Catch helical spring (10) when punch is withdrawn.

6. Rotate elevation index (11) until rear sight assembly base (12) clears upper receiver (8). Catch ball bearing (13) and helical spring (14) as rear sight base clears.

7. Push elevation index (11) out with thumb using slight rotation motion. Catch ball bearing (15) and helical spring (16).

8. Use 1 16 Inch allen wrench to remove index screw (17). Discard Index screw (1 7) Separate elevation index (11) from elevation knob (18) by hand.

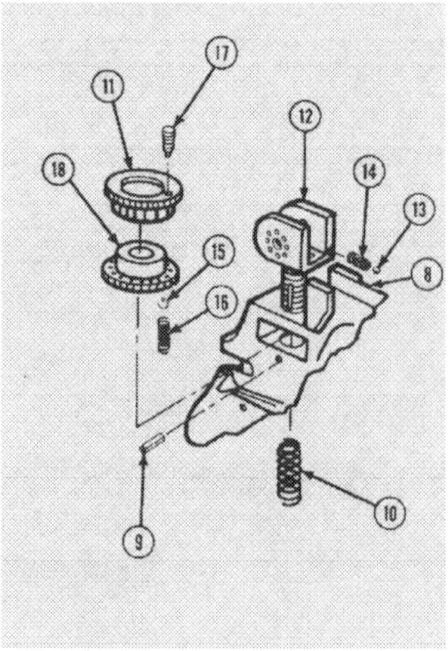

9. Remove pin (1 9) using 3,32 Inch drive pin punch and hand hammer .

WARNING

To avoid Injury to your eyes, use care when removing and installing spring-loaded parts.

10. Remove forward assist assembly (20) and helical spring (21) from upper receiver (8). For further disassembly of forward assist assembly see page 3-59.

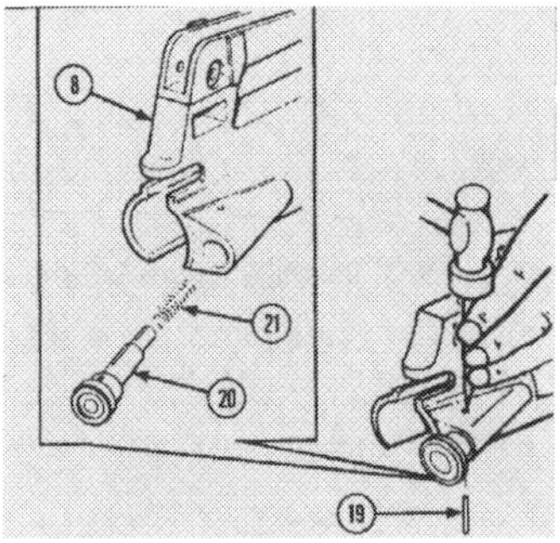

b. INSPECTION

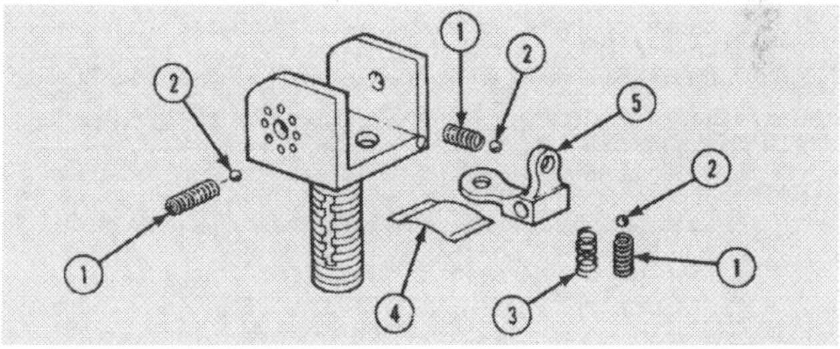

1. Check rear sight parts for serviceability. Inside of apertures should be round and distinct. Replace If defective

2. Visually inspect rear sight assembly helical springs (1) ball bearings (2), and helical spring (3) for breaks, bends, and missing parts Ball bearings should be smooth and round. Replace If defective

3. Check upper receiver for cracks, corrosion, and damage Clear drain hole with a piece of wire. Repair (p 3-29) or replace If defective.

4. Check that flat spring (4) retains sight aperture (5) firmly In either position. Replace flat spring (4) If sight aperture is not firm.

3-11. UPPER RECEIVER ASSEMBLY AND REAR SIGHT ASSEMBLY (CONT).

b. INSPECTION (CONT)

5. Check elevation Index 161 and windage knob (71 for legibility of markings Check underside of windage knobs for cracks. Detent indexing surfaces should be well formed.

6. Check rear sight assembly base (81 for serviceability Clear drain holes for springs Threaded portion of rear sight assembly base and elevation knob should be well formed.

7. Inspect rear sight guards for bends, if bent refer to the following page for repair procedures.

8. Inspect all parts for damage and wear Replace all defective parts.

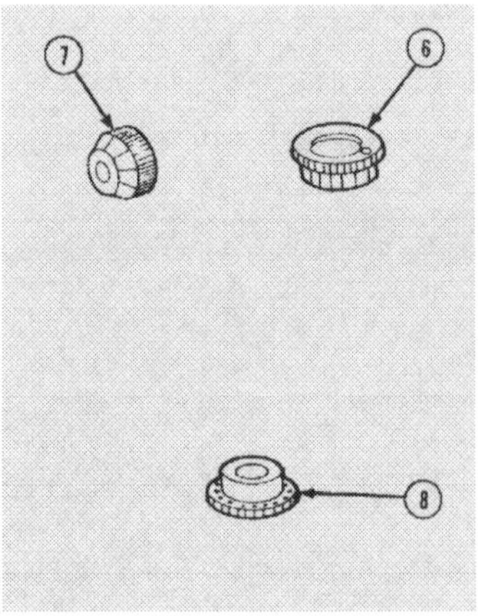

c. REPAIR

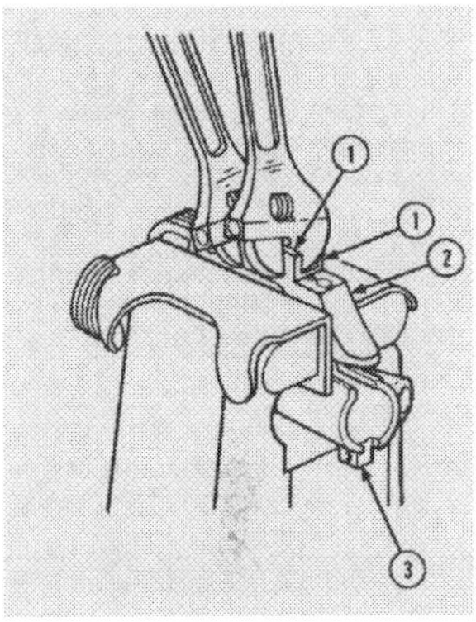

1. To straighten bent rear sight assembly guards (1), remove rear sight assembly components. Place carrying handle (2) In a vise using jaw clamps. Tighten vise to firmly hold upper receiver (3).

2. Using two adjustable wrenches, gradually bend guards (1) to straighten. When bending the guards (1), gradually bend beyond the straight point as the guard will partially return when bending pressure is stopped.

3. After straightening, use a flat file to remove any nicks, kinks, or burrs that remain on the inside of guards (1).

CAUTION
Do not use wire brush on aluminum surfaces

4. Apply solid film lubricant to brightened area for final protective coating.

5. Replace rear sight assembly components and check that rear sight assembly functions properly. If rear sight assembly functions check out, return upper receiver assembly to service.

6. If rear sight guards cannot be straightened utilizing the above procedures, replace the upper receiver

d. LUBRICATION

Lubricate upper receiver assembly and rear sight assembly. Apply CLP to helical springs and ball bearings (three each) and threaded portion of screws before installation. Lubricate helical springs and ball bearings through their respective drain holes.

3-11. UPPER RECEIVER ASSEMBLY AND REAR SIGHT ASSEMBLY (CONT).

e. REASSEMBLY

WARNING
To avoid injury to your eyes, use care when removing and Installing spring-loaded parts.

1. Apply CLP to helical spring (1) and forward assist assembly (2) and Install them into upper receiver (3).

2. Install spring pin (4) using 3, 32 Inch drive pin punch and hand hammer

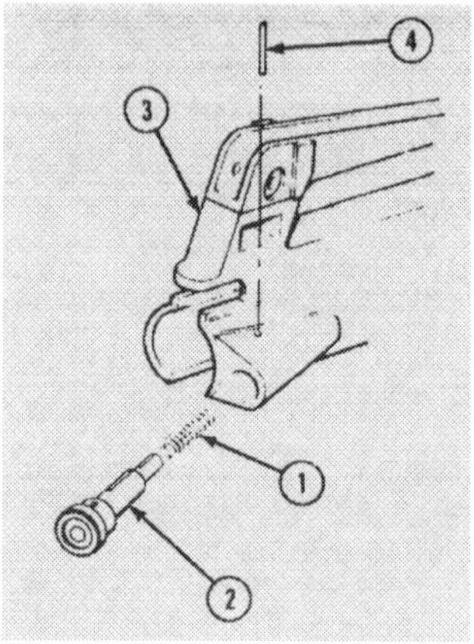

3. Assemble elevation knob (5), elevation index (6), and new index screw (7) using 1 16 Inch alien wrench. Do not overtighten Index screw as scale will require adjustment

4. Install ball bearing (8) and helical spring (9) using needle nose pliers or tweezers

5. Depress ball bearing (8) with a punch Inserted through access hole, and slide elevation knob assembly In upper receiver (3) from the side Center elevation knob assembly

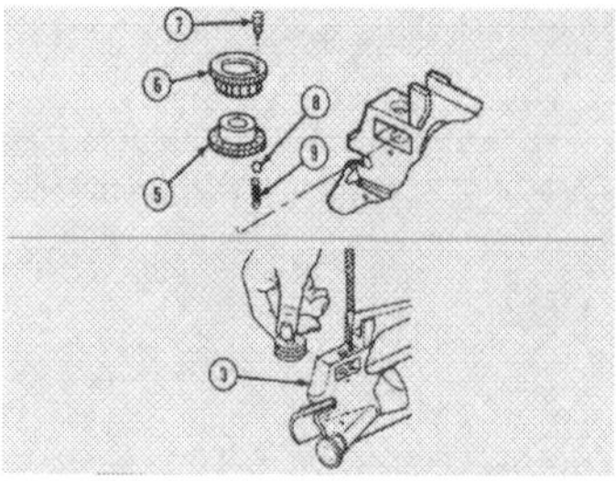

WARNING

To avoid injury to your eyes, use care when removing and installing spring-loaded parts.

NOTE

All springs are identical when new. Once disassembled from the rifle, their free length may vary due to different amounts of compression when installed. If the length of springs varies, use longest as #22, and the shortest as #9

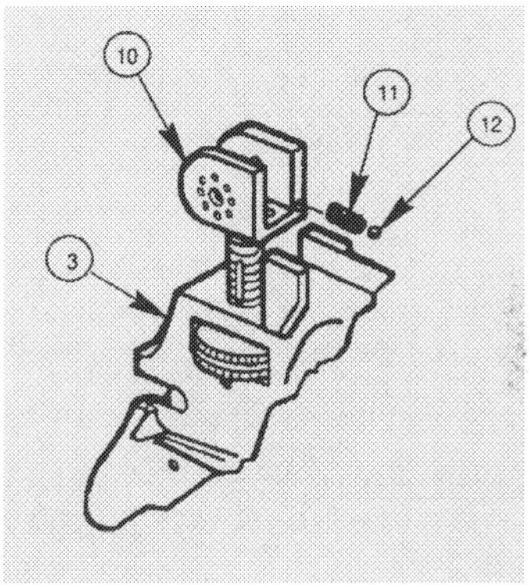

6. Insert threaded portion of rear sight assembly base (10) into upper receiver (3) and rotate elevation knob assembly until threads engage.

7. Insert helical spring (11) and ball bearing (12) in their hole as rear sight assembly base is lowered into upper receiver as elevation knob assembly is further rotated. Rotate elevation knob assembly until rear sight assembly base is all the way down. Then come up 22 clicks before installing spring pin. Check spring action of helical spring (11) on upper receiver.

3-11. UPPER RECEIVER ASSEMBLY AND REAR SIGHT ASSEMBLY (CONT).

e. REASSEMBLY (CONT)

CAUTION

Ensure that spring pin (1 3) passes over helical spring (14), not through its coils

8. Insert helical spring (14) through underside of upper receiver (15). Compress helical spring with a small tip screwdriver (16) to install spring pin (1 3) Spring pin must pass over helical spring, not through its coils Rotate elevation knob (17) until rear sight base (18) Is all the way down.

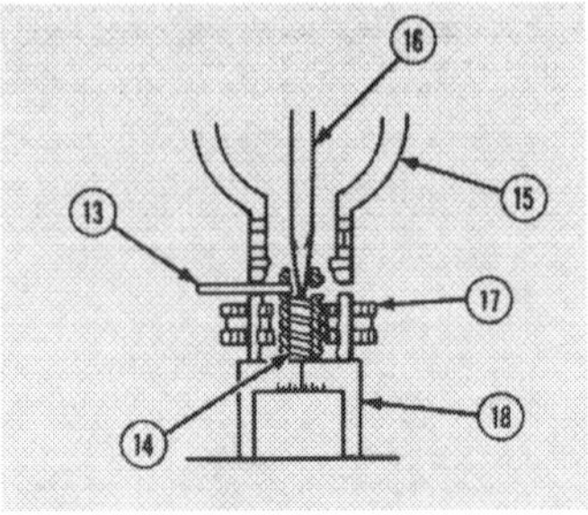

9. Install flat spring (19) and sight aperture (20) In rear sight assembly base (10). Install windage screw (21) with screwdriver.

10. Insert helical spring (22) and ball bearing 123) In windage knob (24)

NOTE

Tilt upper receiver toward wind age knob during positioning to prevent loss of ball bearing.

11. Position windage knob on shaft of windage screw (211 Align holes In windage knob with hole of shaft In windage screw and hold In alignment with a small screwdriver or punch Install spring pin (25).

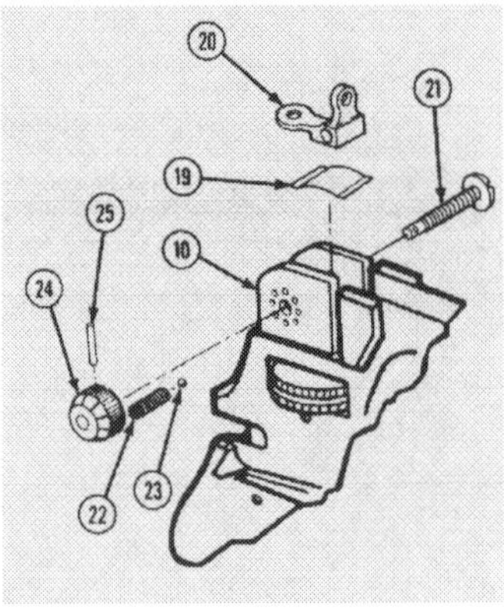

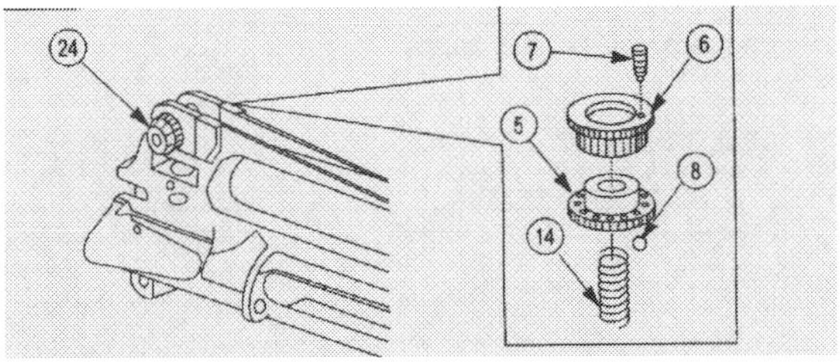

12. Rotate and test elevation index (6) and windage knob (24) for ease of functioning.

13. Inspect elevation knob zero as follows:

(a) Rotate elevation knob (5) counterclockwise until the rear sight assembly is all the way down. If a whole click is not felt as the rear sight assembly stops, the rear sight assembly has bottomed out and will not pivot freely.

(b) Position elevation knob back slightly to its last whole click so the rear sight assembly base is under tension of the ball bearing (8) and helical spring (14). The 300 meter mark should align with the mark on to receiver.

(c) If the 300 meter mark is not aligned with the mark on receiver, slip the range scale in the following manner:

(1) Position the 300 meter mark with the mark on the receiver.

(2) Insert a 1/16 inch allen wrench through the access hole of the rear sight assembly base and into the index screw (7).

(3) Loosen the index screw three turns and leave the wrench in place.

(4) Rotate lower portion of elevation knob

(5) counterclockwise until it stops (range scale should not have moved). Elevation knob should be positioned on its last whole click. (5) Tighten index screw (7) and remove wrench.

(6) Check for proper setting.

NOTE

After the rifle is assembled, the rear sight is centered, and placed at the 300 meter mark, perform the owing check: While looking at a light background, obtain good sight alignment. If the hole in the rear sight aperture appears oval Instead of round, the rear sight base or upper receiver should be replaced. To determine which part requires replacement, replace the rear sight base first. If this does not resolve the problem, replace the upper receiver.

14. Assemble rifle, refer to page 3-77

3-11. UPPER RECEIVER ASSEMBLY AND REAR SIGHT ASSEMBLY (CONT).

f. MEACHANICAL ZERO PROCEDURES (A.F. ONLY)

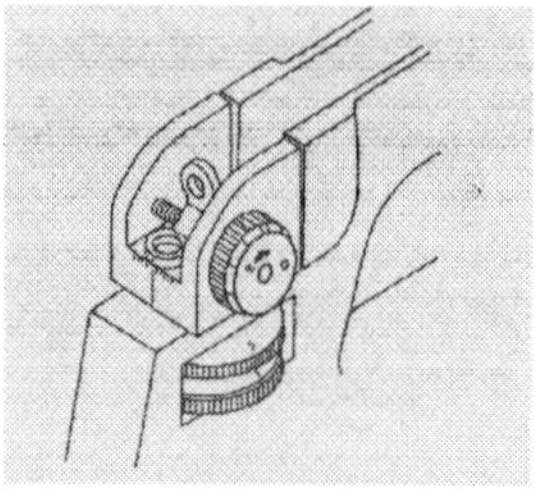

1. Center rear sight by moving windage knob in the appropriate direction.

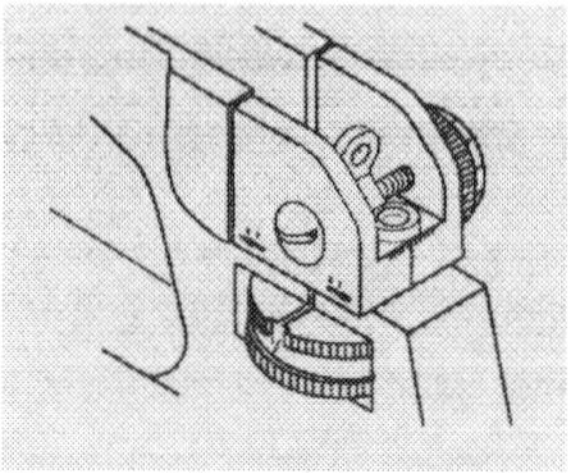

2. Always push in on windage screw head after making rear sight adjustments.

3. Visually check rear sight to ensure it is centered after making adjustments. Also, ensure the rear sight is set in the shortrange (unmarked aperture) position.

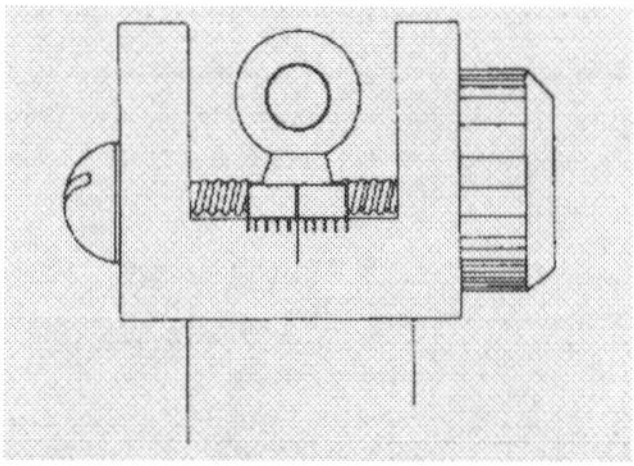

NOTE

This procedure, when used in conjunction with front sight mechanical zero adjustment (p 2-54), will give an approximate battle site zero to most M16A2 rifles. The above steps can also be used before firing a new or newly assigned rifle. Use the procedure to check rifles stored in preferred packaging during routine inspections. This will help ensure people armed with the rifles will stand a better chance of hitting an enemy if the rifles must be used before a live fire zero can be made. Whenever possible, zeroing of the rifle should be accomplished using ball ammunition on a 25 meter zeroing target using the "L" aperture.

After the rifle is assembled, the rear sight is centered, and placed at the 300 meter mark, perform the following check: While looking at a light background sight through the rear sight and obtain good sight alignment with the front sight placed in the center of the rear sight aperture. If while performing this check the hole in the rear sight aperture appears oval instead of round, the rear sight base or upper receiver should be replaced. To determine which part requires replacement, replace the rear sight base first. If this does not resolve the problem, replace the upper receiver.

3-12. FORWARD ASSIST ASSEMBLY.

This task covers:

- **a.** Disassembly
- **b.** Inspection
- **c.** Repair
- **d.** Lubrication
- **e.** Reassembly

INITIAL SETUP:

Tools
(ARMY) Small Arms Repairman Tool Kit (item 3, app B)
Field Maintenance Basic Less Power Small Arms Shop Set (item 1, app B)

Materials/Parts
Cleaner, lubricant, and preservative (CLP) (item 9, app D) | a

Equipment Conditions
3-48 Forward assist assembly removed

General Safety Instructions
To avoid injury to your eyes, use care when removing and installing spring-loaded parts

a. DISASSEMBLY

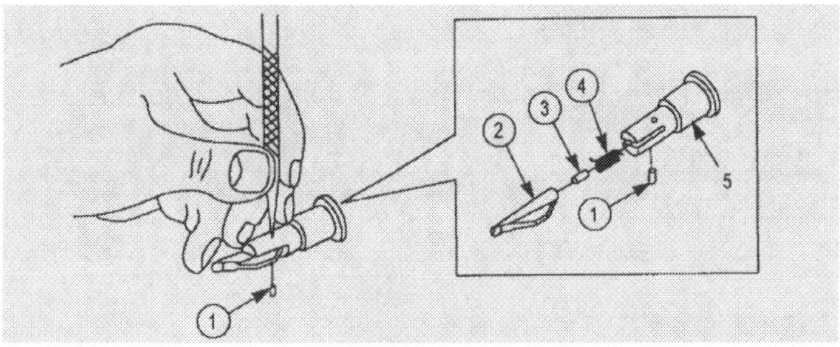

WARNING
To avoid injury to your eyes, use care when removing and installing spring-loaded parts.

1. Drive out spring pin (1) using 1/16 inch drive pin punch and hand hammer
2. Remove forward assist pawl (2), pawl detent (3), and helical spring (4) from plunger assembly (5)

3-12. FORWARD ASSIST ASSEMBLY (CONT)

b. INSPECTION

1. Inspect forward assist pawl (1) for burrs, chips, and cracks. Minor burrs may be removed using fine files or stones, as required. Replace forward assist pawl if defective.

2. Inspect pawl detent (2) for burrs and cracks. Minor burrs may be removed using fine files or stones, as required. Replace pawl detent if defective.

3. Inspect helical spring (3) for kinks, breaks, and wear. Replace helical spring if defective.

4. Inspect plunger assembly (4) for wear, burrs, chips, and breaks. Minor burrs may be removed using fine files or stones, as required. Replace forward assist assembly if defective.

5. Inspect spring pin (5) for wear. Replace if defective.

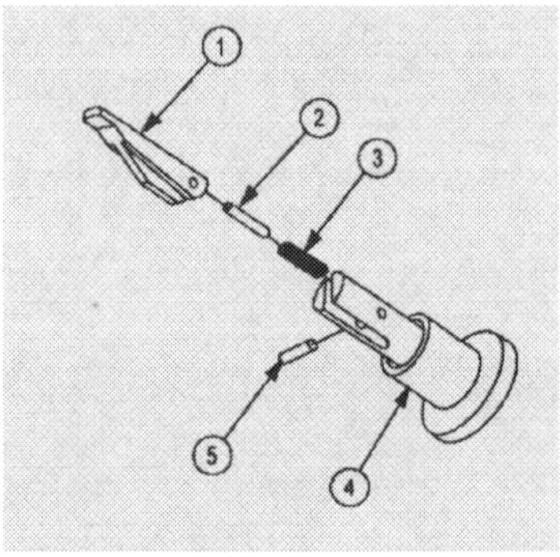

c. REPAIR

1. Repair forward assist pawl using fine files or stones, as required, to smooth burrs. Do not deform forward assist pawl.

2. Repair pawl detent using fine files or stones, as required, to smooth burrs. Do not deform pawl detent.

3. Repair plunger assembly using fine files or stones, as required, to smooth burrs. Do not deform plunger assembly.

d. LUBRICATION

Lubricate helical spring, pawl detent, and forward assist pawl with CLP (p 2-33) before installation.

e. REASSEMBLY

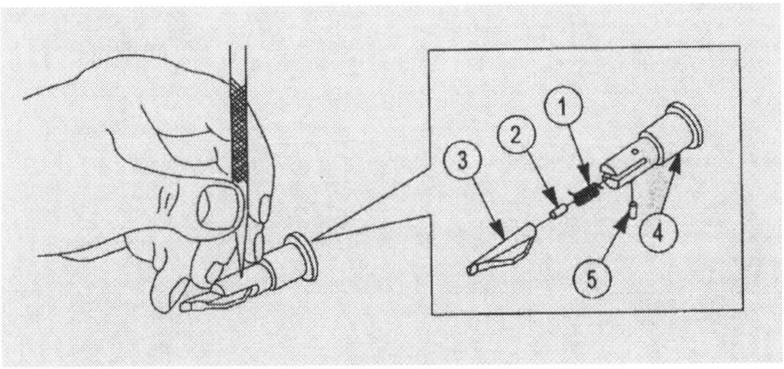

WARNING

To avoid injury to your eyes, use care when removing and installing spring-loaded parts.

1. Install helical spring (1), pawl detent (2), and forward assist pawl (3) into plunger assembly (4).
2. Align holes and install spring pin (5) using 1/16 inch drive pin punch and hand hammer. Spring pin must be flush or slightly below flush after reassembly.
3. Assemble rifle, refer to page 3-77.

3-13. LOWER RECEIVER AND BUTTSTOCK ASSEMBLY.

This task covers:

- **a.** Disassembly
- **b.** Inspection
- **c.** Repair
- **d.** Test
- **e.** Reassembly

INITIAL SETUP

Test Equipment
Tool and Gage Set (item 2, app B)

Tools
(ARMY) Small Arms Repairman Tool Kit (item 3, app B)
Field Maintenance Basic Less Power Small Arms Shop Set (item 1, app B)
Pivot pin removal tool (fig E-3, app E)
Slave pin (fig E-5, app E)

Materials/Parts
Cleaner, lubricant, and preservative (CLP) (item 9, app D)
Dichloromehtane, technical (item 15, app D)
Lubricant, solid film (item 21, app D)
Screw, self-locking (item 6, p C-11)

Equipment Conditions
3-15 Lower receiver and buttstock assembly removed
2-57 Buttstock assembly and pistol grip removed

General Safety Instructions
To avoid injury to your eyes, use care when removing and installing spring-loaded parts.
When using solid film lubricant or dichloro methane, be sure the area is well ventilated.

a. DISASSEMBLY

WARNING

To avoid injury to your eyes, use care when removing and installing spring-loaded parts.

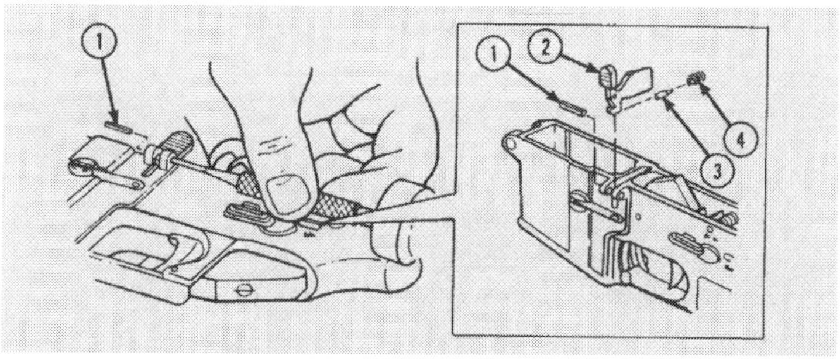

1. Remove spring pin (1) using 3/ 32 inch drive pin punch and hand hammer.

2. Remove bolt catch (2), bolt catch plunger (3), and bolt catch spring (4).

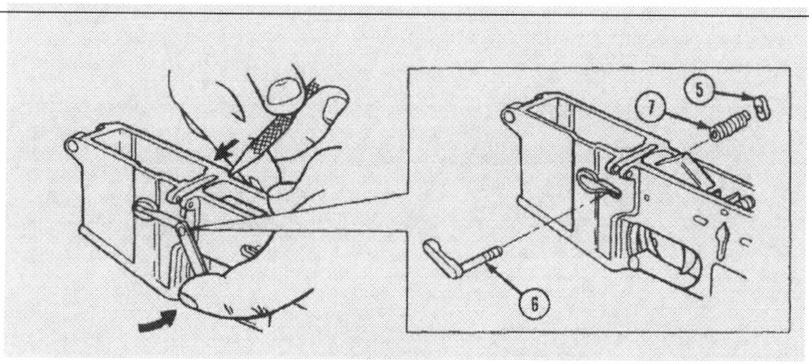

WARNING

To avoid injury to your eyes, use care when removing and installing spring-loaded parts.

3. Using drive pin punch, press in on magazine button (5) and turn magazine catch 16) counterclockwise to unscrew and remove.

4. Remove magazine button (5) and magazine catch spring (7).

3-13. LOWER RECEIVER AND BUTTSTOCK ASSEMBLY (CONT).

a. DISASSEMBLY (CONT)

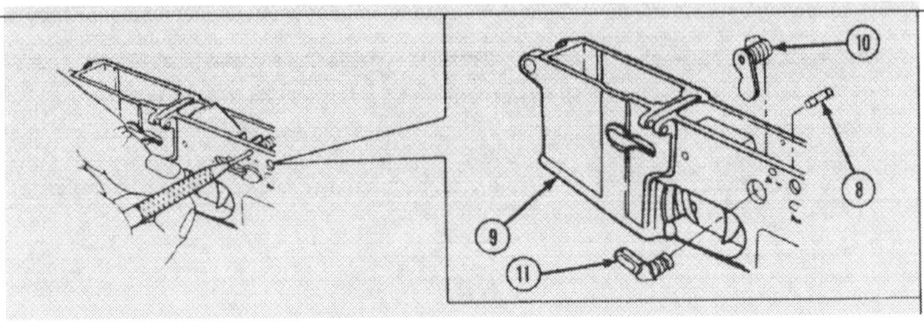

5. Use a rounded tip drive pin punch to push automatic sear pin (8) out of lower receiver (9).

NOTE
To remove automatic sear, selector lever must be positioned to BURST (if installed).

6. Remove automatic sear (10) and selector lever (11).

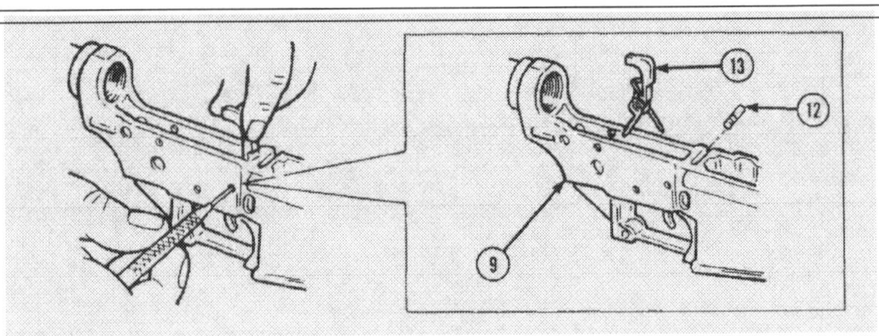

NOTE
To remove (hammer assembly should be forward), place selector lever (if installed) to SEMI position.

7. Use drive pin punch to push hammer pin (12) from lower receiver (9).

8. Remove hammer assembly (13). If further disassembly is required, see page 3-73.

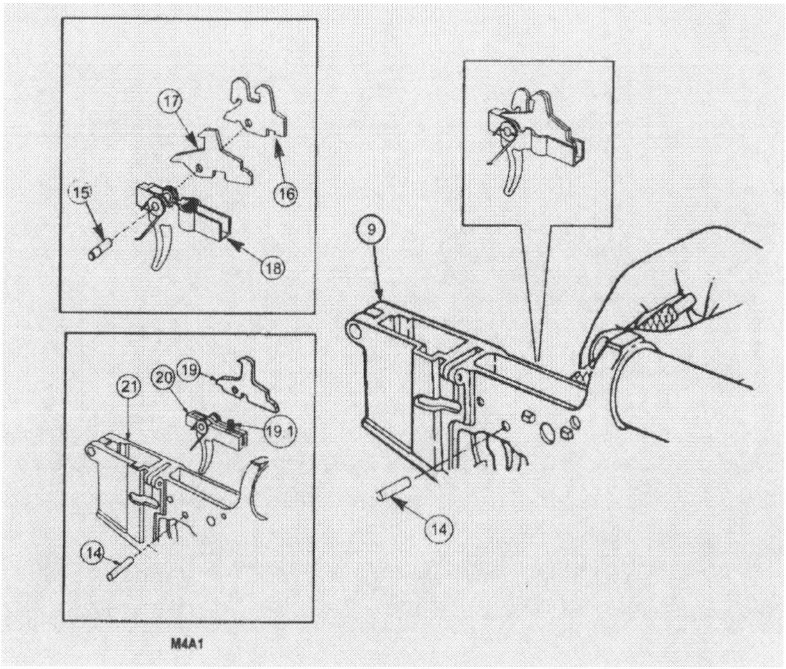

NOTE

Use of fabricated slave pin will allow removal of the following parts as a unit.

9. Remove trigger pin (14) by pushing from the left side of lower receiver (9) with fabricated slave pin (15) and a drive pin punch.

10. Remove semiautomatic disconnector (16), burst disconnector (17), and trigger assembly (18). If further disassembly of trigger assembly is required, see page 3-75.

M4A1 ONLY

11. Remove disconnector (19), disconnector spring (19.1), and trigger assembly (20) from lower receiver and extension subassembly (21).

3-13. LOWER RECEIVER AND BUTTSTOCK ASSEMBLY (CONT).

b. INSPECTION

NOTE

Refer to page 3-35 for repair of corroded surfaces.

Inspect lower receiver and buttstock assembly for corrosion in the lower receiver lobes of the pivot area or hinge pin area. If extensive corrosion appears in these areas, the receiver will not be repaired; turn rifle in for replacement.

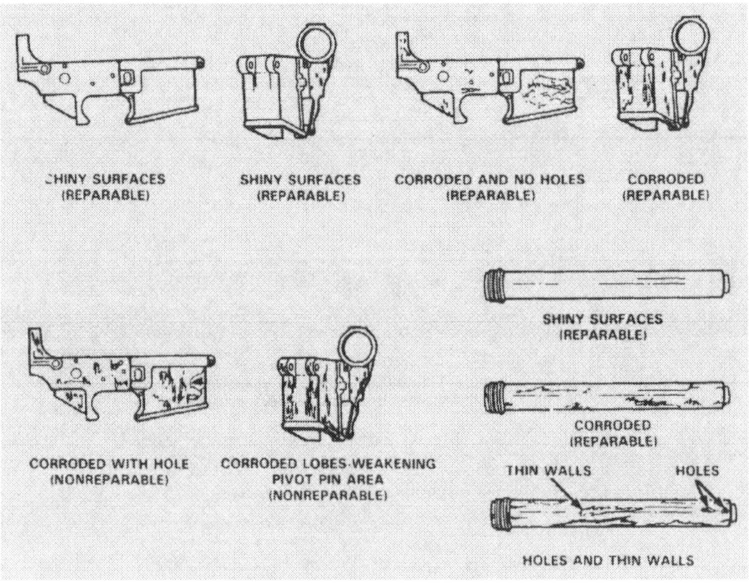

WARNING

When using solid film lubricant or dichloromethane, be sure the area is well ventilated.

CAUTION

Do not use a wire brush on aluminum surfaces.

NOTE

If a weapons lower receiver is missing one third or more of its exterior protective finish, resulting in an unprotected, light reflecting surface, it is candidate for overhaul. This missing finish will be considered a shortcoming. This shortcoming requires action to obtain a replacement weapon. Once a replacement has been received, evacuate the original weapon to depot for overhaul.

Solid Film Lubricant (SFL) is the only authorized touchup for the weapon and may be used on up to one third of the exterior finish of the weapon. **FOR ARMY CONUS USE ONLY AND AIR FORCE TRAINING WEAPONS ONLY:** SFL may be used as a touchup without limitation on the upper receiver and barrel assembly. This is to say that units which DO NOT fall under the category of Divisional Combat Units or rapid deployment type units may have up to 100 percent of the exterior surface of the upper receiver and barrel assembly protected with SFL if necessary.

c. REPAIR

Repair or replace all parts of lower receiver and buttstock assembly if defective.

3-13. LOWER RECEIVER AND BUTTSTOCK ASSEMBLY (CONT).

d. TEST

1. With the upper receiver attached to the lower receiver, and the pivot pin and takedown pins in place, perform the following test:

   (a) Apply hand pressure to push the upper receiver as far to one side as possible.

   (b) Attempt to insert a 0.020 inch thickness gage between the pivot pin lugs of the upper and lower receivers.

   (c) If the thickness gage penetrates to the pivot pin at all accessible locations, repair by replacement of the upper receiver (see (b) below) or replacement of rifle is required.

2. If the rifle fails the above test, remove the upper receiver and install a "NEW" upper receiver and perform the test again.

3. If the rifle now passes the above test, it shall be considered serviceable and continue in use.

4. If the rifle falls the test with a new upper receiver, this failure shall be considered a shortcoming. This shortcoming requires action to obtain a replacement rifle. Once a replacement has been received, evacuate the original rifle to depot for overhaul.

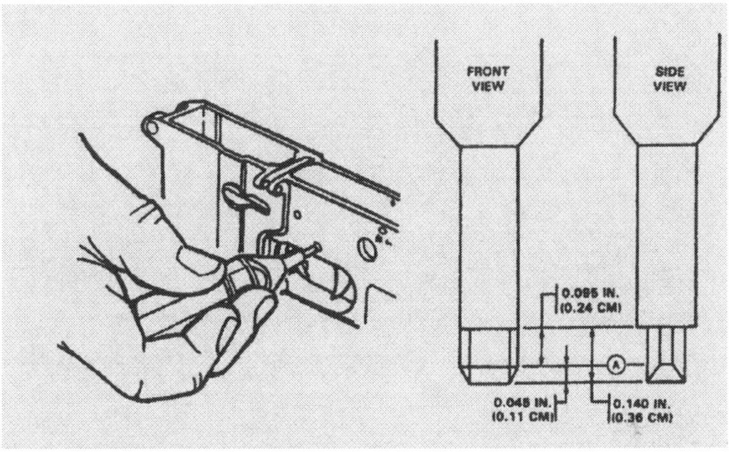

NOTE

If the lower receiver is not disassembled, visually inspect for broken or damaged parts, and to ensure that the hammer and trigger springs are correctly installed before beginning this test. It is not necessary to disassemble the lower receiver for the sole purpose of this visual Inspection. If broken or damaged parts are found, disassemble (p 3-62) and repair as authorized.

5. Test two hammer pin holes and two trigger pin holes using not-go plug gage PN 12006472. This test may be conducted by disassembly of the lower receiver (p 3-62) or by pushing the pin far enough to disengage the end of the pin from the side of the receiver which is being tested. If the lower receiver is not disassembled and the not-go plug gage enters any hole to first shoulder (A), the lower receiver must be disassembled and all four holes must be tested again.

6. Gently insert the not-go plug gage and rotate it 180 degrees. If the not-go plug gage passes through any one of the four pin holes, the rifle is unserviceable and will be turned in for replacement. The gage must extend through the wall thickness to be unserviceable.

7. After completion of gaging operation, visually inspect hammer and trigger springs to ensure proper location of spring legs.

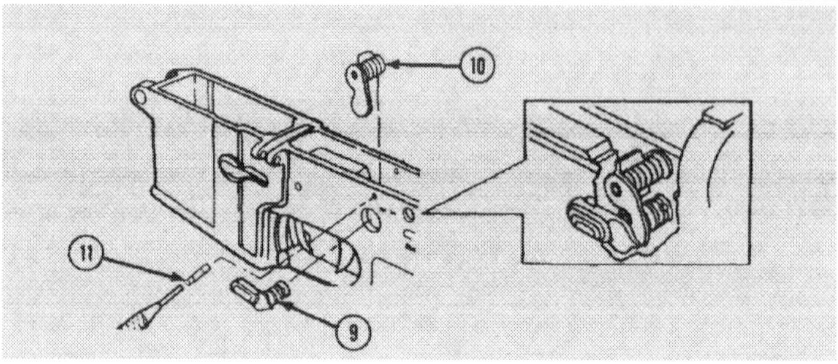

NOTE

Hammer assembly should be cocked prior to installing the selector lever.

Selector lever (9), if installed, must be positioned to BURST. Long leg of automatic sear (10) spring must rest on top of selector lever.

5. Install selector lever (9) and automatic sear (10).

6. Install automatic sear pin (11) (install on the right side) into receiver using drive pin punch. Push in until flush.

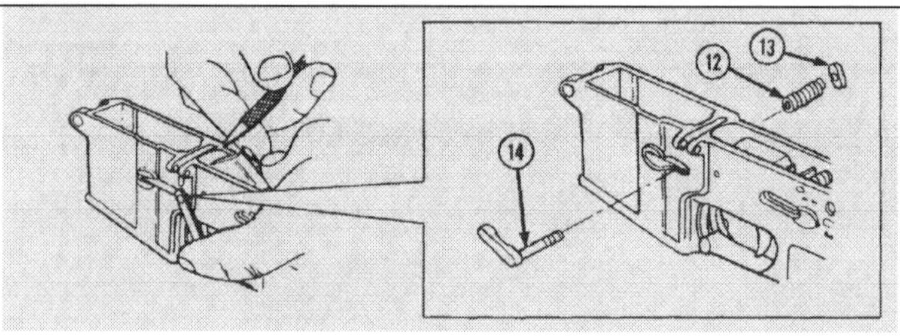

7. Install magazine catch spring (12) and magazine button (13).

NOTE

Drive pin punch should be larger than hole in magazine button.

8. Install magazine catch (14). Push in on magazine button (13) using a drive pin punch and turn magazine catch (14) clockwise until threaded end of magazine catch is flush with magazine button head.

3-13. LOWER RECEIVER AND BUTTSTOCK ASSEMBLY (CONT).

e. REASSEMBLY (CONT)

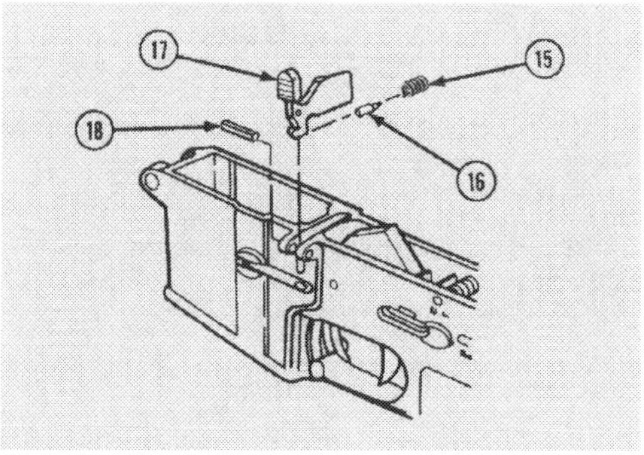

9. Install bolt catch spring (15), bolt catch plunger (16), and bolt catch (17).

10. Secure by installing spring pin (18) using 3/32 inch drive pin punch and hand hammer.

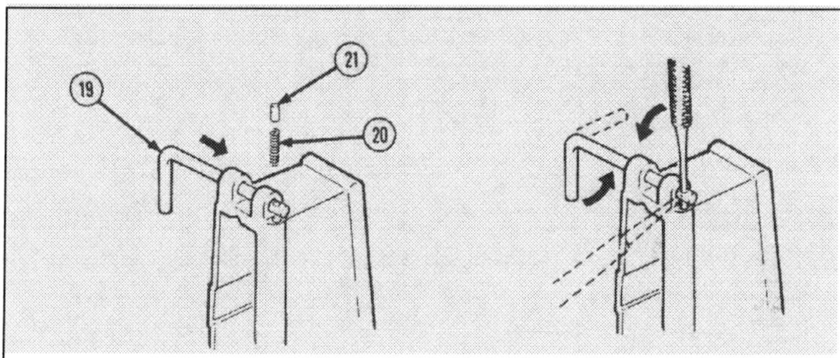

11. Install fabricated pivot pin installation tool (19). Insert helical spring (20) and pivot pin detent 21). Compress pivot pin detent in recess with punch and rotate tool. Remove punch.

WARNING
To avoid injury to your eyes, use care when removing or installing spring-loaded parts.

NOTE
Rounded end of pivot pin detent must be in the groove of the pivot pin (22) when assembly is complete.

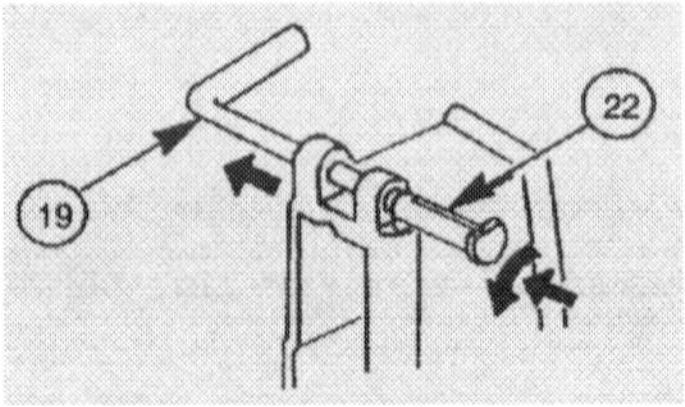

12. Position pivot pin (22) and removing fabricated pivot pin installation tool (19) while maintaining pressure, slide pivot pin (22) into hole. Rotate pivot pin to receive pivot pin detent.

13. Assemble rifle, refer to page 3-82.

3-14. HAMMER ASSEMBLY

This task covers:
a. Disassembly
b. Inspection/Repair
c. Reassembly

INITIAL SETUP

Equipment Conditions
3-62 Hammer assembly removed

a. DISASSEMBLY

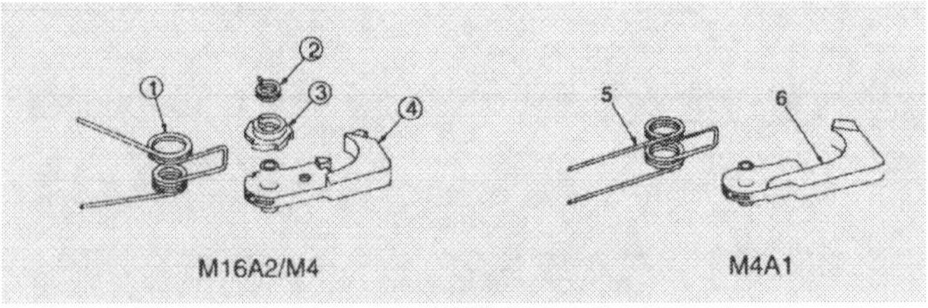

NOTE
M16A2 burst cam is black. M4 burst cam is nickel colored (shiny).

(M16A2/M4) remove hammer spring (1), cam clutch spring (2), and burst cam (3) from hammer and hammer pin retainer assembly (4). **(M4A1)** remove hammer spring (5) from hammer (6).

3-14. HAMMER ASSEMBLY (CONT).

b. INSPECTION/REPAIR

1. Inspect hammer spring for deformities, breaks, and bends. Pay special attention to the large coil.
   Replace hammer spring if defective.

2. Inspect cam clutch spring and burst cam for deformities, breaks, and bends; replace if defective.

3. Inspect hammer and hammer pin retainer assembly for chips and breaks. Hammer pin should click home under strong finger pressure. Install hammer pin into hole in hammer to check spring retention of the hammer pin. Replace hammer and hammer pin retainer assembly if defective.

c. REASSEMBLY

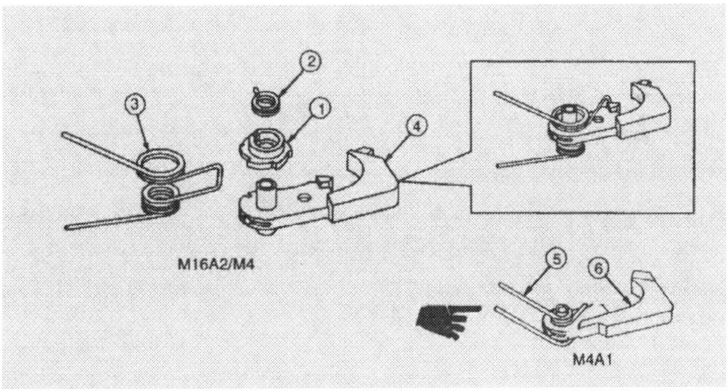

NOTE

M16A2 burst cam is black. M4 burst cam is nickel colored (shiny).

Cam clutch spring should be assembled with bend to the inside. The large loop of the hammer spring should be assembled over the burst cam.

(M16A2/M4) install burst cam (1), cam clutch spring (2), and hammer spring (3) on hammer and hammer pin retainer assembly (4). **(M4A1)** install hammer spring (5) on hammer (6).

3-15. TRIGGER ASSEMBLY AND TRIGGER SUBASSEMBLY (CONT).

b. INSPECTION/REPAIR

1. Inspect trigger spring for kinks, deformities, and weakness. Replace if defective.
2. Inspect disconnector springs for deformities, bends, breaks, and weakness. Replace if defective.
3. Inspect trigger for chips, wear, and cracks. Inspect for damaged searing surface on the trigger nose. Replace if defective.

c. REASSEMBLY

M16A2 AND M4 ONLY
NOTE
Use bolt carrier key tool to install disconnector springs.

1. Use the following procedures to install two disconnector springs (1) using the bolt carrier key tool:

(a) Secure the trigger (2) in a soft vise jaws or similar device.

(b) Place one spring (1) firmly on the tool with large diameter coils outward.

(c) Press spring (1) into recess to solid height.

(d) Hold spring at solid height and slide spring into one of the holes until the punch is flush and perpendicular with the recess wall.

(e) Turn the spring one to two turns opposite of coil winding of the spring.

(f) Discontinue winding when an audible click or snap is heard or felt. This indicates that the spring is seated.

(g) Hold the spring (1) in place when removing the tool to avoid unseating or damaging the spring.

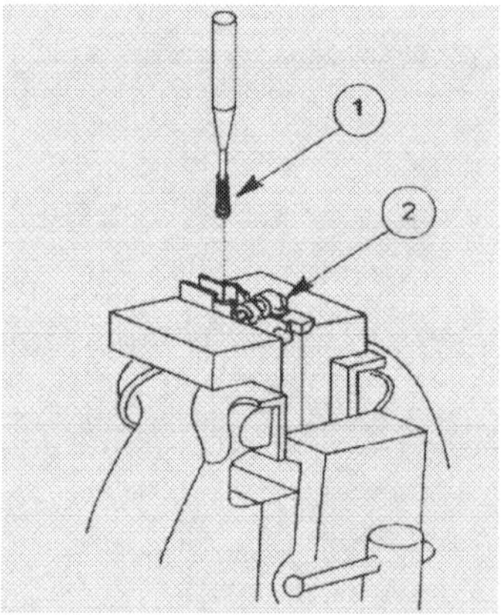

M4 CARBINE ONLY
NOTE
The semi and burst disconnector springs are not the same. The Semi disconnector spring (left side) (figure C-15, item 1B) is black while the burst disconnector spring (right side) (figure C-15, item 1A) is nickel (shiny). Ensure that the correct spring is installed on each side for proper functioning.

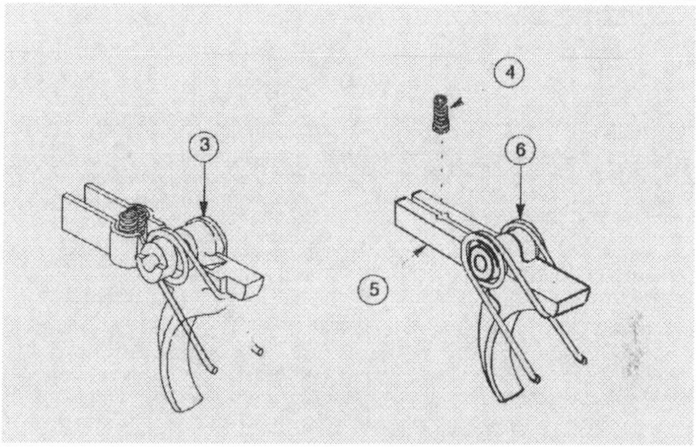

M16A2/M4 M4A1

M16A2 and M4

2. Install trigger spring (3).

M4A1 ONLY

3. Install disconnector spring (4) by inserting large end of spring into trigger (5).

4. Install trigger spring (6) on trigger (5).

3-16. LOWER RECEIVER AND RECEIVER EXTENSION ASSEMBLY.

This task covers:

a. Disassembly
b. Inspection
c. Repair/Modify
d. Reassembly

INITIAL SETUP

Tool Equipment
Tool and Gage Set (item 2, app B)

Tools
(ARMY) Small Arms Repairman Tool Kit (item 3, app B)
Field Maintenance Basic Less Power Small Arms Shop Set (item 1, app B)

Materials/Parts
Cloth, abrasive (item 13, app D)
Grease, molybdenum disulfide (item 19, app D)
Lubricant, solid film (item 21, app D)

Equipment Conditions
3-62 Lower receiver and receiver extension assembly removed

General Safety Instructions
To avoid injury to your eyes, use care when removing and installing spring-loaded parts.
When using solid film lubricant or dichloromethane, be sure the area is well ventilated.

a. DISASSEMBLY

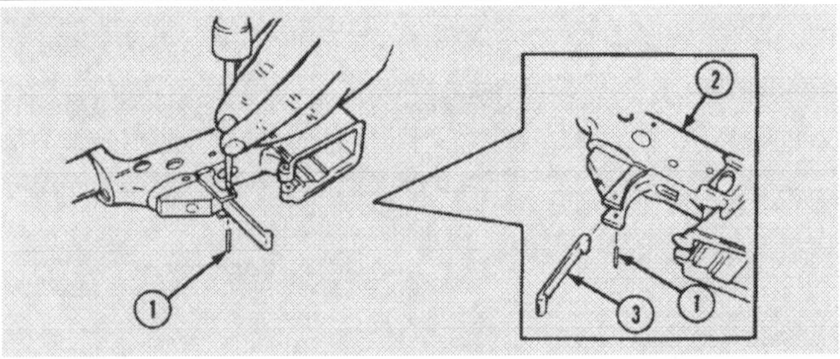

1. Remove spring pin (1) from lower receiver (2) using 1/8 inch drive pin punch and hand hammer.

2. Remove trigger guard (3).

3-16. LOWER RECEIVER AND RECEIVER EXTENSION ASSEMBLY (CONT).

a. DISASSEMBLY (CONT)

3. Use padding between lower receiver and brass vise laws. Grip the solid portion of the lower receiver with brass vise jaws which conform to the shape of the lower receiver in this area.

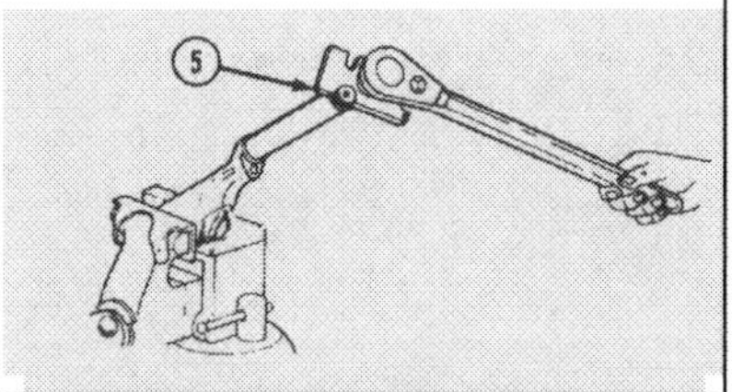

RIFLE ONLY

4. Clamp lower receiver (2) in a machinist's vise using vise jaw caps and tighten on solid portion just tight enough to hold.

WARNING
To avoid injury to your eyes, use care when removing and installing spring-loaded parts.

NOTE
As lower receiver extension is removed, catch buffer retainer and helical spring. Lower receiver is a serial number controlled item.

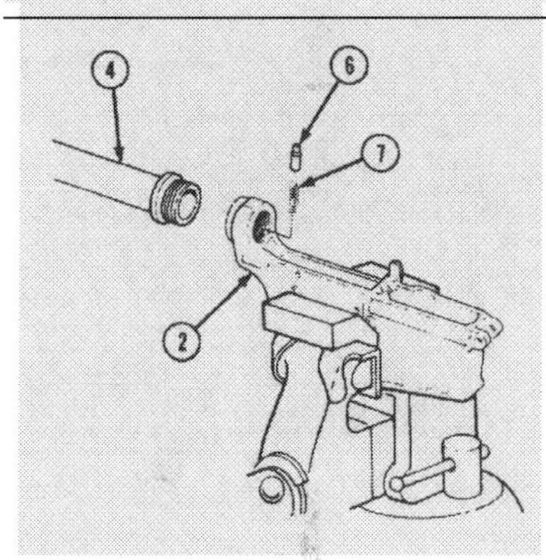

5. Remove lower receiver extension (4) from lower receiver (2) using combination wrench (5) and socket wrench handle. Catch buffer retainer (6) and helical spring (7).

CARBINE ONLY

6. Clamp lower receiver (8) in vise and tighten on solid portion just tight enough to hold.

NOTE
Use wooden vise jaws in place of brass vise jaw caps.

7. Remove the lower receiver extension (9), by loosening the extension locking nut (10) using the special tool (item 12, app C). Catch buffer retainer and spring (11).

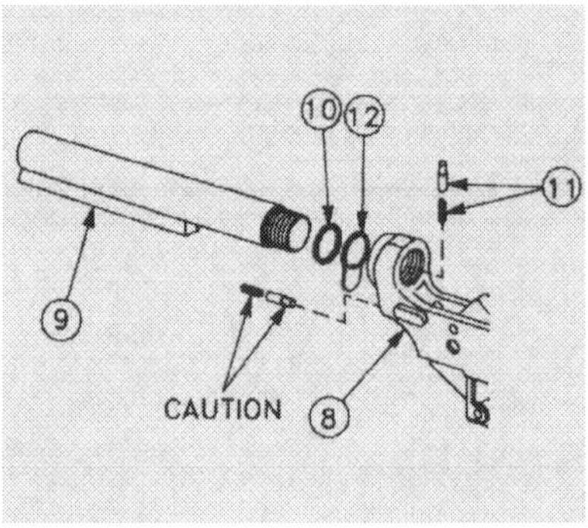

CAUTION
While performing the following step, care should be taken to restrain the pivot pin spring and detent.

8. Loosen the locking nut (10) to allow the receiver end plate (12) to disengage from the lower receiver. Hold the buffer retainer and spring (11) in position with your index finger and unscrew the lower receiver extension (9) from the lower receiver (8).

b. INSPECTION

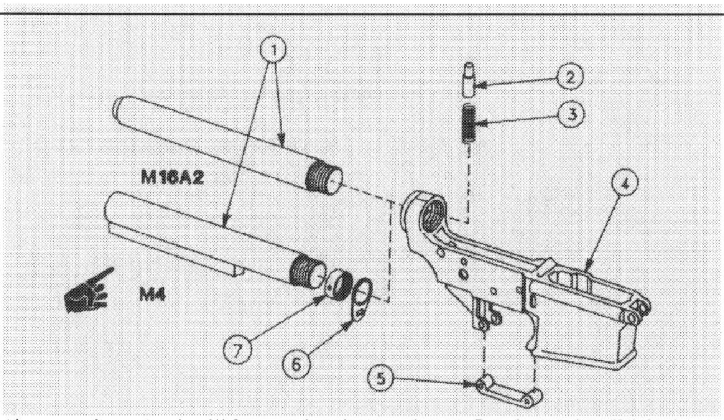

1. Inspect lower receiver extension (1) for corrosion, dents, and wear. Repair or replace if defective.
2. Inspect buffer retainer (2) for wear. Replace if defective.
3. Inspect helical springs (3) for deformities and breaks. Replace if defective.
4. Inspect lower receiver (4). See page 3-66, INSPECTION.
5. Inspect trigger guard (5) for deformities and check operation of plunger and spring. Replace trigger guard if defective.
6. **CARBINE ONLY:** Inspect receiver end plate (6) and locking nut (7) for damage. Replace if damaged.

c. REPAIR/MODIFY

1. Repair lower receiver extension by using abrasive cloth to remove light corrosion. Retouch using solid film lubricant.

NOTE

AIR FORCE ONLY: Only depot maintenance is authorized to restamp the serial number on rifle.

ARMY ONLY: Only direct support level is authorized to restamp serial number.

3-16. LOWER RECEIVER AND RECEIVER EXTENSION ASSEMBLY (CONT).

c. REPAIR/MODIFY (CONT)

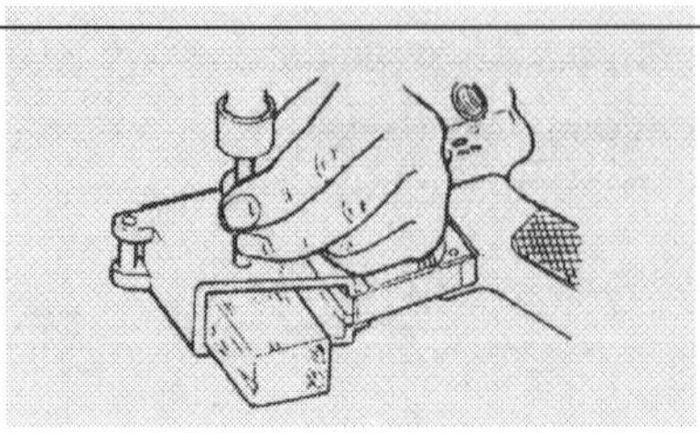

2. **ARMY ONLY:** If serial number is hard to read on rifle, restamp as follows:

   (a) Support the receiver In the stamping area to prevent bending and distortion of the receiver.

   (b) Exercise extreme care to restamp the same serial number as the original.

   (c) Restamp the serial number the same size as the original serial number.

NOTE

Most rifle serial numbers are 1/8 inch (0.31 cm) in height, or close enough that this size is acceptable for such restamping. In the event that a rifle has a serial number that cannot be reproduced by the use of the die sets contained In the Set D Field Maintenance Post, Camp, and Station Small Arms Shop Set, local purchase of an appropriate size die set is authorized.

d. REASSEMBLY

RIFLE ONLY

WARNING

To avoid injury to your eyes, use care when removing and installing spring-loaded parts.

1. Install helical spring (1) and buffer retainer (2) into lower receiver (3).

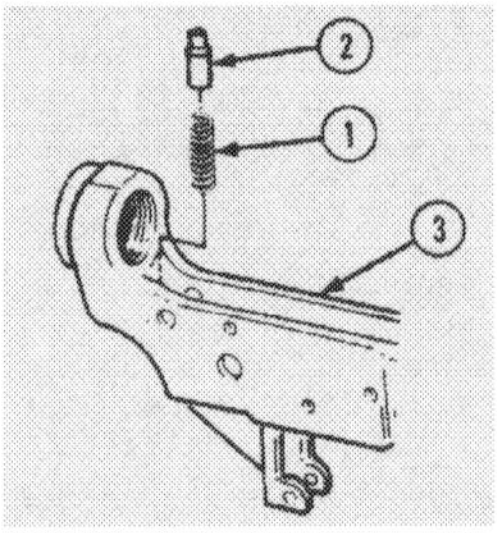

RIFLE ONLY

2. Lubricate threads of lower receiver (3) and lower receiver extension (4) with molybdenum disulfide grease (item 19, app D) before reassembly.

3. Install lower receiver extension (4) into lower receiver (3) while depressing buffer retainer.

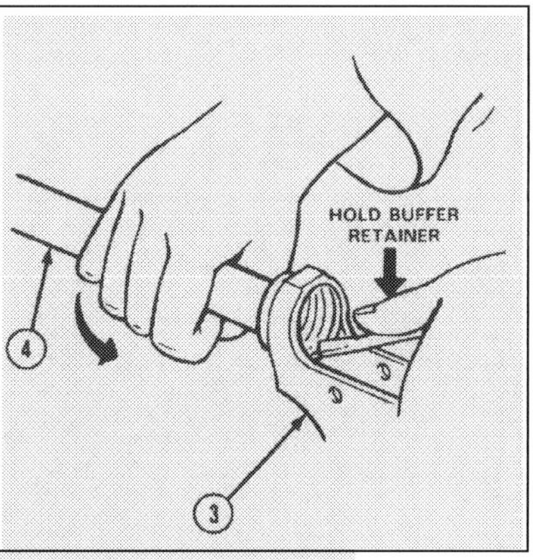

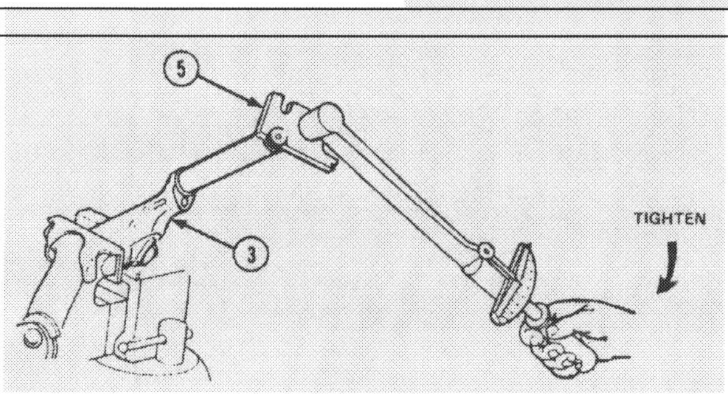

RIFLE ONLY

4. Use padding between lower receiver and brass vise jaws. Use vise jaws in vise and brass vise jaw caps, if available.

5. Clamp solid portion of lower receiver (3) in a machinist's vise using vise jaws. Grip the solid portion of the lower receiver with vise jaws which conform to the shape of the lower receiver in this area.

6. Using combination wrench (5) and torque wrench, torque lower receiver extension to 35-39 ft-lb (47 25 52 65 N•m). Torque is read when both wrenches are used to gether.

3-16. LOWER RECEIVER AND RECEIVER EXTENSION ASSEMBLY (CONT).

d. REASSEMBLY (CONT)

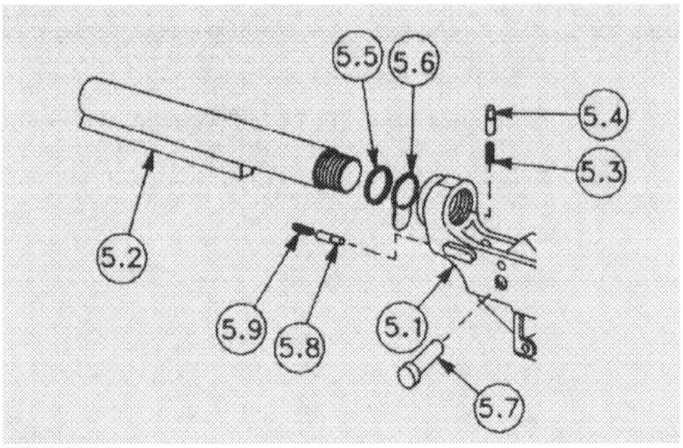

CARBINE ONLY

6A. Lubricate threads of lower receiver (5.1) and lower receiver extension (5.2) with molybdenum disulfide grease (item 19, app D) before reassembly.

6B. Pre-position the spring (5.3) and buffer retainer (5.4) into the retaining hole of the lower receiver (5.1). Screw the locking nut (5.5) onto the lower receiver extension (5.1) with the three notches on the locking ring (5.5) facing forward.

6C. Align the receiver end plate (5.6) onto the lower receiver extension (5.2) with the lug of the receiver end plate (5.6) facing forward.

6D. Pre-position the takedown pin (5.7), detent (5 8), and spring (5.9) in lower receiver assembly (5.1).

6E. Push down on the buffer retainer (5.4) and spring (5.3) and at the same time, screw the lower receiver extension (5.2) in until it retains the buffer retainer (5 4) in position.

6F. Align the lug of the receiver end plate (5.6) into the rear of the lower receiver (5.1). Screw the locking nut (5.5) forward until it contacts the receiver end plate (5.6).

6G. Using the special tool (Item 12, app C) tighten the locking nut (5.5) until snug.

6H. Using the special tool (item 12, app C), and torque wrench, torque locking nut (5.5) to 40 ±2 inch pounds.

6J. Stake the receiver end plate (5.6) in 2 places across from the notches in the locking nut (5.5).

3-16. LOWER RECEIVER AND RECEIVER EXTENSION ASSEMBLY (CONT).

d. REASSEMBLY (CONT)

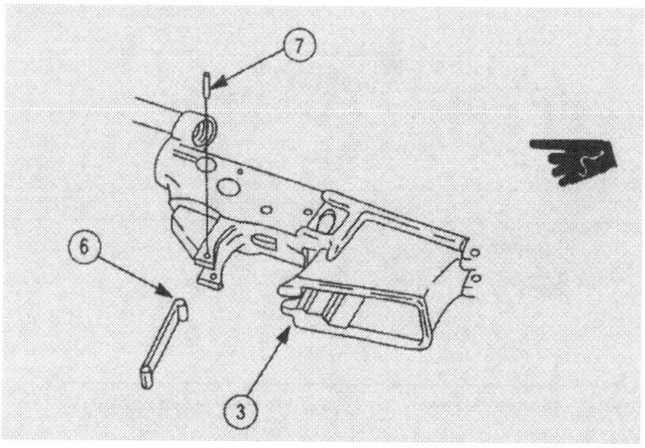

7. Install trigger guard (6) into lower receiver (3).
8. Install spring pin (7) using 1/8 inch drive pin punch and hand hammer.
9. **Deleted.**

3-17. MAJOR COMPONENTS OF M16A2 RIFLE

This task covers:

a. Reassembly b. Test c. Inspection

INITIAL SETUP

Test Equipment
Tool and Gage Set (item 2, app B)

Tools
(ARMY) Small Arms Repairman Tool Kit (item 3, app B)

Reference
TM 9-1005-319-10

General Safety Instructions
To avoid injury to your eyes, use care when removing and installing spring-loaded parts.
Live ammunition must not be near the work area.

a. REASSEMBLY

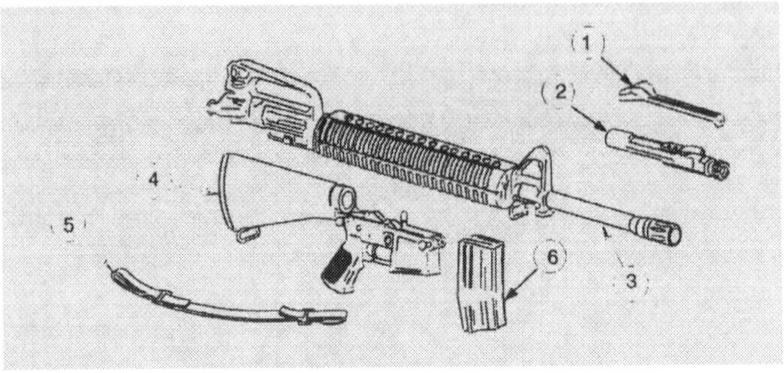

1. Refer to TM 9-1005-319-10.
2. Install charging handle assembly (1) and bolt carrier assembly (2) into upper receiver and barrel assembly (3). Join upper receiver and barrel assembly (3) and lower receiver and buttstock assembly (4).
3. Snap on sling (5) and install magazine (6).

b. TEST

Test trigger pull as follows:

M16A2 and M4 ONLY

(a) Clear the weapon. Place selector to BURST. Pull the trigger and hold it to the rear. Pull the charging handle to the rear and return to the bolt closed position three times. (This will place the BURST disconnector in the deep notch of the BURST cam.)

(b) Release the trigger. Place the selector to SEMI. Hold the rifle in the vertical position. Using trigger pull measuring fixture, 7274758, add weights until hammer trips. Determine weight applied.

(c) Hammer must not trip when 5.5 lbs (2.49 kg) have been applied, hammer must trip on applying 9.5 lbs (4.31 kg).

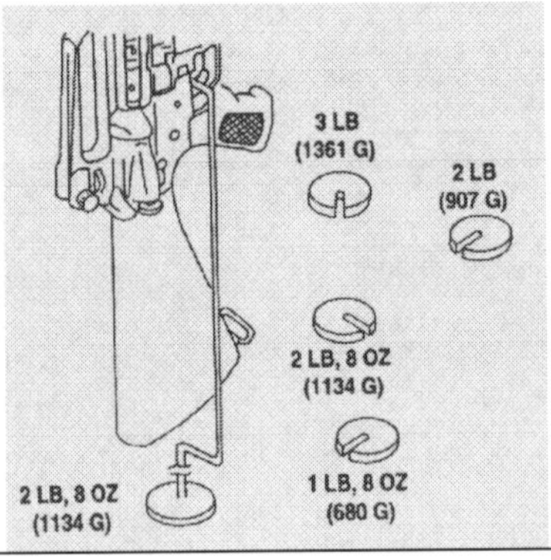

3-17. MAJOR COMPONENTS OF M16A2 RIFLE (CONT)

b. TEST (CONT)

M4A1 ONLY

(d) Clear the weapon. Cock weapon. Place weapon in SEMI position and hold weapon in vertical position.

(e) Using trigger pull measuring fixture 7274758 as shown, add weights until hammer trips. Determine weight applied.

(f) Hammer must not trip when 5.5 lbs (2.49 kg) have been applied. Hammer must trip on applying 8.5 lbs (3.86 kg).

ALL WEAPONS

(g) If weapon fails trigger pull test or excessive creep is present, replace trigger and/or hammer.

NOTE

Always gage trigger and hammer pin holes with not go plug gage 12006472 before replacing parts.

c. INSPECTION

Perform final inspection procedures below.

3-18. M16A2 RIFLE FINAL INSPECTION.

This task covers:

a. Final inspection
b. Test
c. Functional theory of three-round burst control

INITIAL SETUP

Test Equipment
Tool and Gage Set (item 2, app B)

References
TM 9-1005-319-10

General Safety Instructions
Live ammunition must not be near the work area.

Air Force rifles will be inspected in accordance with AFM 36-2227, Vol. 1.

a. FINAL INSPECTION I

1 Visually inspect general appearance of rifle should look almost new. All metal surfaces are to have a dull, rustor corrosion-resistant finish with no burrs or deep scratches.

2 Visually inspect barrel for serviceability. Barrels must be straight, clean, free of rust, powder fouling, and free of bulges and rings. Fine pitting is allowable.

Using moderate hand pressure, check for rotational movement of the front sight in relation to the barrel. Also, using moderate hand pressure, check for rotational movement of the barrel in relation to the upper receiver. If movement between the front sight and the barrel exist, the barrel assembly must be replaced. If movement between the barrel and the upper receiver exist, the barrel assembly must be aligned and tightened (page 3-29).

3 Visually inspect rifle for missing parts. All parts must be attached and all modifications must be applies. Steel parts must be rust free. Spring pins must be secure and screws must be tight.

4 Functionally inspect key and bolt carrier assembly and gas tube alignment. Refer to TM 9-1005-319-10 and use the following procedures:

(a) Disengage the takedown pin and open the receiver.

(b) Remove bolt carrier assembly.

(c) Remove bolt assembly from bolt carrier assembly.

(d) Insert key and bolt carrier assembly into upper receiver and barrel assembly. The bolt assembly must not be installed while performing test.

(e) Slide key and bolt carrier assembly forward to detect binding between key and bolt carrier assembly and gas tube by feel. Badly bent gas tube could cause damage to both the key and bolt carrier assembly and the gas tube. A slightly bent gas tube will cause unnecessary wear of the key and bolt carrier assembly and gas tube.

(f) Correct slight binding by removing handguard assemblies and slightly bending gas tube in the handguard area while repeating step (e) above until no binding is detected. Badly bent gas tubes will be replaced and realigned.

(g) Remove key and bolt carrier assembly from upper receiver and barrel assembly.

(h) Reassemble bolt assembly into key and bolt carrier assembly.

(i) Reinstall key and bolt carrier assembly into upper receiver and barrel assembly.

5 Make a functional check on an assembled weapon with selector lever in SAFE, SEMI, and BURST/AUTO positions Any portion of this check may be used alone to determine the operational condition of any specific firing position selected

3-18 M16A2 RIFLE FINAL INSPECTION (CONT)

a. FINAL INSPECTION (CONT)I

6 Check rear sight assembly as follows:

(a) Rotate elevation knob counterclockwise until the rear sight assembly is all the way down. If a whole click Is not felt as the rear sight assembly stops, the rear sight assembly has bottomed out and will not pivot freely.

(b) Position elevation knob back slightly to its last whole click as the rear sight assembly base is under tension of the ball bearing and helical spring. The 300 meter mark should align with the mark on the receiver (c) If the 300 meter mark is not aligned with the mark on receiver, slip the range scale in the following manner:

(1) Position the 300 meter mark with the mark on the receiver.

(2) Insert a 1/16 inch alien wrench through the access hole of the rear sight assembly base and into the index screw.

(3) Loosen the index screw three turns and leave the wrench in place.

(4) Rotate lower portion of elevation knob counterclockwise until it stops (range scale should not have moved). Elevation knob should be positioned on its last whole click.

(5) Tighten index screw and remove wrench.

(6) Check for proper setting.

7 Pull charging handle assembly to rear. Check that chamber is clear. Leave hammer in cocked position.

8. Place selector lever is SAFE position and pull trigger. Hammer should not fall.

WARNING

If rifle falls any of the following inspections, continued use of the rifle could result In injury to, or death of, personnel.

NOTE

For the purpose of the following tests, "SLOW" is defined as 1/4 to 1/2 the normal rate of trigger release

9. Place selector lever in SEMI position.

10 Pull trigger. Hammer should fall.

11 Hold trigger to the rear, charge rifle and release the trigger with a slow, smooth motion without hesitations or stops, until the trigger is fully forward; an audible click should be heard. Hammer should not fall.

12 Repeat the above selector lever SEMI position test (steps 10 and 11) five times. The rifle must not malfunction during any of these five repetitions. If the rifle malfunctions during any of these five tests, refer to page 3-7.

M16A2 and M4 ONLY

13 Place selector lever in BURST position. Charge rifle and pull trigger. Hammer should fall.

14 While holding the trigger to the rear, pull charging handle assembly to the rear and release it three times. Hammer should not fall the third time the charging handle assembly is released. It may or may not fall the first and second time the charging handle assembly is released. When the burst disconnector reached the deep notch of the burst cam, the burst disconnector should have held the hammer to the rear while the trigger was in the pulled position. Release the trigger with a slow, smooth motion without hesitations or stops, until the trigger is fully forward. The hammer should not fall.

15 Pull trigger. Hammer should fall. This would be the first round of a three-round burst.

NOTE

A detailed explanation of the three-round burst control can be found in the following pages.

M4A1 ONLY

15.1 Place selector lever in AUTO position. Charge weapon and pull trigger. Hammer should fall.

15.2 Hold trigger to the rear, charge weapon, and release trigger. Pull trigger. Hammer should not fall. Automatic assembly sear should have released the hammer as the bolt closed.

ALL WEAPONS

16 With the hammer in the forward position, using moderate finger/thumb pressure, attempt to place the selector lever in the SAFE position. Selector lever should not go in the SAFE position.

17 Perform the following additional functional checks and adjustments on assembled weapon:

(a) Press magazine catch button. Make sure it functions properly.

(b) Press bolt catch. Make certain it operates smoothly and holds bolt in open position.

(c) Inspect front sight and rear sight assembly. Make certain they can be adjusted properly.

(d) Actuate forward assist assembly It must work freely.

3-18 M16A2 RIFLE FINAL INSPECTION (CONT)

a. FINAL INSPECTION (CONT)

(e) Inspect upper receiver and barrel assembly. Barrel assembly should not rotate within upper receiver assembly.

(f) Check that third or middle slot of compensator is straight up (TDC).

b. TEST

1 Check headspace using headspace gage PN 7799734. See page 3-45, TEST.

2 Check firing pin protrusion using firing pin protrusion gage PN 7799735. See page 3-45, TEST.

3 Check extent of barrel erosion using barrel erosion gage PN 8448496. See page 3-45, TEST.

4 Check barrel straightness using barrel straightness gage PN 8448202. See page 3-45.

c. FUNCTIONAL THEORY OF THREE-ROUND BURST CONTROL (M16A2 AND M4 ONLY).

NOTE

First become familiar with the functioning of the firing mechanism especially when in the SAFE and SEMI positions. You should also understand the role that the automatic sear plays when firing in the BURST position. Functioning of the mechanism is explained below In a step by step manner. This actually will seem to complicate something that is really very simple and happens in less than 1 second. The diagrams below and on the following pages do not show the associated springs for the sake of simplicity The positioning of the burst cam is shown in detail

Functional check of three-round burst is as follows:

NOTE
Assume the rifle is fully loaded with a live round in the chamber and the selector lever on BURST.

(a) Hammer is cocked

(b) Front hook of burst lever is in stop notch

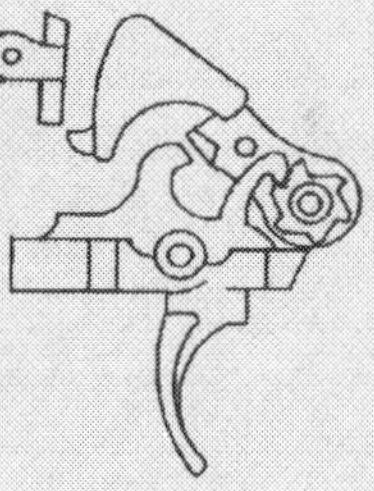

(c) Trigger Is pulled.

(d) Trigger nose drops and hammer falls firing the FIRST ROUND.

(e) Front hook of burst disconnector holds burst cam In place as hammer falls.

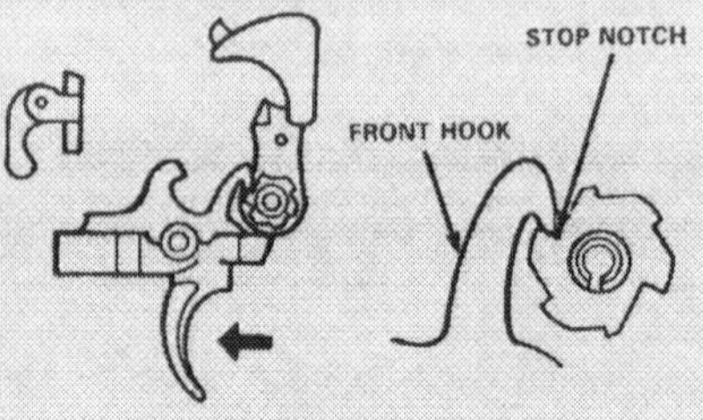

NOTE

Anytime the hammer falls forward, the clutch spring releases the burst cam and allows the front hook of the burst disconnector to keep It In place.

(f) As the key and bolt carrier assembly moves to the rear, the hammer is forced to the rear.

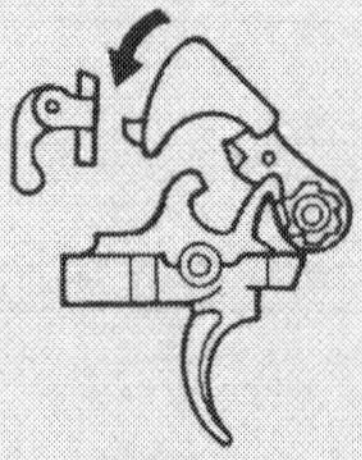

(g) The clutch spring of the burst cam clutches the burst cam and causes It to rotate one notch as the hammer is forced back.

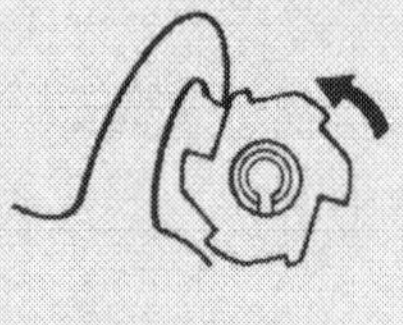

(h) When hammer is fully to the rear, the automatic sear catches It.

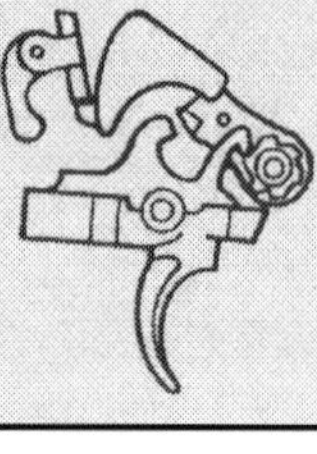

3-18 M16A2 RIFLE FINAL INSPECTION (CONT).

c. FUNCTIONAL THEORY OF THREE-ROUND BURST CONTROL (CONT)

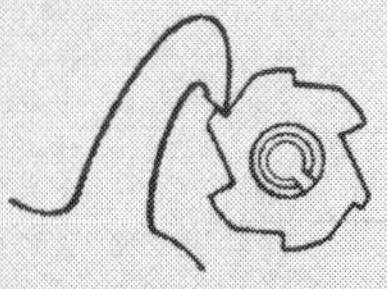

(i) The front hook of the burst disconnector is now fully In the second notch.

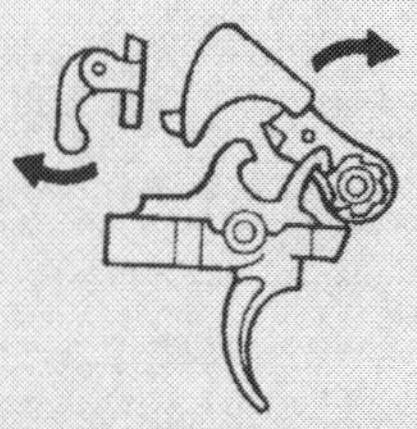

(j) As the key and bolt carrier assembly travels forward, the automatic sear releases the hammer and the hammer falls.

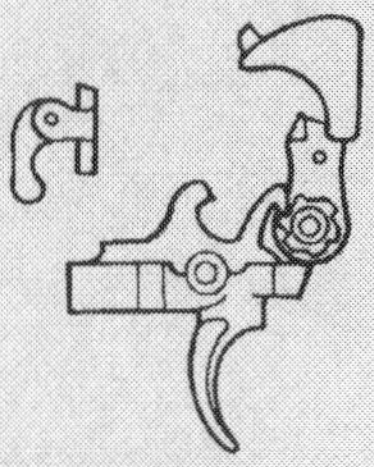

(k) When the hammer falls, the SECOND ROUND is fired.

(l) As the key and bolt carrier assembly moves to the rear, the hammer is forced back to the rear.

(m) The clutch spring of the burst cam clutches against the burst cam and causes It to rotate one notch as the haiimmer is forced back

(n) When the hammer is fully to the rear, the automatic sear catches It

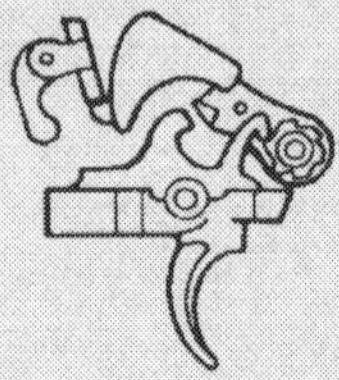

(o) The front hook of the burst disconnector IS now fully In the third notch

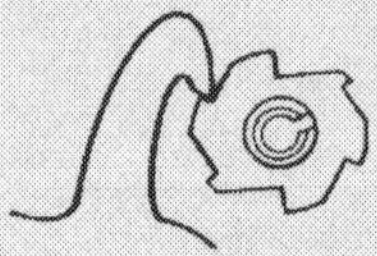

(p) As the key and bolt carrier assembly) travels forward, the automatic sear releases the hammer and the hammer falls

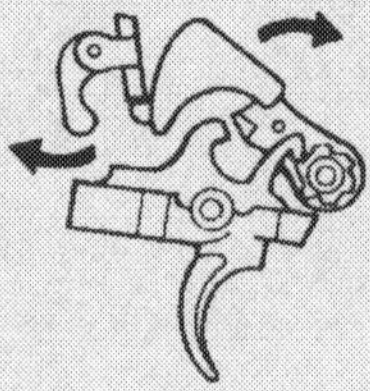

3-18. M16A2 RIFLE FINAL INSPECTION (CONT).

c. FUNCTIONAL THEORY OF THREE-ROUND BURST CONTROL (CONT)

(q) When the hammer falls, the THIRD ROUND Is fired

(r) As the key and bolt carrier assembly moves to the rear, the hammer is forced back to the rear

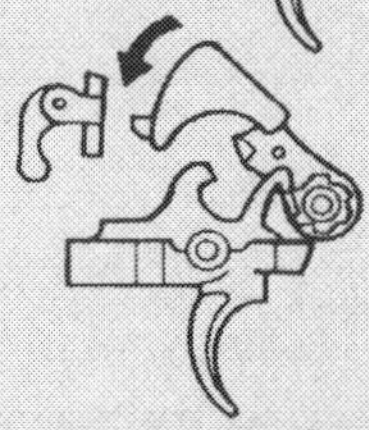

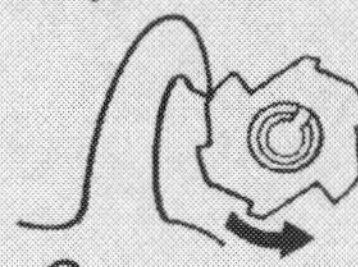

(s) The clutch spring of the burst cam clutches against the burst cam and causes it to rotate one notch as the hammer is forced back.

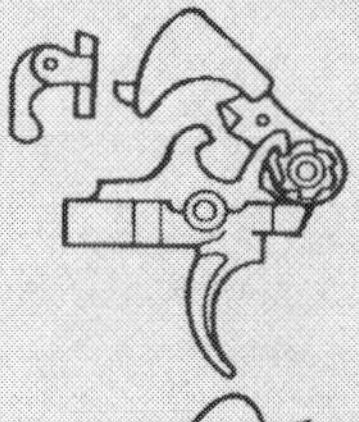

(t) When the hammer Is fully to the rear, It Is initially caught the, ;automatic sear However, the front hook of the first disconnector Is now fully in the next stop notch which is deeper than the others

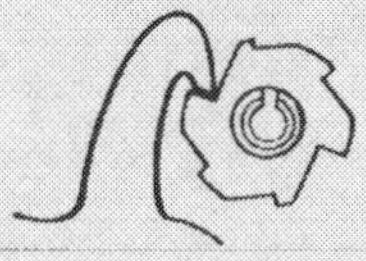

(u) Because a stop notch is deeper than the others, It allows the front hook of the burst disconnector further forward than before. This allows the rear hook of the burst disconnector to latch on the rear hammer notch. This holds the hammer fully to the rear even though the trigger is still to the rear. This happens when the burst Is over and the firing Is stopped.

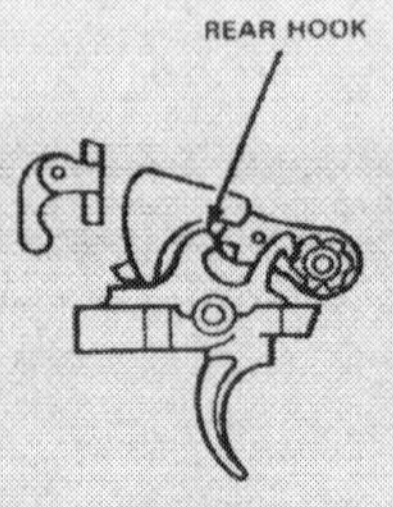

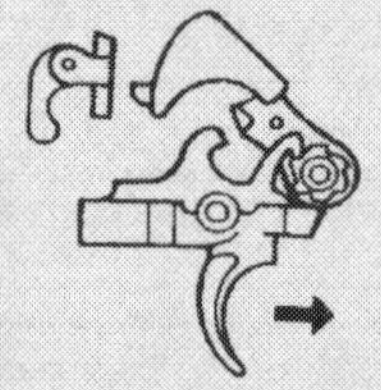

(v) Once the trigger is released, the trigger nose comes up and holds the hammer back.

NOTE

Pulling the trigger to the rear and holding it back again will fire another three round burst This will continue until the magazine is empty However the trigger must be released between each burst

3-19 MI 6A2 RIFLE AND M4/M4A1 CARBINE ANNUAL GAGING REQUIREMENTS.

This task covers.

a Inspection

b Gaging

INITIAL SETUP

Test Equipment
Tool and Gage Set (item 2, app B)

Tools
(ARMY) Small Arms Repairman Tool Kit (item 3, app B)
Field Maintenance Basic Less Power Small Arms Shop Equipment (item 1, app B)

References
DA PAM 738-750
TM 9-1005-319-10
AFR 50-36
AFTO Form 105

General Safety Instructions
- To avoid injury to your eyes, use care when removing and installing spring-loaded parts
- Initial gaging is required 1 year from receipt of the weapons.
- All rifles and carbines must be gaged at least once annually for safety.
- All Army Reserve and Army National Guard M16A2 rifles must be inspected and gaged at least once every 2 years, after the initial inspection/gaging procedures have been accomplished. This 2 year interval may be maintained unless preventive maintenance checks and services (PMCS), or other physical evidence, indicates that an individual unit's rifles require inspection/ gaging at a more frequent interval. If it is determined that a yearly inspection is necessary for an individual unit, only that unit will be affected. This will not affect other units in regard to the interval of inspection.
- It is recommended that training units inspect/ gage all rifles at the end of each training cycle. Training units will inspect/gage all rifles at least once annually.
- Air Force M16 rifles will be inspected in accordance with AFR 50-36, Vol 1, Chapter 5.

a. INSPECTION

1 Visually inspect general appearance of weapon Overall appearance will be approximately that of a new weapon. For inspection criteria refer to final inspection, page 3-84 All visual and functional Inspection requirements must be met.

NOTE

To perform the following tests, disassemble weapon only as far as allowed in chapter 2 (Unit Maintenance Instructions). unless a deficiency is uncovered

2 Perform a general inspection of weapon per section III of this chapter Repair as required and authorized

b. GAGING

1	Gage bolt carrier assembly for firing pin protrusion using firing pin protrusion gage PN 7799735. See page 3-20, TEST.
2.	Gage bolt carrier assembly for firing pin hole wear using not-go plug gage PN 12620101. See page 3-24, TEST.
3	Inspect chamber in upper receiver and barrel assembly using chamber reflector tool PN 8448201 See page 3-33, INSPECTION/CLEANING.
4	Gage barrel in upper receiver and barrel assembly using barrel erosion gage PN 8448496 and bore straightness gage PN 8448202. See page 3-45, TEST.
5	Check headspace in upper receiver and barrel assembly by inserting headspace gage PN 7799734 in chamber. See page 3-45, TEST.
6	Gage pivot pin lug area clearance in lower receiver assembly using 0.020 thickness gage. See page 3-68.
7	Gage hammer and trigger pin holes in lower receiver assembly using taper plug gage PN 12006472. See page 3-68, TEST.
8	Gage trigger pull using trigger pull measuring fixture PN 7274758. See page 3-83, TEST.
9	Document inspection with DA PAM 738-750, AFTO Form 105, or NAVMC 11003 when completed.

Section IV. PREEMBARKATION INSPECTION OF MATERIEL IN UNITS ALERTED FOR OVERSEAS MOVEMENT

NOTE

This section refers to all the weapons (MI 16A2 rifle, M4, and M4A1) unless otherwise stated.

3-20 PURPOSE: This section establishes standards for overseas shipment (pre-embarkation in section criteria) for all weapons. These standards are provided to ensure that the user is furnished equipment which will perform its mission without early failure or major maintenance problems.

3-21 SCOPE.

a. The standards prescribed provide for a high percentage of remaining life in affected rifles; therefore, rifles designated for overseas shipment must qualify under the standards contained in the following paragraph, table, and in referenced DA publications, before they can be approved for shipping action.

b. Provisions of this standard apply to all US Army agencies/activities selecting or preparing rifles for shipment to US troops overseas. It also applies to CONUS troops preparing rifles for shipment overseas. Provisions do not apply either to rifles being prepared for shipment to MAP/MAS recipients unless specifically prescribed by MAP/MAS transaction for the materiel or to rifle being returned to CONUS from overseas. The maintenance instructions and standards contained herein do not apply to rifles once the material has arrived at the overseas destination. At that time, maintenance instructions contained in the applicable TM's will be used.

c. This applies to rifles which are the logistic responsibility of the US Army Armament, Munitions and Chemical Command.

3-22. GENERAL.

a. Only rifles which have been classified as serviceable condition code A, B, or C under AR 725-50 will be considered for overseas shipment. All items of equipment for which equipment serviceability criteria have been published must, as a minimum, be rated green under the ESC as a prerequisite to overseas shipment. In addition to the condition code standard, as enumerated above, and the required ESC rating prescribed herein, the rifle being considered for overseas shipment must meet the requirements of this section. The ESC will be discontinued as new operator manuals are revised which will be used to determine serviceability condition of rifle.

b. Waivers to provisions can only be granted by the gaining command of any particular end Item being considered for issue, deployment, or shipment. The issuing services may recommend issue or shipment of rifles not meeting the provisions when all the following conditions exist:

(1) Repair parts in required quantities cannot be obtained from the supply system prior to delivery of the end item.

(2) The gaining command concurs in the receipt of the end item for storage until required repair parts become available. The gaining command must also state that capability, facilities, and funds are available to perform the necessary work when parts become available.

(3) Department of the Army approval is obtained on a case-by-case basis.

(4) Required repair parts are requisitioned by the issuing command for delivery to the Framing command.

c. All Department of the Army MWO's applicable to the specific rifle being considered for shipment overseas must have been applied.

d. Refer to SB 746-1 for pertinent publications relating to equipment processing and marking information

e. Refer to AMC-P 310-9 for publications containing applicable overhaul standards.

3-23. SHIPMENT OR ISSUE.

a. **Organizational Repair Parts, Tools, and Equipment**. Rifles must be complete with all Items required by applicable Department of the Army publications, including those In the basic Issue items list of the appropriate operator's manual.

b. **Publications**. Operator publications applicable to the equipment log book must accompany the equipment. All log book entries must be complete and up-to-date including those covering any repairs, replacements, or adjustments made to the rifle In complying with this section.

c. **Documentation**. Prepare DA Form 2408-9 (Equipment Control Record) at time of overseas shipment or Issue to another stock record or property book account, in accordance with the provisions of DA PAM 738-750.

d. **Preparation**. Process rifles for shipment as required by shipping documents and pertinent regulations.

3-24. DISPOSITION. Disqualified rifles which do not qualify for shipment will either be redistributed within the camp, post, or station, be repaired, or become candidates for overhaul, cannibalization, or other disposition as required by existing regulations.

3-25. GENERAL INSPECTION CRITERIA.

WARNING

Before starting an Inspection, be sure to clear the rifle Do not actuate the trigger until the rifle has been cleared. Inspect the bore and chamber to ensure that It is empty and free from obstructions, and check to see that no ammunition Is In position to be Introduced.

a. Before Inspection, the materiel must be thoroughly cleaned of all grease, dirt, or other foreign matter that might Interfere with Its proper function or the use of gages and tools during inspection.

b. Materiel must be free of burrs, rust, or corrosion on functional surfaces.

3-25. GENERAL INSPECTION CRITERIA (CONT).

c. Parts must not be cracked, bent, distorted, or damaged and must be free of detrimental wear or looseness

d. Minor defects In metal components do not normally affect their acceptability For example, scratches and tool marks are ordinarily of no Importance

e. Inspect finish of metal surface

(1) General. Satisfactory metal surfaces for rifles range from black to light gray. A worn shiny metal surface is objectionable only when It is capable of reflecting light No rifle will be reelected unless exterior parts have a shine All rear sights must have a dull gray or black finish of all surfaces that would cause a glare

(2) M16A2 Rifle. Minor loss of finish (shiny spots, nicks, scratches) on exterior surfaces of the barrel and flash suppressor shall not be cause for rejection of M16A2 rifles located In hands of troops at training centers Large shiny surfaces, nicks, scratches, etc., can be restored by the use of solid film lubricant (Item 21, app D) Rifles (small arms) missing in excess of one-third or more of the exterior finish resulting In an unprotected, light reflecting surface, are considered candidates for overhaul The only authorized level of maintenance to phosphate finish small arms Is depot.

f. Plastic components must not be cracked or damaged In such a way as to Interfere with their structural strength Surface cracks, bruises, or dents that do not affect their strength will not be cause for rejection. Cracks will be cause for rejection. Criteria for determining which cracks are repairable are on page 2-1 4.

g. Barrels must be clean and free of corrosion such as that caused by moisture and powder fouling Standards of serviceability are indicated In (1) through (9).

(1) Pits In the chamber are allowable If they do not cause extraction difficulties

(2) Pits as wide as a land and 3'8 Inch (O 95 cm) or less in length are allowable for 5 56-mm barrels Pits not greater than the width of a land and less than 3 8-inch (O 95 cm) long are permissible

(3) Scattered or uniformly fine pits, or fine pits In a densely pitted area are allowable

(4) Tool marks are acceptable regardless of length They will appear as lines running laterally In the grooves, or may run spirally across the top of lands

(5) Ringed bores or bores ringed sufficiently to bulge the outside surface of the barrel are cause for rejection. However, faint rings or shadowy depressions do not indicate an unserviceable barrel and will not be cause for rejection. Gap in lined barrels will not be classified as a ringed bore.

(6) Lands that appear dark due to coating of gilding metal from projectiles will not be cause for rejection.

(7) Breech bore diameter will be checked on unlined barrels using the appropriate breech bore gage.

(8) Barrel erosion gages are provided for lined barrels. Bore wear will be checked using barrel erosion gage for the M16A2. For detailed instructions in the use of the above gage and for serviceability limits, refer to page 3-46.

(9) Flaking or checking (fine cracks) of chromium plate in barrels or chambers will not be cause for rejection, unless accompanied by pitting to the degree that extraction difficulties are encountered or accuracy is unacceptable.

h. Springs must be free of distortion and broken coils. Springs must have sufficient tension to perform their intended function.

i. Screw heads must be in serviceable condition and threads must not be stripped. Internal threads must not be stripped.

j. The sear, hammer, and/or cocking notches must be in good condition. Chipped engaging corners will be cause for rejection. Slight wear on functional surfaces, including engaging corners, shall be acceptable, providing the minimum trigger pull requirements and selector lever checks are met in accordance with instructions on page 3-83.

k. Chips, flat spots, or bent striker points on firing pins will be cause for rejection.

l. The cartridge engaging surfaces on extractors must not be chipped or deformed.

m. Evidence of any damage to sights will be cause for a sight alignment check. Rear sight bases should have no movement.

n. Rear sight elevating and windage mechanisms must operate with distinct clicks, without binding. Sights must have sufficient tension to retain their setting during firing. Graduations and numerals must be legible. Graduation filler Is not required.

o. Safeties must positively position in both the ON and OFF position. When in the ON or safe position, the rifle must not fire when the trigger is squeezed; when in the OFF or fire position, the rifle must fire when the trigger is squeezed.

3-25. GENERAL INSPECTION CRITERIA (CONT).

p. All locking devices such as latches, magazine latches, or detents must be positive in action and must not become disengaged due to normal handling and firing. Retaining pins and similar devices must not be subject to accidental loss during use or transportation.

q. Each rifle must be hand functioned to check for unusual binding, positive cocking action, and general operation. Dummy ammunition must be used to assure positive feeding, chambering, extraction, and ejection action.

Table 3-1. 5.56MM Rifle M16A2

Item	Standard
RIFLE:	
General	Clear rifle of any ammunition and Inspect In accordance with paragraph 3-25.
Barrel and barrel extension	Check barrel erosion. Use barrel erosion gage 8448496 for chrome lined barrels. Stripping of lands and grooves shall not be cause for rejection unless so determined by barrel erosion gage. Visually inspect, using chamber reflector tool 8448201. Pits 1/8 inch (0.31 cm) In length and those pits large enough to extend from the body of the chamber into the shoulder stop area and forcing cone area are cause for rejection. Large pits are defined as those 1/8 inch (0.31 cm) or more in diameter as determined by visual Inspection. Only closed flash suppressors are acceptable. Check barrel for straightness using bore straightness gage 8448202. Gage must pass freely through the bore to be acceptable, either dropped from the muzzle or chamber end.
Front sight and gas tube	Inspect gas tube for proper alignment with carrier key. Gas tube must not bind when mating with the key. Evidence of gas leaks around the front sight connection of the gas tube shall be cause for rejection until rifle has been function fired to determine if the loss of gas is sufficient to cause malfunction. If function firing malfunctions occur, repairs are necessary. Inspect front sight for damage.

Table 3-1. 5.56MM Rifle M16A2, M4/M4A1 Carbine - Cont

Item	Standard
Bolt carrier group	Inspect bolt for elongated or oversized firing pin hole using plain cylinder gage 12620101. Firing pinholes which permit the plain cylinder plug gage to fully penetrate at any position on the circumference will be rejected. Bolt face with a cluster of pits which are touching or tightly grouped, covering an area measuring approximately 1/8 inch (0.31 cm) across will be rejected. Bolts which contain pits extending into the firing pinhole will not be rejected unless firing pin hole gaging check determines rejection. Bolts which contain individual pits or scattered pits will not be cause for rejection. Only phosphated bolt carriers are acceptable. Both phosphated and chrome plated bolts are acceptable.
Bolt locking lugs and bolt cam pin hole	Inspect for cracks in the locking lugs and cam pin hole area. Use a black light, if available; otherwise, use a glass of no more than 3X magnification or use inspection penetrant (item 25, app D). Use instructions contained in kit for application. If cracks are detected, the bolts will be replaced. **NOTE** Particular attention must be given to the area where the lugs meet the bolt body and around the side walls of the cam pin hole. Bolt rings must not be broken. Ring gaps must be properly spaced approximately 1/3 turn apart and not in line. Firing pin protrusion must be not less than 0.028 inch (0.071 cm) or more than 0.036 inch (0.091 cm). (Use firing pin protrusion gage 7799735.) Socket head capscrews must be staked. Carrier key must not be dented where end mates with gas tube. Repair or replace damaged carrier keys.
Headspace	Inspect headspace using headspace gage 7799734. Excessive headspace will be cause for rejection.
Trigger pull	Inspect trigger pull using trigger measuring fixture 7274758. Trigger pull must be minimum 5.5 pounds (2.49 kg), maximum for M16A2 and M4 is 9.5 pounds (4.31 kg) and maximum for M4A1 is 8.5 pounds (3.86 kg). Test trigger pull, refer to page 3-83, steps a, b, c, and d.

3-25. GENERAL INSPECTION CRITERIA (CONT).

Table 3-1. 5.56MM Rifle M16A2, M4/M4A1 Carbine - Cont

Item	Standard
Lower receiver group	Inspect hammer and trigger pin holes using plain cylinder plug gage 12006472 Penetration of the gage in any one or more of the four holes will be cause for rejection. Inspect for cracks, corrosion, or mutilation which would affect functioning. Small dents or gouges will not be cause for rejection. Inspect receiver for corrosion in the lobes of the pivot or hinge pin area. Width between lobes shall not exceed 0 515 inch (1.30 cm). Inspect receiver for break through of metal. Inspect receiver and receiver extension for initial loss of protective coating.
Action spring	Free length of spring shall be between 11-3/4 and 13-1/2 inches (29.84 and 34.29 cm) **M16A2 ONLY** and 10-1/16 and 11-1/4 inches (25.56 and 28.58 cm) **M4/M4A1 ONLY**.
Handguard	Inspect handguard assembly for breaks, separations, and cracks. Breaks and separations of material which prevent proper retention or interfere with functioning of the weapon will be cause for handguard rejection and replacement. Cracks up to 1 inch (2.54 cm) in length are acceptable provided they do not extend into the handguard retaining flange (1) (critical area). **M16A2 ONLY** Each handguard assembly may have up to two of the three front retaining tabs (2) missing. If all three front tabs are missing, the handguard assembly must be replaced. **M4/M4A1 ONLY** No tabs can be missing. **ALL WEAPONS** Replace severely cracked handguards. Handguards that have a heat shield loose enough to rattle when installed on the weapon must be replaced.
Stock assembly	Inspect buttstock assembly for dents, cracks, and chips. Check for breaks and separation of material which could prevent proper functioning of weapon. M4/M4A1 inspect for proper functioning. Repair as required

Table 3-1. 5.56MM Rifle M16A2, M4/M4A1 Carbine - Cont

Item	Standard
	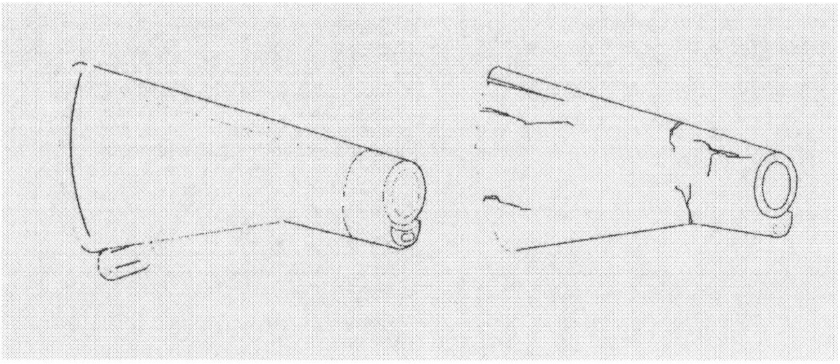
	M16A2 ONLY Under the following conditions, hairline cracks originating from buttplate end of buttstock are acceptable. No chipped away material is allowed. a. One hairline crack, not to exceed 1 inch (2.54 cm) in length, per side of buttstock. b. Two additional hairline cracks up to 0.22 inch (0.55 cm) in length, per side of buttstock. Buttstocks with unauthorized markings stamped into their surfaces will be replaced. Unauthorized markings, scratched, etched, carved, etc., are acceptable if they do not extend into the fiber of the buttstock which may weaken it. These marks may lie at any location on the buttstock. Cracks in the critical area at the front end of the buttstock are not acceptable and these buttstocks must be replaced.

CHAPTER 4

MAINTENANCE OF AUXILIARY EQUIPMENT

CHAPTER OVERVIEW

This chapter contains information and instructions to keep auxiliary equipment used with your weapon in good repair.

Section I. AUXILIARY EQUIPMENT REPAIR

4-1 GENERAL.

a The following items of auxiliary equipment are used in conjunction with the Weapons:

(1) 40mm Grenade Launcher M203, NSN 1010-00-179-6447, (M16A2 ONLY)

(2) Lock Plate, NSN 1005-00-233-9031.

(3) Top Sling Adapter, NSN 1005-00-406-1570.

(4) Blank Firing Attachment M15A2, NSN 1005-00-118-6192, (M16A2 ONLY), Blank Firing Attachment M23, NSN 1005-01-361-8208 (M4 ONLY)

Change 4**(5)** ARMY ONLY: Conversion Kit, M261 (caliber .22 rimfire adapter), NSN 1005-01-010-1561.

(6) Bayonet-Knife M7, NSN 1005-00-073-9238.

(7) Bayonet-Knife Scabbard M10, NSN 1095-00-223-7164.

(8) M9 Multi-Purpose Bayonet System, NSN 1005-01-227-1739.

(9) Night Vision Sight, Individual Served Weapon, AN/PVS-4, NSN 5855-00-629-5334 (M16A2 ONLY)

(10) M2 Practice Bolt, NSN 1005-01-184-4041 (M16A2 ONLY)

b Refer to TM 9-1010-221-23&P for unit maintenance for the Grenade Launcher M203.

c ARMY ONLY: Refer to TM 9-6920-363-12&P for unit maintenance for the M261 Conversion Kit (caliber rimfire adapter) M16, M16A1, and M16A2 Rifles.

d Refer to TM 9-1005-237-23&P for repair instructions and repair parts for Bayonet-Knife M7 and Bayonet Knife Scabbard M10 and M9 Multi-Purpose Bayonet System.

e Refer to TM 11-5855-213-23&P for maintenance for the Night Vision Sight, Individual Served Weapon, AN/PVS-4.

f Refer to TM 9-6920-746-12&P for unit maintenance for the M2 Practice Bolt.

4-2. LOCK PLATE

This task covers:

a Installation
b Removal
c Inspection

INITIAL SETUP

Tools
(ARMY) Small Arms Repairman Tool Kit
(item 3, app B)

WARNING
The lock plate prevents the selector lever from being placed in BURST and will be installed at the discretion of the unit commander. It is mandatory for use in civil disturbance (riot control). Do not keep live ammunition near work area.

a. INSTALLATION

1 Using a screwdriver, reach inside pistol grip (1) and remove screw (2) and lockwasher (3).

WARNING
To avoid injury to your eyes, use care when removing and installing spring-loaded parts.

2 Carefully remove pistol grip (1). Hold detent helical spring (4) in place.

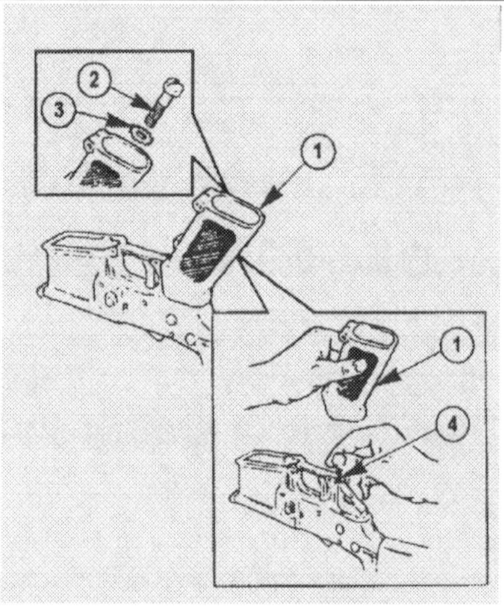

3 Install lock plate (5) with the detent helical spring (4) passing through the hole In the right side of the lock plate and the arm on the outside of the receiver The selector lever just point to the SAFE position

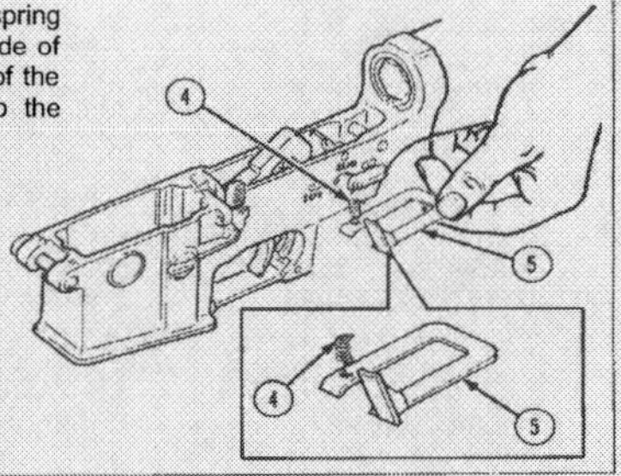

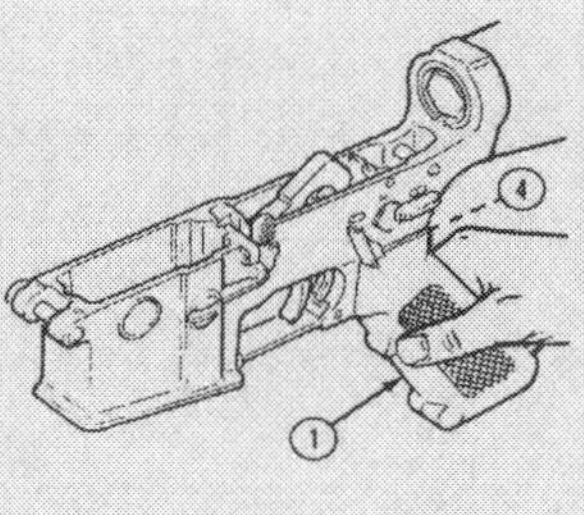

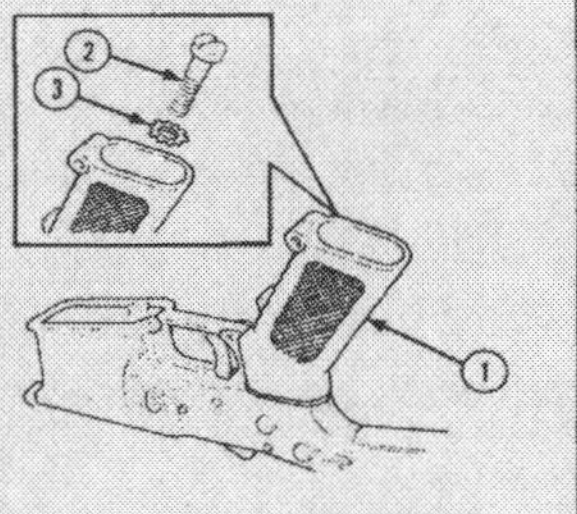

4 Carefully compress detent helical grip (1)

5. Using a screwdriver secure pistol by Installing lockwasher 131 spring (4) and position pistol grip (1) and screw 12)

4-2. LOCK PLATE (CONT).

b REMOVAL

1. Using a screwdriver, reach inside pistol grip (l) and remove screw (2) and lockwasher (3)

WARNING
To avoid Injury to your eyes, use care when removing and installing spring loaded parts

2 Carefully remove pistol grip (1) Hold detent helical spring (4) In place

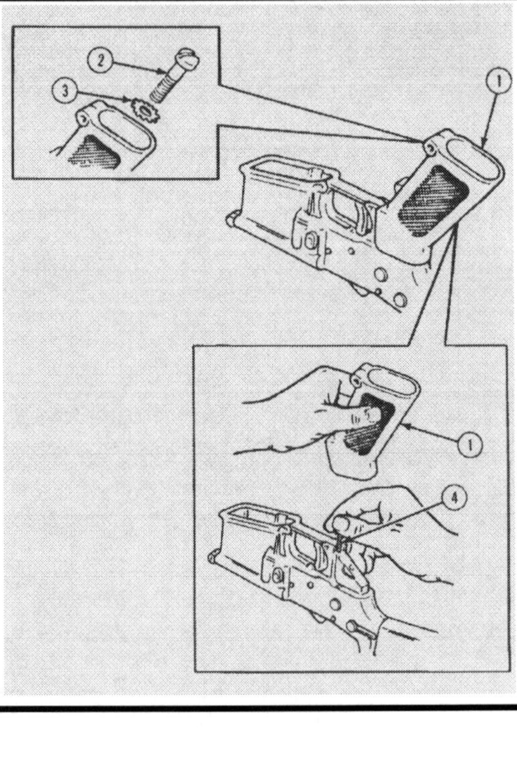

3 Remove lock plate (51 while continuing to hold detent helical spring (4) In place The selector lever must point to the SAFE position

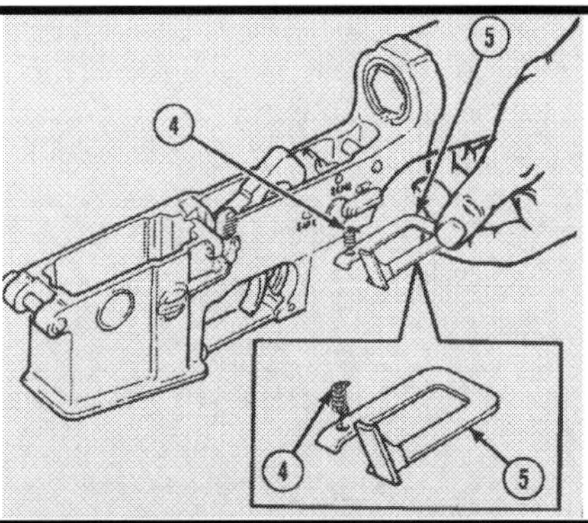

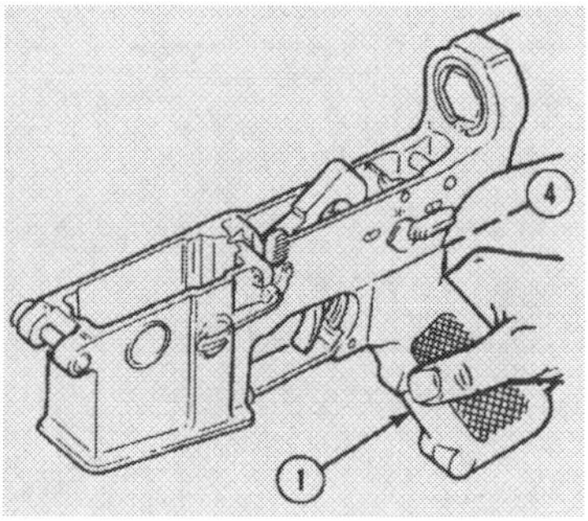

4 Carefully compress detent helical grip (I)

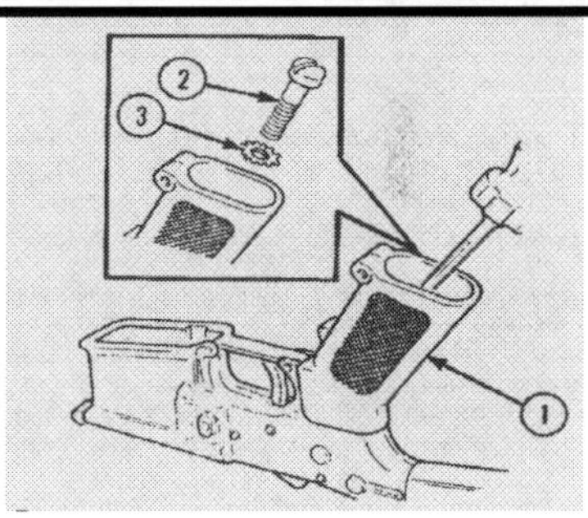

5. Using a screwdriver, secure pistol by installing lockwasher (3) spring (4) and position pistol grip (1) and screw (2)

c. INSPECTION

Inspect lock plate for serviceability and broken arm Replace If unserviceable or if arm is broken off

4-3. TOP SLING ADAPTER

This task covers:

a Installation
b Removal
c Inspection

INITIAL SETUP

Materials/Parts
Top sling adapter kit PN 8448471

References
TM 9-1005-319-10

a. INSTALLATION I

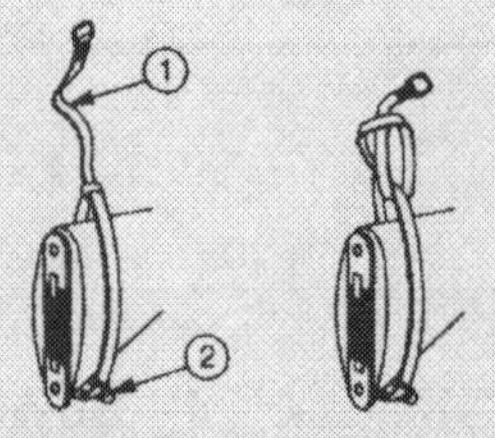

RIFLE

1 Refer to TM 9-1005-319-10 to remove sling.

RIFLE ONLY

2 Install top sling adapter strap (1) through sling swivel (2) and tie.

CARBINE ONLY

2.1. Install top sling adapter strap (1) through sling opening (2.1) and tie.

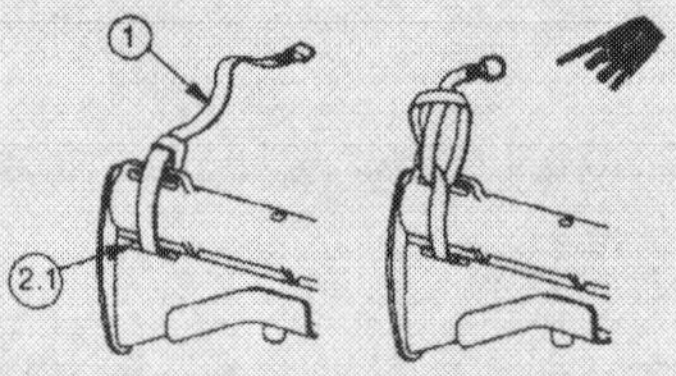

CARBINE

3 Remove upper handguard assembly.

4 Use pliers to install clamp (3) on front sight base (4).

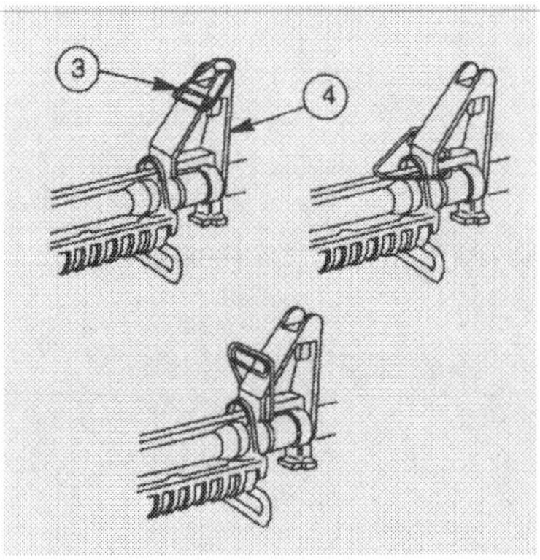

5 Attach sling (5) to top sling adapter strap (1) and to clamp (3).

6 Reinstall upper handguard assembly (6).

7 Adjust sling (5).

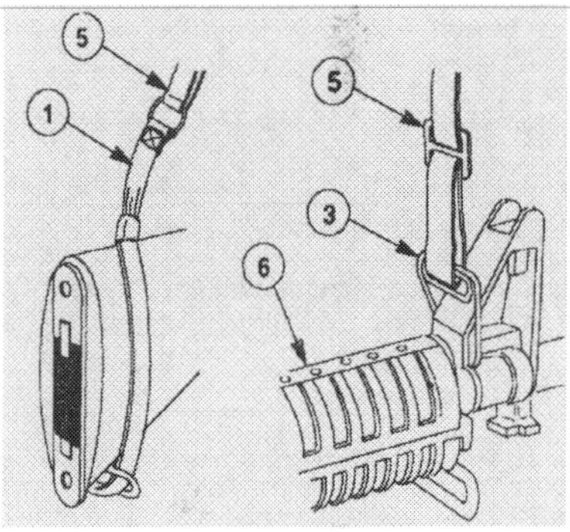

4-3. TOP SLING ADAPTER (CONT).

b. REMOVAL

1 Remove upper handguard assembly.

2 Remove sling (1) from clamp (2) and top sling adapter strap (3).

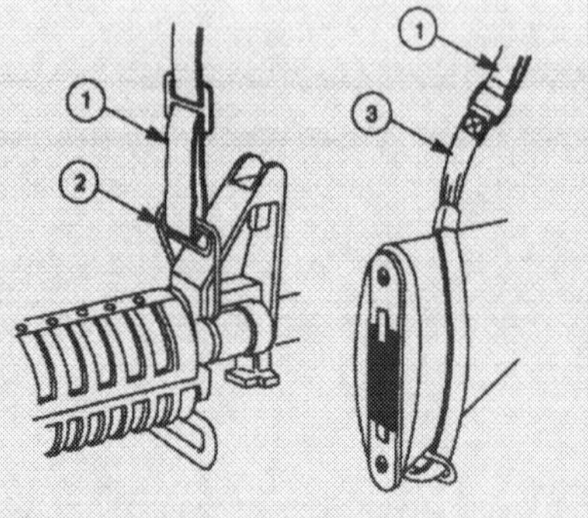

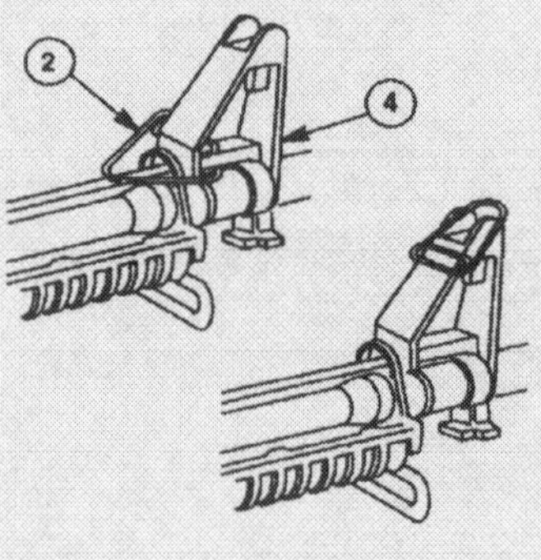

3 Use pliers to remove clamp (2) from front sight base (4). (5).

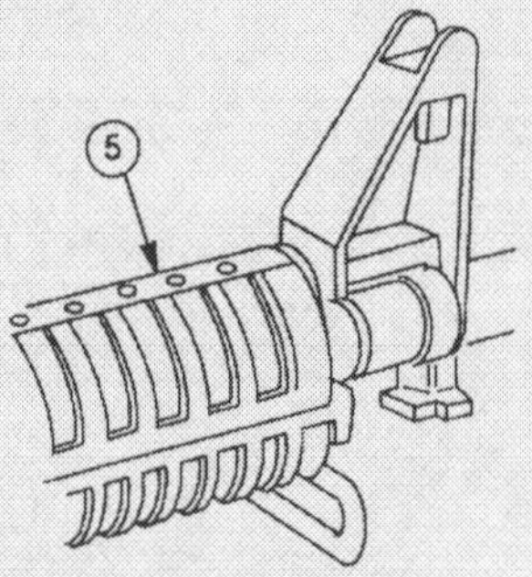

4 Reinstall upper handguard assembly

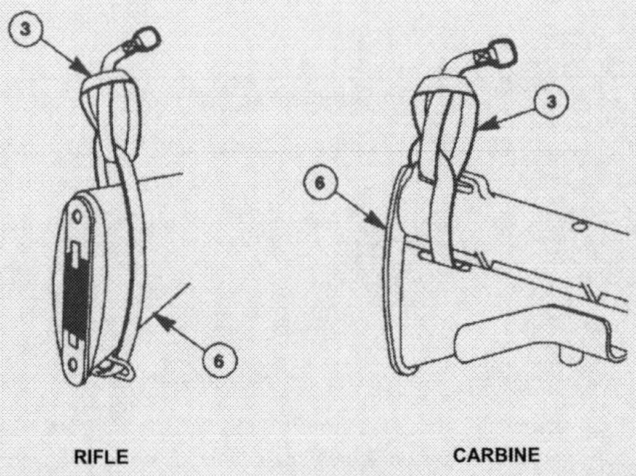

5 Untie top sling adapter strap (3) and remove from buttstock (6).

6 Refer to TM 9-1005-319-10 and install sling.

4-3. TOP SLING ADAPTER (CONT)

c. INSPECTION

Visually Inspect top sling adapter strap and replace If it is badly worn or damaged

4-4. BLANK FIRING ATTACHMENTS M15A2 (M16A2) AND M23 (M4/M4A1).

This task covers:

a Installation
b Removal
c Cleaning
d Inspection
e Repainting
f Replacement

INITIAL SETUP

Materials/Parts
Cleaner, lubricant, and preservative (CLP) (item 9, app D)
Coating compound, enamel (Red - M16A2) (item 23.2, app D)
Coating compound, enamel (Yellow - M4/M4A1) (item 23.3, app D)

General Safety Instructions
Do not keep live ammunition near the work area.

Only blank cartridge M200 is to be used when the blank firing attachment is attached to the weapon.
Do not fire blank ammunition at a representative enemy at distances of less than 20 feet (6.10m).
The unburned propellant grains can cause injury within this distance.

a. INSTALLATION

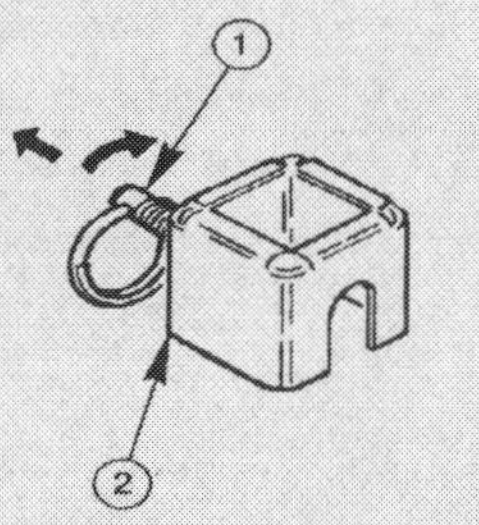

NOTE
M23 blank firing attachment is stamped "M4 Carbine Only" painted yellow and may be used on the M4 and M4A1 carbines. M15A2 BFA is painted red and is used on the M16A2 rifle.

1 Unscrew and pull slide (1) all the way out on blank firing attachment (2)

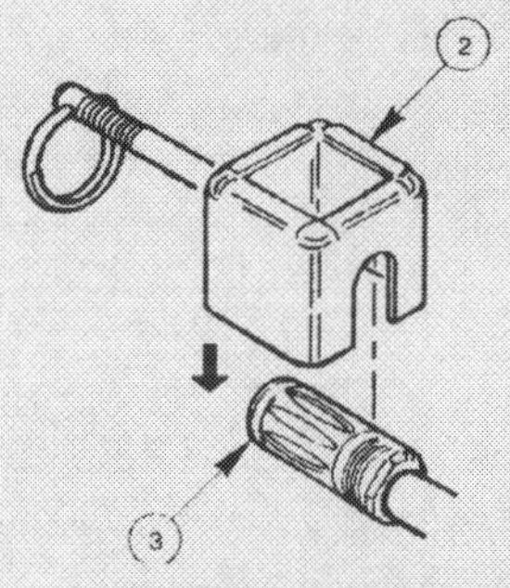

2 Hook blank firing attachment (2) behind the first groove of the compensator (3).

4-4 BLANK FIRING ATTACHMENTS M15A2 (M16A2) AND M23 (M4/M4A1)(CONT).

a. INSTALLATION (CONT)

CAUTION

Do not use tools to tighten the blank firing attachment. **USE HANDS ONLY.**

3 Push slide (2) into compensator (3) and hand tighten.

NOTE

Check for tightness after firing approximately 50 blank rounds.

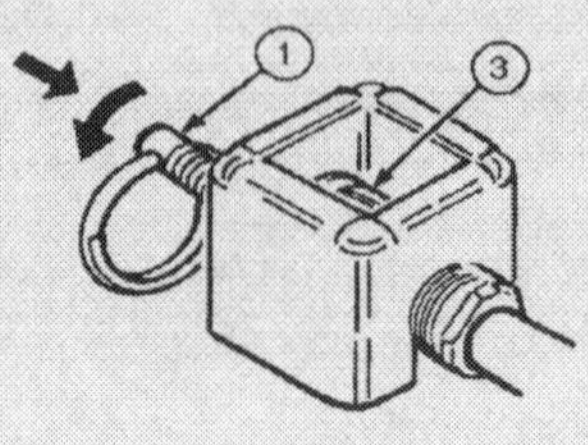

b. REMOVAL

CAUTION

Do not use tools to tighten the blank firing attachment. USE HANDS ONLY.

1 Unscrew slide (1) to remove from compensator (2).

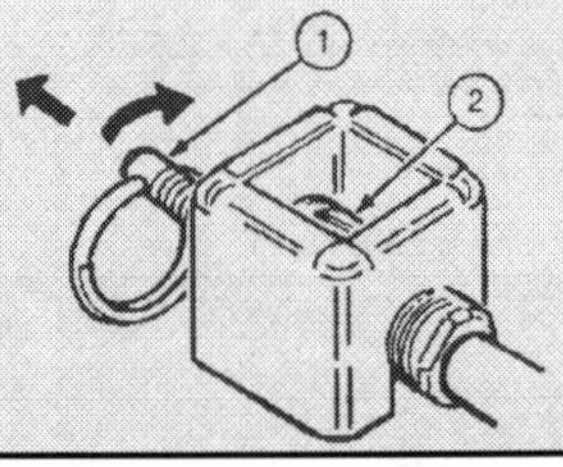

2. Unhook blank firing attachment (3) from behind the first groove of the compensator (2)

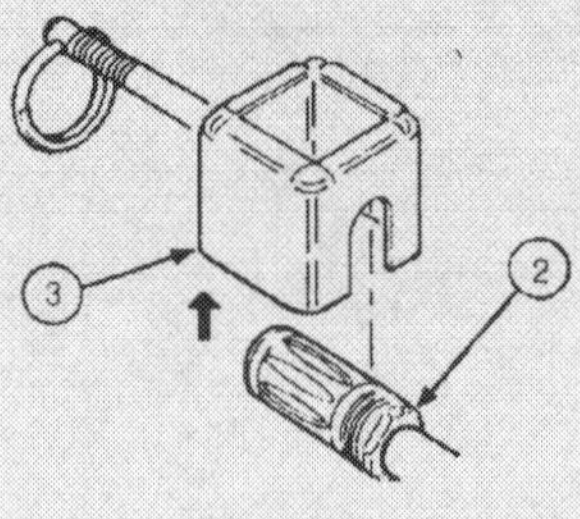

3 Screw slide (1) all the way in on blank firing attachment (3).

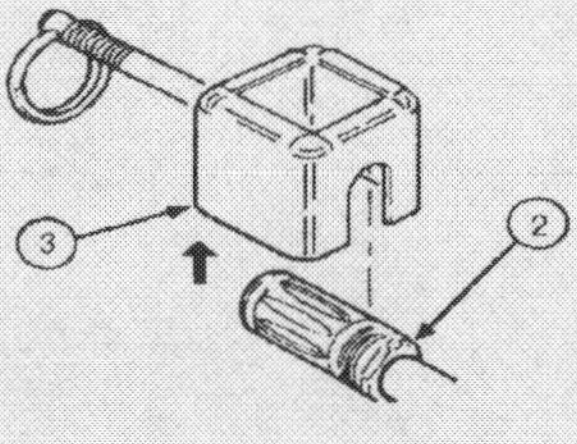

c. CLEANING

Clean blank firing attachment with CLP, wipe dry, and coat with CLP.

d. INSPECTION

Inspect blank firing attachment for cracks or distortion. Be sure the parts in the slide are clear and clean. If blank firing attachment is cracked or distorted, it is unserviceable.

e. REPAINTING

Repaint blank firing attachment using enamel coating compound (Red for M16A2 rifle or Yellow for M4/M4A1 carbine). Painting is the only repair authorized.

f. REPLACEMENT

Replace blank firing attachment if unserviceable.

APPENDIX A
REFERENCES

A-1. TECHNICAL BULLETINSINSTRUCTIONS/MANUALS/ORDERS.

TB 9-1000-247-34	Standards for Overseas Shipment or Domestic Issue of Small Arms, Aircraft Armament, Toward Howitzers, Mortars, Recoilless Rifles, Rocker Launchers and Associated Fire Control Equipment
TB 43-0002-73	Maintenance Expenditures Limits for FSC Group 10; FSC Classes 1005, 1010, 1015, 1030, 1055, 1090, and 1095
TI 05538A-15/2A	Early Warning Indications of Malfunctions
TI 05538A-35/12A	Checking Muzzle and Breech Erosion in Barrel
TI 8005-24/20	Prefire Inspection, Small Arms Weapons, Ordnance Material
TM 9-1005-237-23&P	Organizational and Direct Support Maintenance Manual (Including Repair Parts and Special Tools List) for Bayonet-Knife M6 and M7, with Bayonet-Knife Scabbard M10 and M9 Multipurpose Bayonet System
TM 9-1005-319-10	Operator's Manual for 5.56mm, M16A2 Rifle and M4/M4A1 Carbine
TM 9-1010-221-23&P	Organizational, Direct Support and General Support Maintenance Manual (Including Repair Parts and Special Tools List) for Launcher, Grenade: 40mm M203 W/E (NSN 1010-00-179-6447)
TM 9-6920-363-12&P	M261 Conversion Kit
TM 9-6920-746-12&P	M2 Practice Bolt
TM 11-5855-213-23&P	Night Vision Sight, Individual Served Weapon, AN/PVS-4
TM 750-244-7	Procedures for Destruction of Equipment in Federal Supply Classifications 1000, 1005, 1015, 1020, 1025, 1030, 1055, 1090, and 1095 to Prevent Enemy Use
TM 4700-15/1	Equipment Record Procedures
TO 00-35D-54	Materiel Deficiency Reporting and Investigating System
TO 11 A-13-10-7	Storage and Maintenance Procedures for Small Arms Ammunition
TO 11W-1-10	Historical Data Recording of Inspection, Maintenance, and Firing Data for Ground Weapons
TO 33K-1-100	(TMDE) Interval Calibration and Repair Reference Guide and Work Unit Code Manual

A-2. FIELD MANUALS.

FM 21-11 .... First Aid for Soldiers

FM 23-9 .... M16A1 and M16A2 Rifle and Rifle Marksmanship

A-3 PAMPHLETS.

DA PAM 25-30 .... Consolidated Index of Army Publications and Blank Forms

DA PAM 738-750 .... The Army Maintenance Management System (TAMMS)

A-4 RELATED PUBLICATIONS.

AFM 36-2227, Vol 1 .... Combat Arms Training Management and Range Operation

AMC-P 310-9 .... Equipment Publications Listing

AR 725-50 .... Requisitioning, Receipt, and Issue System

AR 750-1 .... Army Materiel Policies

CTA 8-100 .... Army Medical Department Expendable/Durable Items

CTA 50-970 .... Expendable/Durable Items (Except: Medical, Class V, Repair Parts and Heraldic Items)

DOD 4160.21-M-1 .... Defense Demilitarization Manual

FMFM 1-3A .... Field Firing Techniques

SB 746-1 .... Publications for Packing Army General Supplies

SPI 00-856-6885 .... Special Packaging Instructions for M16 Rifle

A-5 FORMS.

AFTO Form 22 .... Technical Order System Publications Improvement Report

AFTO Form 105 .... Inspection, Maintenance, and Firing Data for Ground Weapons

DA Form 2028 .... Recommended Changes to Publications and Blank Forms

DA Form 2028-2 .... Recommended Changes to Equipment Technical Manuals

DA Form 2404 .... Equipment Inspection and Maintenance Worksheet

DA Form 2408-9 .... Equipment Control Record

DD Form 314 .... Preventive Maintenance Schedule and Record

NAVMC 10772..Recommended Changes to Technical Publications

NAVMC 11003..Ordnance Serialized Items Subsidiary Records

SF 364...Report of Discrepancy (ROD)

SF 368...Product Quality Deficiency Report

A-6 SUPPLY CATALOGS/SUPPLY LISTS.

SC 4933-95-CL-A11................................Shop Set, Small Arms Field Maintenance

SC 5180-95-CL-A07................................Tool Kit, Small Arms Repair

APPENDIX B

MAINTENANCE ALLOCATION CHART

Section I. INTRODUCTION

B-1. GENERAL.

a. This section provides a general explanation of all maintenance and repair functions authorized at various maintenance levels

b. The Maintenance Allocation Chart (MAC) In section II designates overall authority and responsibility for the performance of maintenance functions on the identified end Item or component. The application of the maintenance functions to the end Item or component will be consistent with the capacities and capabilities of the designated maintenance levels

c. Section III lists the tools and test equipment (both special tools and common tool sets) required for each maintenance function as referenced from section II.

d. Section IV contains supplemental instructions and explanatory notes for a particular maintenance function.

B-2. MAINTENANCE FUNCTIONS.

Maintenance functions will be limited to and defined as follows: (except for ammunition MAC')

a. Inspect. To determine the serviceability of an Item by comparing Its physical, mechanical, and or electrical characteristics with established standards through examination leg , by sight, sound, or feel).

b. Test. To verify serviceability by measuring the mechanical, pneumatic, hydraulic, or electrical characteristics of an Item and comparing those characteristics with prescribed standards

c. Service. Operations required periodically to keep an Item In proper operating condition, I e to clean (includes decontaminate, when required), to preserve, to drain, to paint, or to replenish fuel, lubricants, chemical fluids. or gases.

d. Adjust. To maintain or regulate, within prescribed limits, by bringing Into proper or exact position, or by setting the operating characteristics to specified parameters

'Exception is authorized for ammunition MAC to permit the redesignation/redefinition of maintenance function headings to more adequately identify ammunition maintenance functions The heading designations and definitions will be Included In the appropriate technical manual for each category of ammunition

B-2. MAINTENANCE FUNCTIONS (CONT).

e. Align. To adjust specified variable elements of an item to bring about optimum or desired performance

NOTE

Air Force Precision Measurement Equipment Laboratory (PMEL) person will calibrate small arms Inspection gages In accordance with TO 33K-1-100

f. Calibrate. To determine and cause corrections to be made or to be adjusted on Instruments or test, measuring, and diagnostic equipments used In precision measurement Consists of comparisons of two Instruments, one of which Is a certified standard of known accuracy, to detect and adjust any discrepancy in the accuracy of the Instrument being compared

g. Remove/Install. To remove and Install the same item when required to perform service or other maintenance functions. Install may be the act of emplacing, seating, or fixing Into position a spare, repair part, or module (component or assembly) In a manner to allow the proper functioning of an equipment or system.

h. Replace. To remove an unserviceable Item and Install a serviceable counterpart In Its place "Replace"' is authorized by the MAC and is shown as the 3d position code of the SMR code.

i. Repair. The application of maintenance services2, including fault location troubleshooting3, removal 'installation, and disassembly'assembly4 procedures, and maintenance actions" to Identify troubles and restore serviceability to an Item by correcting specific damage, fault, malfunction, or failure In a part, subassembly, module (component or assembly), end Item, or system.

j. Overhaul. That maintenance effort (service/action) prescribed to restore an item to a completely serviceable operational condition as required by maintenance standards In appropriate technical publications (i.e , DMWR). Overhaul is normally the highest degree of maintenance performed by the Army Overhaul does not normally return an Item to like new condition

k. Rebuild. Consists of those services/actions necessary for the restoration of unserviceable equipment to a like new condition In accordance with original manufacturing standards Rebuild is the highest degree of materiel maintenance applied to Army equipment The rebuild operation Includes the act of returning to zero those age measurements (hours miles, etc) considered In classifying Army equipment/components.

---

[2] Servlces--Inspect, test, service, adjust, align, calibrate, and or replace
[3] Fault locate troubleshoot The process of investigating and detecting the cause of equipment malfunctioning; the act of isolating a fault within a system or unit under test (UUT).
[4] Disassemble assemble encompasses the step-by-step taking apart (or breakdown) of a spare/functional group coded Item to the level of Its least competency identified as maintenance significant (.e , assigned an SMR code) for the level of maintenance under consideration '
[5] Actionswelding, grinding, riveting, straightening, facing, remachinery, and or resurfacing

B-3. EXPLANATION OF COLUMNS IN THE MAC, SECTION II.

a. Column 1, Group Number. Column 1 lists functional group code numbers, the purpose of which is to identify maintenance significant components, assemblies, subassemblies, and modules with the next higher a assembly. End item group number shall be "OO."

b. Column 2, Component Assembly. Column 2 contains the names of components. Assemblies, subassemblies, and modules forwhich maintenance is authorized.

c. Column 3, Maintenance Function. Column 3 lists the functions to be performed on the it em listed in Column 21. (for detailes explanation of these functions, see paragraph B-2.)

d. Column 4, Maintenance Level. Column 4 specifies, by the list ing of a work time figure in the appropriate subcolumn(s), the level of maintenance. If the number of complexity of the t asks within the listed maintenance function vary at different maintenance levels, appropriate work time figures will be shown for each level. The work time figure represents the average time required to restore an item (assembly, subassembly, component, module, end it4m, or system) or a serviceable condition under typicalfield operating conditions. This time includes preparation time (including any necessary disassembly assembly time), troubleshooting fault location time, and quality assurance quality control time in addition to the time required to perform the specific tasks identified for the maintenance functions authorized i the maintenance allocation chart. The symbol designations for the varies maintenance levels are as follows:

C..Operator or Crew
O..Unit Maintenance
F..Direct Support Maintenance
H..General Support Maintenance
L..Specialized Repair Activity (SRA)[6]
D..Depot Maintenance

e. Column 5, Tools and Equipment. Column 5 specifies, by code, those common tool sets (not individual tools) and special tools., TMDE, and support equipment required to perform the designated function.

f. Column 6, Remarks. This column shall, when applicable, contain a lett4er code, in alphabetic order, which shall be keyed to the remarks contained in Section IV.

[6]This maintenance level is not included in Section II, column (4) of the Maintenance Allocation Chart. To identify functions to this level of maintenance, enter a work time figure in the "H" column of section II, column (4), and use an associated reference code in the Remarks column (6). Key the c ode to Section IV . Remarks, and explain the SRA complete repair application there. The explanatory remarks)s) shall reference the specific Repair Parts and Special Tools List (RPSTL) TM which contains additional SRA criteria and the authorized spare/repair parts.

B-4. EXPLANATION OF COLUMNS IN TOOL AND TEST EQUIPMENT REQUIREMENTS, SECTION III.

a. Column 1, Reference Code. The tool and test equipment reference code correlates with a code used In the MAC, Section II, Column 5.

b. Column 2, Maintenance Level. The lowest level of maintenance authorized to use the tool or test equipment.

c. Column 3, Nomenclature. Name or identification of the tool or test equipment.

d. Column 4, National Stock Number. The National stock number of the tool or test equipment.

e. Column 5, Tool Number. The manufacturer's part number.

B-5. EXPLANATION OF COLUMNS IN REMARKS, SECTION IV.

a. Column 1, Reference Code. The code recorded In column 6, Section II.

b. Column 2, Remarks. This column lists information pertinent to the maintenance function being performed as indicated In the MAC, Section II.

Section II. MAINTENANCE ALLOCATION CHART
FOR
M16A2 RIFLE AND M4/M4A1 CARBINES

(1) GROUP NUMBER	(2) COMPONENT ASSEMBLY	(3) MAINTENANCE FUNCTION	(4) MAINTENANCE LEVEL UNIT C	UNIT O	DS F	DS H	DEPOT D	(5) TOOLS AND EQPT	(6) REMARKS CODE
00	M16A2 5.56MM RIFLE, AND M4/ M4A1 5.56MM CARBINES	Inspect	0.1	0.3	0.7			2	
		Test			0.2			2,3	
		Service	0.2	0.3	0.3				
		Replace		0.1					
		Overhaul							
00.1	CARRYING HANDLE ASSEMBLY M4/M4A1	Inspect	0.1	0.1	0.1				
		Service	0.1						
		Remove/ Install	0.1	0.1	0.1				
		Replace			0.3			1,3	
		Repair		0.1	0.2			1,3	
00.101	Rear Sight Assembly M4/ M4A1	Inspect			0.1				
		Replace			0.3			1,3	
		Repair			0.2			1,3	
01	BOLT CARRIER ASSEMBLY	Inspect	0.1	0.1	0.1				
		Service	0.1	0.1					
		Remove/ Install	0.1						
		Replace			0.1			1,2	
		Repair	01	0.1				3	
0101	Bolt Assembly	Inspect	0.1	0.1	0.1				
		Test			0.1			2	
		Service	0.1						
		Remove/ Install	0.1						
		Replace			0.1			2	
		Repair		0.1	0.1			3	

*'Worktimes are included in DMWR 9-1005-249

MAINTENANCE ALLOCATION CHART (CONT)

(1) GROUP NUMBER	(2) COMPONENT ASSEMBLY	(3) MAINTENANCE FUNCTION	(4) MAINTENANCE LEVEL UNIT C	UNIT O	DS F	DS H	DEPOT D	(5) TOOLS AND EQPT	(6) REMARKS CODE
0102	Key and Bolt Carrier Assembly	Inspect	0.1		0.1				
		Service	0.1						
		Remove/ Install	0.1						
		Replace			0.1			1	
		Repair		0.1	0.1			1,3	
02	HANDLE ASSEMBLY	Inspect	0.1						
		Service	0.1						
		Remove/ Install	0.1						
		Replace			0.1		**		
		Repair			0.1			3	
03	UPPER RECEIVER AND BARREL ASSEMBLY	Inspect	0.1	0.2	0.1				
		Test			0.2			2	
		Service	0.2		0.1			1	
		Remove/ Install	0.1						
		Replace			0.5			1,2	
		Repair		0.2	0.5			1,2,3	A,B
0301	M16A2 Barrel Assembly and M4/ M4A1 Replacement Barrel	Inspect		0.1	0.1				
		Replace			0.3			1,2	
		Repair	0.1					3	
0302	Upper Receiver Assembly	Inspect			0.1				
		Replace			0 5			1.2	
		Repair			0.3			1.3	
030201	Forward Assist Assembly	Inspect			0.1				
		Replace			0.1				
		Repair			0 1			1.3	
030202	Rear Sight Assembly M16A2	Inspect			0 1				
		Replace			0.3			1,3	
		Repair			0 2			1,3	

MAINTENANCE ALLOCATION CHART (CONT)

(1) GROUP NUMBER	(2) COMPONENT ASSEMBLY	(3) MAINTENANCE FUNCTION	(4) MAINTENANCE LEVEL UNIT C	UNIT O	DS F	DS H	DEPOT D	(5) TOOLS AND EQPT	(6) REMARKS CODE
04	LOWER RECEIVER AND BUTTSTOCK ASSEMBLY	Inspect	0.1	0.1	0.1				
		Test			0.1			2	
		Service	0.2	0.2					
		Repair		0.2	0.3			1,2,3	C,D,E
0401and 0401A	Buttstock Assembly	Inspect		0.1					
		Remove/ Install		0.1				3	
		Replace		0.1				3	
		Repair		0.1				3	
0402	Hammer Assembly	Inspect			0.1				
		Remove/ Install			0.1				
		Replace			0.1				
		Repair			0.1				
0403	Trigger Assembly	Inspect			0.1				
		Remove/ Install			0.1			1,3	
		Replace			0.1			1,3	
		Repair			0.1			1,3	
040301	Trigger Subassembly M16A2 and M4	Inspect			0.1				
		Replace			0.1			1,3	
		Repair			0.1			1,3	
0404 and 0404A	Lower Receiver and Receiver Extension Assembly	Inspect			0.1				
		Test			0.1			2	
		Repair			0.2			1,3	

Section III. TOOL AND TEST EQUIPMENT REQUIREMENTS FOR M16A2 RIFLE AND M4/M4A1 CARBINES

TOOL OR TEST EQUIPMENT REF CODE	(2) MAINTENANCE CATEGORY	(3) NOMENCLATURE	(4) NATIONAL/NATO STOCK NUMBER	(5) TOOL NUMBER
1 (ARMY ONLY)	F	Shop Set, Small	4933-00-754-0664	SC 4933-95-CL-A11
2 (ARMY ONLY)	F	Tool and gage Set, DS/GS Maintenance for 5.56mm Rifle, M16 Series	4933-00-056-7106	8426685
3 (ARMY ONLY)	O	Tool Kit, Small Arms Repairman	5180-00-357-7770	SC 5180-95-CL-A07
4 (A.F ONLY)	F	Torque Wrench, ft-lb	5120-00-640-6365	A-A-411
5 (A.F ONLY)	F in.-lb	Torque Wrench,	5120-00-230-6380	T-E-12A
6 A.F ONLY)	F	Trigger Weights	4933-00-647-3696	7274758

Section IV. REMARKS

Reference Code	Remarks
A	Tool, Front Sight Post Removal and Installation
B	Depressor, Front Sight Detent
C	Tool, Pivot Pin Removal
D	Tool, Pivot Pin Installation
E	Pin, Slave

APPENDIX C

REPAIR PARTS AND SPECIAL TOOLS LIST

Section I. INTRODUCTION

C-1. SCOPE.

This RPSTL lists and authorizes spares and repair parts; special tools; special test, measurement, and diagnostic equipment (TMDE); and other special support equipment required for performance of unit and direct support maintenance of the M16A2 rifle and M4/M4A1 carbines. It authorizes the requisitioning, issue, and disposition of spares, repair parts, and special tools as indicated by the source, maintenance, and recoverability (SMR) codes.

C-2. GENERAL.

In addition to Section I, Introduction, this Repair Parts and Special Tools List is divided into the following sections:

a. Section II Repair Parts List. A list of spares and repair parts authorized by this RPSTL for use in the performance of maintenance. The list also includes parts which must be removed for replacement of the authorized parts. Parts lists are composed of functional groups in ascending alphanumeric sequence, with the parts in each group listed in ascending figure and item number sequence. Bulk materials are listed by item name in FIG BULK at the end of the section. Repair parts kits or sets are listed separately in their own functional group within Section II. Repair parts for repairable special tools are also listed in the section.

b. Section III Special Tools List. A list of special tools, special TMDE, and other special support equipment authorized by this RPSTL (as indicated by Basis of Issue (BOI) information in DESCRIPTION AND USABLE ON CODE (UOC) column) for the performance of maintenance.

c. Section IV Cross-Reference Indexes. A list, in National item identification number (NIIN) sequence, of all National stock numbered items appearing in the listings, followed by a list in alphanumeric sequence of all part numbers appearing in the listing. National stock numbers and part numbers are cross-referenced to each illustration figure and item number in alphanumeric sequence and cross-references NSN, CAGEC, and part number.

C-3. EXPLANATION OF COLUMNS (SECTIONS II AND III).

a. ITEM NO. (Column (1)). Indicates the number used to identify items called out in the illustration.

b. SMR CODE (Column (2)). The Source, Maintenance, and Recoverability (SMR) code is a 5-position code containing supply/requisitioning information, maintenance level authorization criteria, and disposition instruction, as shown In the following breakout:

C-3. EXPLANATION OF COLUMNS (SECTION II AND III) (CONT).

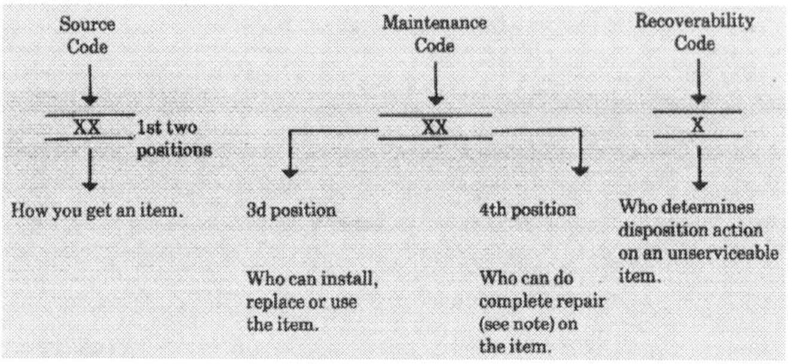

*Complete Repair: Maintenance capacity, capability, and authority to perform all corrective maintenance tasks of the "Repair" function in a use/user environment In order to restore serviceability to a failed item.

(1) Source Code. The source code tells you how to get an item needed for maintenance, repair, or overhaul of an end item/equipment. Explanations of source codes follow:

Code	Explanation
PA PB PC* * PD PE PF PG	Stocked items; use the applicable NSN to request/requisition Items with these source codes They are authorized to the level indicated by the code entered In the 3d position of the SMR code. * *NOTE: Items coded PC are subject to deterioration.
KD KF KB	Items with these codes are not to be requested/requisitoned Individually They are part of a kit which is author->zed to the maintenance level indicated In the 3d positon of the SMR code The complete kit must be requisitioned and applied.
MO-(Made at Unit/ AVUM Level) MF- (Made at DS/ AVIM Level) MH-(Made at GS Level) ML- (Made at Specialized Repair Act (SRA)) MD-(Made at Depot)	Items with these codes are not to be requested/requisitioned Individually They must be made from bulk material which Is identified by the part number In the DESCRIPTION AND USABLE ON CODE (UOC) column and listed In the Bulk Material group of the repair parts list in this RPSTL If the Item Is authorized to you by the 3d position code of the SMR code, but the source code Indicates It is made at a higher level, order the Item from the higher level of maintenance.

Code	Explanation
AO- (Assembled by unit/AVUM Level) AF- (Assembled by DS/AVIM Level) AH- (Assembled by GS Level) AL- (Assembled by SRA) AD- (Assembled by Depot)	Items with these codes are not to be requested/requisitioned Individually. The parts that make up the assembled Item must be requisitioned or fabricated and assembled at the level of maintenance indicated by the source code. If the 3d position code of the SMR code authorizes you to replace the Item, but the source code Indicates the Item Is assembled at a higher level, order the Item from the higher level of maintenance.

XA-Do not requisition an "XA"-coded item. Order Its next higher assembly. (Also, refer to the NOTE below.)

XB--If an "XB" item is not available from salvage, order It using the CAGEC and part number given.

XC-installation drawing, diagram, instruction sheet, field service drawing, that is identified by manufacturer's part number.

XD-Item is not stocked. Order an "XD"-coded Item through normal supply channels using the CAGEC and part number given, If no NSN Is available.

NOTE

Cannibalization or controlled exchange, when authorized, may be used as a source of supply for items with the above source codes, except for those source coded "XA" or those aircraft support Items restricted by requirements of AR 750-1.

(2) Maintenance Code. Maintenance codes tell you the level(s) of maintenance authorized to USE and REPAIR support items. The maintenance codes are entered In the third and fourth positions of the SMR code as follows:

(a) The maintenance code entered in the third position tells you the lowest maintenance level authorized to remove, replace, and use an Item. The maintenance code entered In the third position will indicate authorization to one of the following levels of maintenance

Code	Application/Explanation
C	-Crew or operator maintenance done within unit or aviation unit maintenance.
O	-Unit or aviation unit level can remove, replace, and use the Item.
F	-Direct support or aviation intermediate level can remove, replace, and use the Item.
H	-General support level can remove, replace, and use the Item.
L	-Specialized repair activity can remove, replace, and use the Item.
D	-Depot level can remove, replace, and use the Item

C-3. EXPLANATION OF COLUMNS (SECTIONS II AND III) (CONT).

(b) The maintenance code entered In the fourth position tells whether or not the Item is to be repaired and Identifies the lowest maintenance level with the capability to do complete repair (I e, perform all authorized repair functions). (NOTE: Some limited repair may be done on the Item at a lower level of maintenance, If authorized by the Maintenance Allocation Chart (MAC) and SMR codes I This position will contain one of the following maintenance codes

Code	Application/Explanation
O	-Unit or aviation unit is the lowest level that can do complete repair of the item
F	-Direct support or aviation Intermediate is the lowest level that can do complete repair of the Item
H	-General support is the lowest level that can do complete repair of the item .
L	-Specialized repair activity is the lowest level that can do complete repair of the item
D	-Depot is the lowest level that can do complete repair of the Item
Z	-Nonreparable. No repair Is authorized
B	-No repair is authorized. (No parts or special tools are authorized for the maintenance of a "B" coded item) However, the Item may be reconditioned by adjusting, lubricating, etc , at the user level

(3) Recoverability Code. Recoverability codes are assigned to Items to Indicate the disposition action on unserviceable items The recoverability code is entered In the fifth position of the SMR code as follows.

Recoverability Codes	Application/Explanation
Z	Nonreparable item unserviceable, condemn and dispose of the item at the level of maintenance shown In 3d position) of SMR code
O	Reparable Item. When uneconomically reparable condemn and dispose of the item at unit or aviation unit level
F	Reparable Item. When uneconomically reparable, condemn and dispose of the Item at the direct support or aviation level
H	Reparable Item. When uneconomically reparable, condemn and dispose of the Item at the .general support level

Recoverability Codes	Application/Explanation
D	-Reparable item. When beyond lower level repair capability, return to depot. Condemnation and disposal of item not authorized below depot level.
L	-Reparable item. Condemnation and disposal not authorized below Specialized Repair Activity (SRA).
A	-Item requires special handling or condemnation procedures because of specific reasons (e.g., precious metal content, high dollar value, critical material, or hazardous material). Refer to appropriate manuals/directives for specific instructions.

c. NSN (Column (3)). Indicates the National stock number (NSN) assigned to the item and which will be used for requisitioning.

d. CAGEC (Column (4)). The Contractor and Government Entity Code (CAGEC) is a 5-digit numeric code which is used to identify the manufacturer, distributor, or Government agency, etc., that supplies the Item.

e. PART NUMBER (Column (5)). Indicates the primary number used by the manufacturer (individual, company, firm, corporation, or Government activity), which controls the design and characteristics of the item by means of its engineering drawings, specifications standards, and inspection requirements to identify an item or range of items.

NOTE

When you use an NSN to requisition an item, the item you receive may have a different part number from the part ordered.

f. DESCRIPTION AND USABLE ON CODE (UOC) (Column (6)). This column includes the following Information:

(1) The Federal item name and, when required, a minimum description to identify the Item.

(2) The physical security classification of the item is indicated by the parenthetical entry which is a physical security classification abbreviation, e.g., Phy Sec C1 (C) Confidential, Phy Sec C1 (S) Secret, Phy Sec C1 (T) Top Secret.

(3) Items that are included in kits and sets are listed below the name of the kit or set.

(4) Spare/repair parts that make up an assembled item are listed immediately following the assembled item line entry.

(5) Part numbers for bulk materials are referenced in this column In the line item entry for the Item to be manufacture red/fabricated.

(6) When the item is not used with all serial numbers of the same model, the effective serial numbers are shown on the last line(s) of the description (before UOC).

(7) The usable on code, when applicable (see paragraph 5, Special Information).

C-3. EXPLANATION OF COLUMNS (SECTION II AND III) (CONT).

(8) In the Special Tools List section, the basis of issue (BOI) appears as the last line(s) in the entry for each special tool, special TMDE, and other special support equipment. When density of equipment supported exceeds density spread indicated in the basis of Issue, the total authorization is increased proportionately.

(9) The statement "END OF FIGURE" appears just below the last item description in Column 5 for a given figure In both Section II and Section III.

g. Oty (Column (7)) . The Qty (quantity per figure column) indicates the quantity of the item used in the breakout shown on the illustration figure, which is prepared for a functional group, subfunctional group, or an assembly. A "V" appearing In this column in lieu of a quantity indicates that the quantity is variable and the quantity may vary from application to application.

C-4. EXPLANATION OF COLUMNS (SECTION IV).

a. NATIONAL STOCK NUMBER (NSN) INDEX.

(1) STOCK NUMBER column. This column lists the NSN by National Item Identification Number (NIIN) sequence. The NIIN consists of the last nine digits of the NSN.

NSN
(i.e., 5305-01-674-1467). When using this column to locate an item, ignore the first 4 digits NIIN of the NSN. However, the complete NSN should be used when ordering items by stock number.

(2) FIG. column. This column lists the number of the figure where the item Is Identified/located. The figures are in numerical order in Section II and Section III.

(3) ITEM column. The item number identifies the item associated with the figure listed in the adjacent FIG. column. This item is also identified by the NSN listed on the same line.

b. PART NUMBER INDEX. Part numbers in this index are listed by part number in ascending alphanumeric sequence (i.e., vertical arrangement of letter and number combination which places the first letter or digit of each group In order A through Z, followed by the numbers 0 through 9 and each following letter or digit in like order).

(1) CAGEC column. The Contractor and Government Entity Code (CAGEC) is a 5-digit numeric code used to identify the manufacturer, distributor, or Government agency, etc., that supplies the item.

(2) PART NUMBER column. Indicates the primary number used by the manufacturer (individual, firm, corporation, or Government activity), which controls the design and characteristics of the item by means of Its engineering drawings, specifications standards, and inspection requirements to identify an Item or range of Items.

(3) STOCK NUMBER column. This column lists the NSN for the associated part number and manufacturer identified In the PART NUMBER and CAGEC columns to the left.

(4) FIG. column. This column lists the number of the figure where the item is identified/located in Sections II and III.

(5) ITEM column. The item number is that number assigned to the item as it appears in the figure referenced in the adjacent figure number column.

C-5. SPECIAL INFORMATION.

a. ASSEMBLY INSTRUCTION. Detailed assembly instructions for items source coded to be assembled from component spare/repair parts are found in Chapter 3. Items that make up the assembly are listed immediately following the assembly item entry or reference is made to an applicable figure.

b. ASSOCIATED PUBLICATIONS. The publications listed below pertain to the M16A2 rifle, M4/M4A1 carbine, and its components:

Publication	Short Title
TM 9-1005-319-10	M16A2 Rifle and M4/M4A1 Carbine

c. USABLE ON CODE. The usable on code appears in the lower left corner of the DESCRIPTION column heading. Usable on codes are shown as "UOC..." in DESCRIPTION column (justified left) on the first line applicable item description/nomenclature. Uncoded items are applicable to all models. Identification of the usable on codes used in the RPSTL are:

Code	Used On
AR8	M16A2 Rifle
AS1	M4 Carbine
AY6	M4A1 Carbine

C-6. HOW TO LOCATE REPAIR PARTS.

a. When National Stock Number or Part Number is Not Known.

(1) First. Using the table of contents, determine the assembly group or subassembly group to which the item belongs. This is necessary since figures are prepared for assembly groups and subassembly groups, and listings are divided into the same groups.

C-6. HOW TO LOCATE REPAIR PARTS (CONT).

(2) Second. Find the figure covering the assembly group or subassembly group to which the item belongs.

(3) Third. Identify the item on the figure and use the Figure and Item Number Index to find the NSN.

b. When National Stock Number or Part Number is Known.

(1) First. Using the National Stock Number or the Part Number Index, find the pertinent National Stock Number or Part Number. The NSN index is in National Item Identification Number (NIIN) sequence. (See C4.a.(1).) The pat numbers in the Part Number Index are listed in ascending alpha numeric sequence. (See C4.b.) Both indexes cross-reference you to the illustration/figure and item number of the item you are looking for.

(2) Second. Turn to the figure and item number, verify that the item is the one you're looking for, then locate the item number in the repair parts list for the figure.

C-7. ABBREVIATIONS.

Not applicable.

(1) ITEM NO QTY	(2) SMR CODE	(3) NSN	(4) CAGEC	(5) PART NUMBER	(6) DESCRIPTION AND USABLE	(7) ON CODE (UOC)
					GROUP 00 (M16A2) RIFLE 5.56MM 9349000,(M4) CARBINE 5.56MM 9390000, AND (M4A1) CARBINE 5.56MM 12972700 FIG. C-1. (M16A2) RIFLE, (M4) AND CM4A1) CARBINES	
1	PAOOO	1005000179546	19204	8448517	HANDLE ASSEMBLY,CHA CHARGING................................1	
2	AFFFF		19204	8448501	BOLT CARRIER ASSEMBLY (M16A2)..................................1 UOC:AR8	
2A	AFFFF	19200	12011849	BOLT	AND BOLT CARRI ASSEMBLY (M4.....................................1 AND M4A1).. UOC:AS1,AY6	
3	AFFFF	1005011879716	19200	9349050	UPPER RECEIVER AND BARREL ASSEMBLY ....1 (M16A2).. UOC:AR8	
3A	AFFFF	19200	12972680	UPPER	RECEIVER AND BARREL ASSEMBLY.................................1 (M4 AND M4A1).. UOC:AS1,AY6	
4	PACZZ	1005009215004	19200	8448670	MAGAZINE,CARTRIDGE (30 ROUND).................................1	
5	PACZZ	1005012164510	19204	12624561	SLING,SMALL ARMS (M16A2)..1 UOC:AR8	
5A	PACZZ	1005013689852	19200	12011996	SLING,SMALL ARMS (M4 AND M4A1) 1 UOC:AS1,AY6	
6	XAFFA	19200	9349100	LOWER	RECEIVER AND BUTTSTOCK ....1 ASSEMBLY (M16A2)... UOC:AR8	
6A	XAFFA		19200	9390011	LOWER RECEIVER AND BUTTSTOCK ....1 ASSEMBLY (4)... UOC:AS1	
6B	XAFFA		19200	12972690	LOWER RECEIVER AND BUTTSTOCK ....1 ASSEMBLY (M4A1).. UOC:AY6	
7	AFFFF		19200	12951011	CARRYING HANDLE ASS (M4 AND M4A1).........................1 UOC:AS1,AY6	

END OF FIGURE

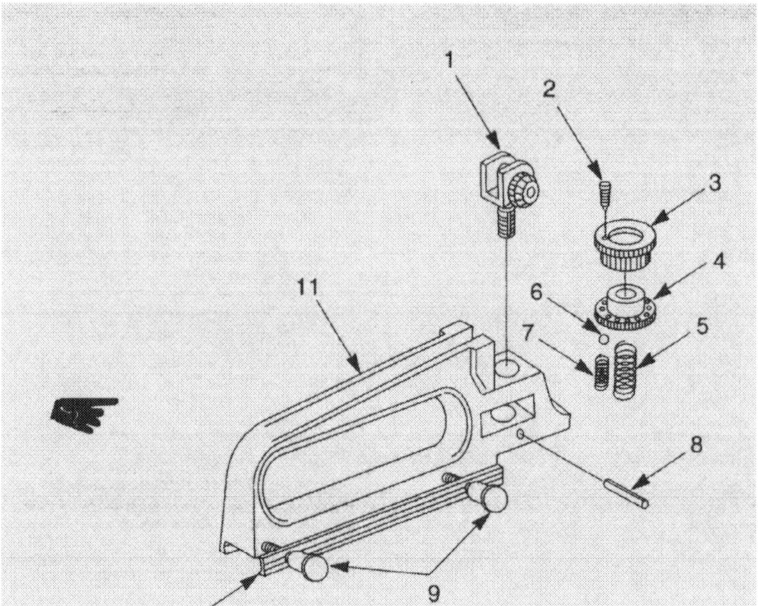

Figure C-1A. (M4 and M4A1) Carrying Handle Assembly 12951011.

(1) ITEM NO	(2) SMR CODE	(3) NSN	(4) CAGEC	(5) PART NUMBER	(6) DESCRIPTION AND USABLE ON CODE (UOC)	(7) QTY
					GROUP 00.1 (M4 AND M4A1) CARRYING HANDLE ASSEMBLY 12951011 FIG. C-1A. (M4 AND M4A1) CARRYING HANDLE ASSEMBLY	
1	AFFFF		19200	12951026	REAR SIGHT ASSY UOC:AS1,AY6	1
2	PAFZZ	5305011343622	19200	9349065	SCREW,INDEX	1
3	PAFZZ	1005013827089	19200	12951018	ELEVATING MECHANISM UOC:AS1 ,AY6	1
4	PAFZZ	5355013826801	19200	12951019	KNOB UOC:AS1,AY6	1
5	PAFZZ	5360011343710	19200	9349070	SPRING ,HELICAL,COMP	1
6	PAFZZ	3110001839175	96906	MS19060-4808	BALL,BEARING UOC:AS1,AY6	1
7	PAFZZ	5360013826802	19200	12951020	SPRING,HELICAL,COMP UOC:AS1,AY6	1
8	PAFZZ	5315008403812	96906	MS16562-121	PIN,SPRING	1
9	PAOZZ	5310013826793	19200	12951023	NUT,PLAIN,KNURLED UOC:AS1,AY6	2
10	PAOZZ	5340013823201	19200	12951017	BRACKET,MOUNTING UOC:AS1,AY6	1
11	PAFZZ	1005013827083	19200	12951021	HANDLE,GUN CARRYING UOC:AS1,AY6	1

END OF FIGURE

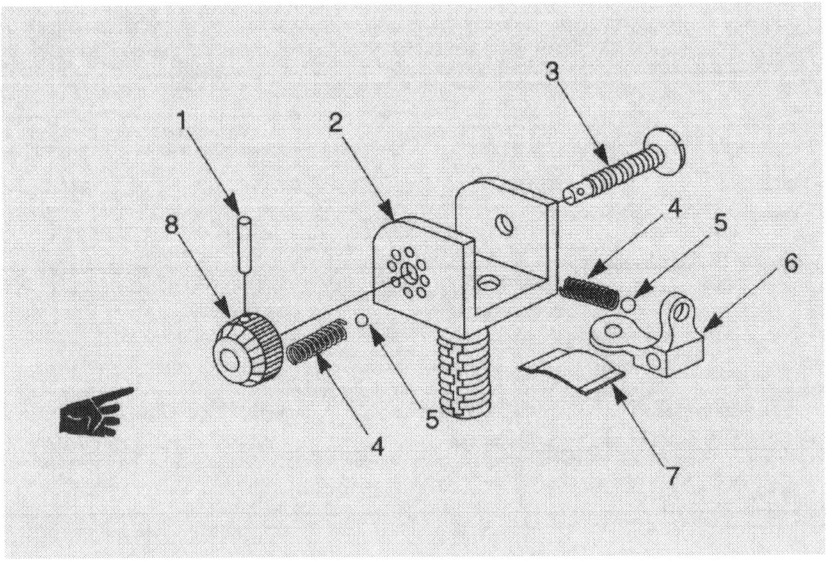

Figure C-1B. (M4 and M4A1) Rear Sight Assembly 12951026.

(1) ITEM NO	(2) SMR CODE	(3) NSN	(4) CAGEC	(5) PART NUMBER	(6) DESCRIPTION AND USABLE ON CODE (UOC)	(7) QTY
					GROUP 00.101 (M4 AND M4A1) REAR SIGHT ASSEMBLY 12951026 FIG. C-18. (M4 AND M4A1) REAR SIGHT ASSEMBLY	
1	PAFZZ	5315000586678	96906	MS16562-103	PIN,SPRING	1
2	PAFZZ	1005013827086	19200	12951028	BASE,REAR SIGHT UOC:AS1,AY6	1
3	PAFZZ	5305011441490	19200	9349076	SCREW,EXTERNALLY RE	1
4	PAFZZ	5360011481751	19200	9349069	SPRING,HELICAL,COMP UOC:AS1,AY6	2
5	PAFZZ	3110001839175	96906	MS19060-4808	BALL,BEARING UOC:AS1,AY6	2
6	PAFZZ	1005011353697	19200	9349075	APERTURE,SIGHT	1
7	PAFZZ	5360013816183	19200	12011987	SPRING,FLAT	1
8	PAFZZ	5355011343627	19200	9349077	KNOB,WINDAGE	1

END OF FIGURE

SECTION II

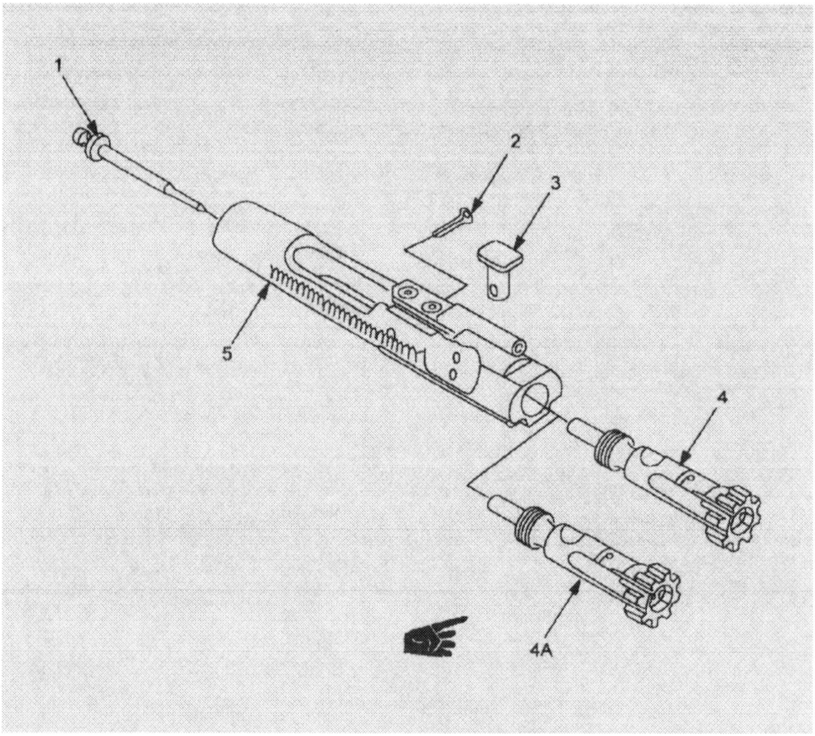

Figure C-2. Bolt Carrier Assembly (M16A2) 8448501 and (M4 and M4A1) 12011849.

(1) ITEM NO	(2) SMR CODE	(3) NSN	(4) CAGEC	(5) PART NUMBER	(6) DESCRIPTION AND USABLE ON CODE (UOC)	(7) QTY
					GROUP 01 BOLT CARRIER ASSEMBLY (M16A2) 8448501 AND (M4 AND M4A1) 12011849 FIG. C-2. BOLT CARRIER ASSEMBLY	
1	PAFZZ	1005000179547	19204	8448503	PIN,FIRING	1
2	PAOZZ	1005009991509	19204	8448504	PIN,FIRING PIN RETA RETAINING	1
3	PAOZZ	5315009927294	19204	8448502	PIN,GROOVED,HEADED BOLT CAM	1
4	PAFFF	1005009927285	19200	8448509	BOLT ASSEMBLY (M16A2).......... UOC:AR8	1
4A	PAFFF	1005014223770	19200	12972691	BOLT,BREECH ASSEMBLY (M4 AND M4A1).......... UOC:AS1,AY6	1
5	AFFFF		19204	8448505	KEY AND BOLT..........	1

END OF FIGURE

SECTION II

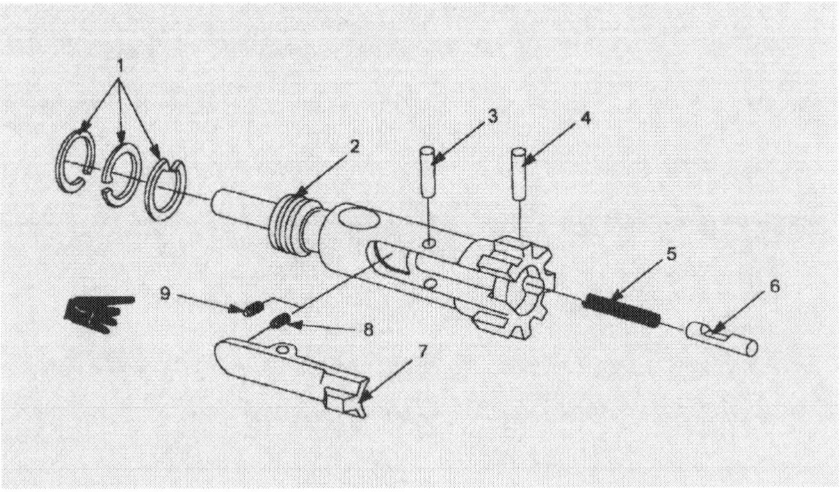

Figure C-3. Bolt Assembly (M16A2) 8448509 and (M4 and M4A1) 12972691.

(1) ITEM NO	(2) SMR CODE	(3) NSN	(4) CAGEC	(5) PART NUMBER	(6) DESCRIPTION AND USABLE ON CODE (UOC)	(7) QTY
					GROUP 0101 BOLT ASSEMBLY (M16A2) 8448509 AND (M4 AND M4A1) 12972691 FIG. C-3. BOLT ASSEMBLY	
1	PAFZZ	1005009927287	19204	8448511	RING,BOLT	3
2	XAFZZ		19204	8448510	BOLT	1
3	PAOZZ	1005009927290	19204	8448513	PIN,EXTRACTOR	1
4	PAOZZ	5315005975086	96906	MS16562-9B	PIN,SPRING EJECTOR	1
5	PAOZZ	5360009927292	19204	8448516	SPRING,HELICAL,COMP COMPRESSION, EJECTOR	1
6	PAOZZ	1005009927291	19204	8448515	EJECTOR,CARTRIDGE	1
7	PAOZZ	1005009927288	19204	8448512	EXTRACTOR,CARTRIDGE	1
8	PAOZZ	1005007603768	19200	8448755	SPRING ASSEMBLY,EXT EXTRACTOR (M16A2) UOC:AR8	1
9	PAOZZ	1005014245899	19200	12972692	SPRING ASSEMBLY,EXT EXTRACTOR (M4 AND M4A1)(BLACK) UOC:AS1 ,AY6	1

END OF FIGURE

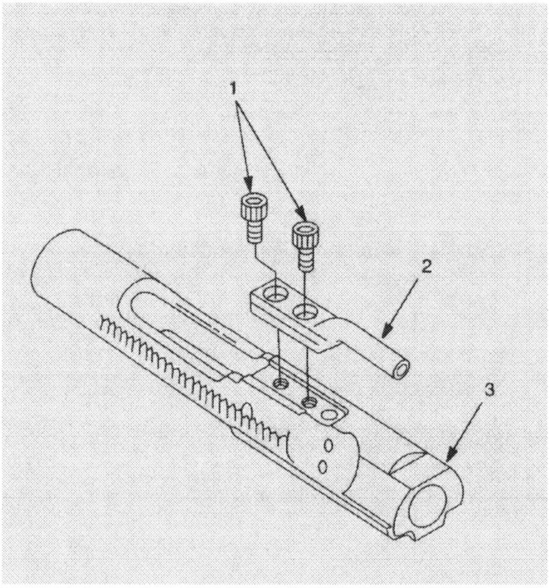

Figure C-4. Key and Bolt Carrier Assembly 8448505.

(1) ITEM NO	(2) SMR CODE	(3) NSN	(4) CAGEC	(5) PART NUMBER	(6) DESCRIPTION AND USABLE ON CODE (UOC)	(7) QTY
					GROUP 0102 KEY AND BOLT CARRIER ASSEMBLY 8448505 FIG. C-4. KEY AND BOLT CARRIER ASSEMBLY	
1	PAFZZ	5305009927284	19204	8448508	SCREW,CARRIER AND K KEY	2
2	PAFZZ	1005009927283	19200	8448506	KEY,BOLT CARRIER	1
3	PAFZZ		19200	8448507	CARRIER,BOLT	1

END OF FIGURE

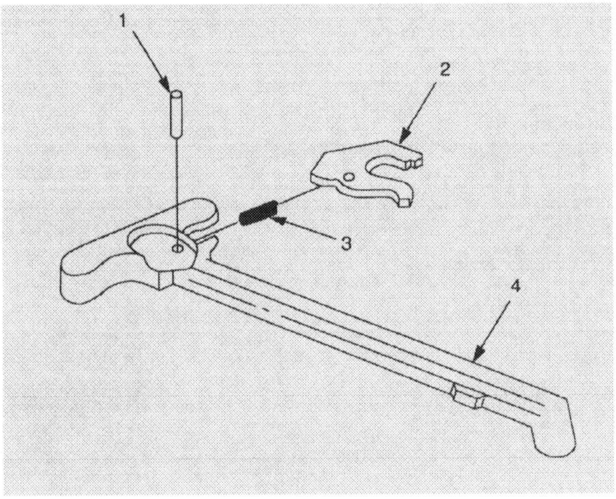

Figure C-5. Handle Assembly 8448517.

(1) ITEM NO	(2) SMR CODE	(3) NSN	(4) CAGEC	(5) PART NUMBER	(6) DESCRIPTION AND USABLE ON CODE (UOC)	(7) QTY
					GROUP 02 HANDLE ASSEMBLY 8448517 FIG. C-5. HANDLE ASSEMBLY	
1	PAOZZ	5315010489372	19204	8448521-2	PIN,SPRING CHARGING HANDLE.......................................	1
2	PAOZZ	5340009990405	19200	8448519	LATCH,CHARGING HAND HANDLE.....................................	1
3	PAOZZ	5360009990404	19204	8448520	SPRING,HELICAL,COMP COMPRESSION, ... CHARGING HANDLE...	1
4	XAOZZ		19204	8448518	HANDLE ,CHARGING...	1

END OF FIGURE

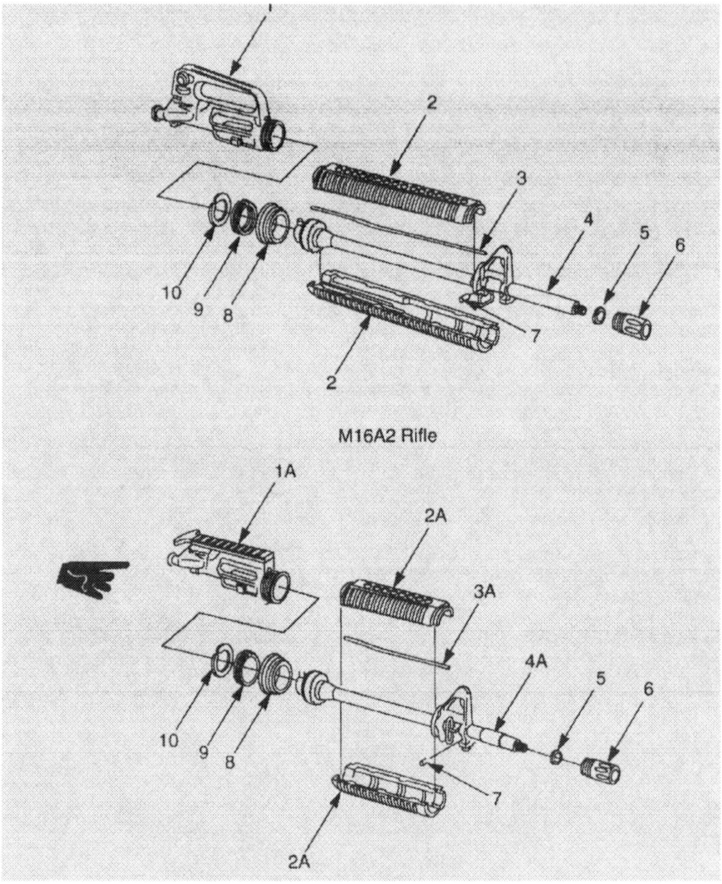

M4/M4A1 Carbine

Figure C-6. Upper Receiver and Barrel Assembly (M16A2) 9349050, and (M4 and M4A1) 12972680.

(1) ITEM NO	(2) SMR CODE	(3) NSN	(4) CAGEC	(5) PART NUMBER	(6) DESCRIPTION AND USABLE ON CODE (UOC)	(7) QTY
					GROUP 03 UPPER RECEIVER AND BARREL ASSEMBLY (M16A2) 9349050, (M4 AND M4A1] 12972680 FIG. C-6. UPPER RECEIVER AND BARREL ASSEMBLY	
1	AFFFF	19200	9349062	UPPER	RECEIVER ASSEMBLY (M16A2) UOC:AR8	1
1A	AFFFF	19200	12972675	UPPER	RECEIVER ASSY (M4 AND M4A1) UOC:AS1,AY6	1
2	PAOZZ	1005011343629	19200	9349059	HANDGUARD ASSEMBLY (M16A2) UOC:AR8	2
2A	PAOZZ	1005012342297	19200	9390003	GUARD,HAND,GUN (M4 AND M4A1) UOC:AS1,AY6	2
3	PAFZZ	4710009781038	19200	8448567	TUBE,BENT,METALLIC (M16A2) UOC:AR8	1
3A	PAFZZ	4710012338637	19200	9390016	TUBE,BENT,METALLIC (M4 AND M4A1) UOC:AS1,AY6	1
4	PAFFF	1005011467684	19200	9349124	BARREL ASSEMBLY (M16A2) UOC:AR8	1
4A	PAFFF	1005012338529	19200	9390007	BARREL AND FRONT SI SIGHT ASSEMBLY, REPLACEMENT (M4 AND M4A1) UOC:AS1,AY6	1
5	PAFZZ	5365011441496	19200	9349052	SHIM COMPENSATOR	1
6	PAFZZ	1005011343633	19200	9349051	COMPENSATOR	1
7	PAFZZ	5315000586044	96906	MS16562-106	PIN,SPRING ,GAS TUBE	1
8	PAFZZ	1005000878998	19204	8448712	RING,SLIP,HAND GUAR GUARD	1
9	PAFZZ	5360009781036	19204	8448555	SPRING,SLIP RING,HA HANDGUARD UPPER RECEIVER	1
10	PAFZZ	5325009990863	96906	M5S16626-3137	RING,RETAINING END OF FIGURE	1

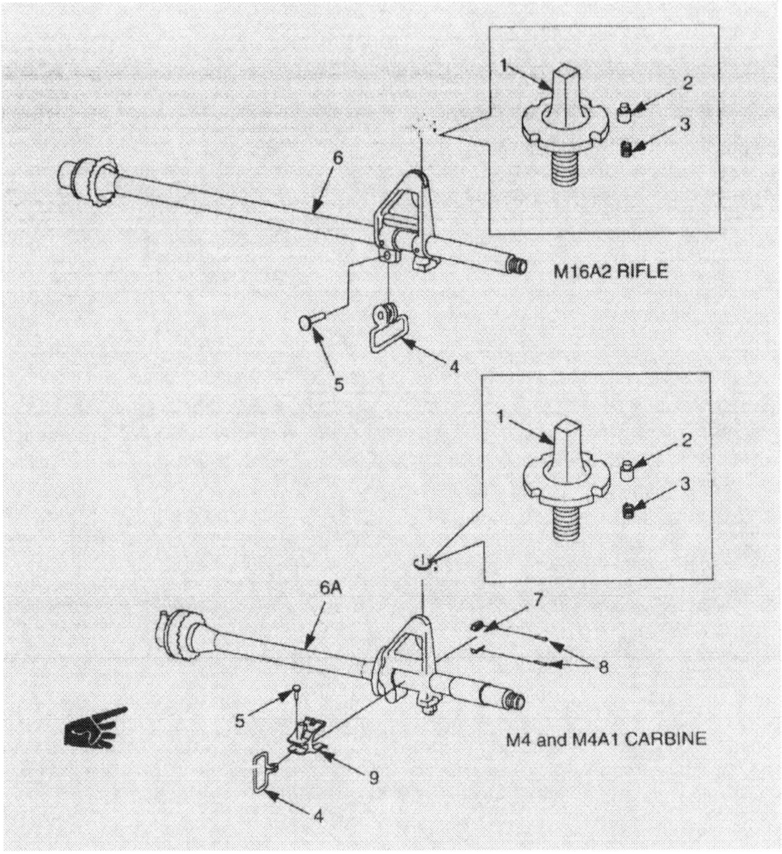

Figure C-7. (M16A2) Barrel Assembly 9349124 and (M4 and M4A1) Replacement Barrel and Front Sight Assembly 9390007.

(1) ITEM NO	(2) SMR CODE	(3) NSN	(4) CAGEC	(5) PART NUMBER	(6) DESCRIPTION AND USABLE ON CODE (UOC)	(7) QTY
					GROUP 0301 (M16A2) BARREL ASSEMBLY 9349124, AND (M4 AND M4A1) REPLACEMENT BARREL AND FRONT SIGHT ASSEMBLY 9390007 FIG. C-7. (M16A2) BARREL ASSEMBLY, AND (M4 AND M4A1) REPLACEMENT BARREL AND FRONT SIGHT ASSEMBLY	
1	PAOZZ	1005011343625	19200	9349056	POST,FRONT SIGHT	1
2	PAOZZ	5315009793930	19204	8448573	PIN,SHOULDER,HEADLE HEADLESS (DETENT, FRONT SIGHT)	1
3	PAOZZ	5360009793931	19204	8448574	SPRING,HELICAL,COMP ,COMPRESSION, FRONT SIGHT	1
4	PAOZZ	1005000179543	19204	8448571	SWIVEL,SLING,SMALL	1
5	PAOZZ	5320010637635	19204	8448697	RIVET,TUBULAR	1
6	XAFZZ		19200	9349054	BARREL AND BARREL E EXTENSION ASSEMBLY (M16A2) UOC:AR8	1
6A	XAFZZ		19200	9390009	BARREL AND BARREL EXTENSION ASSEMBLY (M4 AND M4A1) UOC:AS1,AY6	1
7	PAOZZ	5340012646530	19200	12598618	CLAMP,RIM CLENCHING (M4 AND M4A1) UOC:AS1,AY6	1
8	PAOZZ	5315006900544	96906	MS39086-93	PIN,SPRING (M4 AND M4A1) UOC:AS1,AY6	2
9	PAOZZ	1010012646517	19200	12598617	MOUNT,SWIVEL (M4 AND M4A1) UOC:AS1,AY6	1

END OF FIGURE

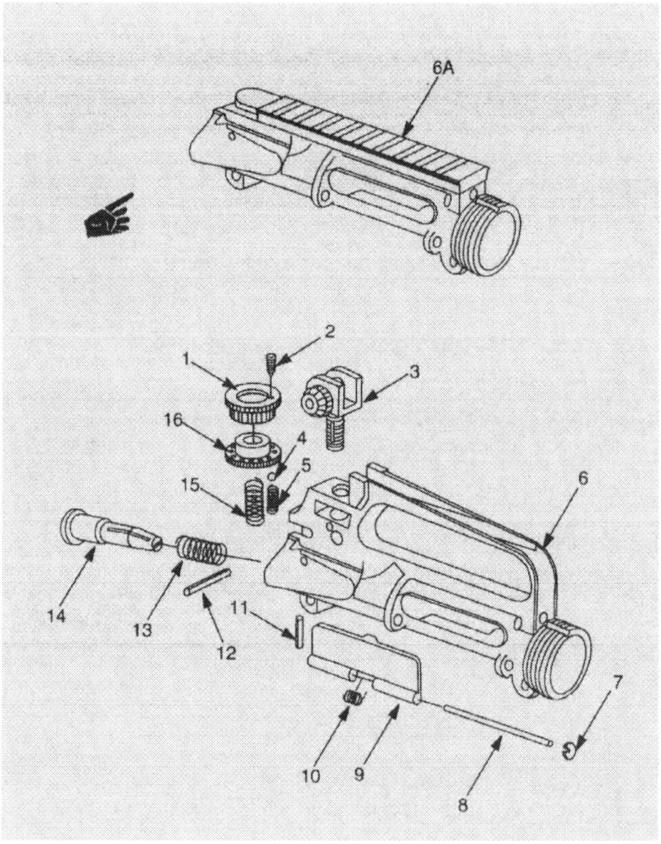

Figure C-8. Upper Receiver Assembly (M16A2) 9349062, and (M4 and M4A1) 12972675.

(1) ITEM NO	(2) SMR CODE	(3) NSN	(4) CAGEC	(5) PART NUMBER	(6) DESCRIPTION AND USABLE ON CODE (UOC)	(7) QTY
					GROUP 0302 UPPER RECEIVER ASSEMBLY (M16A2) 9349062, (M4 AND M4A1) 12972675 FIG. C-8. UPPER RECEIVER ASSEMBLY	
1	PAFZZ	1005011343621	19200	9349066	INDEX,ELEVATION (M16A2) UOC:AR8	1
2	PAFZZ	5305011343622	19200	9349065	SCREW,INDEX (M16A2)	1
3	AFFFF	19200	9349072	REAR	SIGHT ASSEMBLY (M16A2) UOC:AR8	1
4	PAFZZ	3110001839175	96906	MS19060-4808	BALL,BEARING (M16A2) UOC:AR8	1
5	PAFZZ	5360011481751	19200	9349069	SPRING,HELICAL,COMP COMPRESSION, ... INDEX (M16A2) UOC:AR8	1
6	PAFZZ	1005011343701	19200	9349063	RECEIVER,UPPER (M16A2) UOC:AR8	1
6A	PAFZZ	1005013826795	19200	12972670	RECEIVER,CARTRIDGE (M4 AND M4A1) UOC:AS1,AY6	1
7	PAOZZ	5325009990864	96906	MS16632-3012	RING,RETAINING ,COVER	1
8	PAOZZ	5315009781023	19204	8448533	PIN,GROOVED,HEADLES COVER	1
9	PAOZZ	1005009781022	19204	8448525	COVER,EJECTION PORT	1
10	PAOZZ	5360009781025	19204	8448532	SPRING,HELICAL,TORS ,COVER	1
11	PAFZZ	5315008403812	96906	M5S16562-121	PIN,SPRING FORWARD ASSIST	1
12	PAFZZ	5315008403812	96906	M5S16562-121	PIN,SPRING (M16A2)	1
13	PAFZZ	5360000179541	19200	8448540	SPRING,HELICAL,COMP COMPRESSION .... FORWARD ASSIST	1
14	AFFFF		19204	9349086	FORWARD ASSIST ASSY ASSEMBLY	1
15	PAFZZ	5360011343710	19200	9349070	SPRING,HELICAL,COMP COMPRESSION, ... ELEVATION (M16A2)	1
16	PAFZZ	5355011354972	19200	9349067	KNOB,ELEVATION (M16A2) UOC:AR8	1

END OF FIGURE

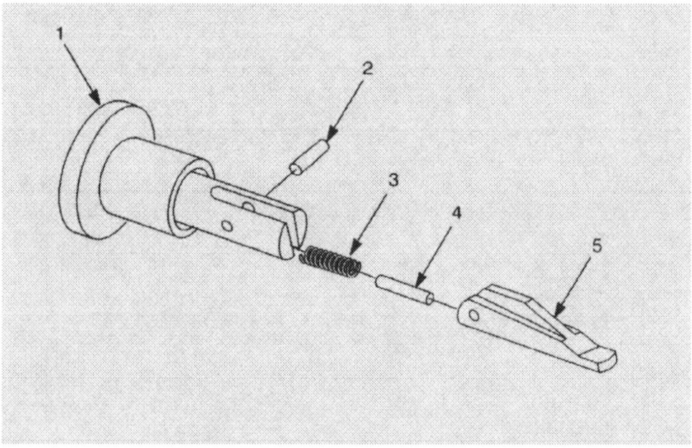

Figure C-9. Forward Assist Assembly 9349086.

(1) ITEM NO	(2) SMR CODE	(3) NSN	(4) CAGEC	(5) PART NUMBER	(6) DESCRIPTION AND USABLE ON CODE (UOC)	(7) QTY
					GROUP 030201 FORWARD ASSIST ASSEMBLY 9349086	
					FIG. C-9. FORWARD ASSIST ASSEMBLY	
1	PAFZZ	1005011441468	19200	9349085	PLUNGER ASSEMBLY	1
2	PAFZZ	5315010489372	19204	8448521-2	PIN,SPRING ,PAWL	1
3	PAFZZ	5360005238084	19200	8448542	SPRING,HELICAL,COMP COMPRESSION PAWL	1
4	PAFZZ	1005000179540	19204	8448544	DETENT,PAWL	1
5	PAFZZ	3040000179539	19204	8448543	PAWL,FORWARD ASSIST	1

END OF FIGURE

SECTION II

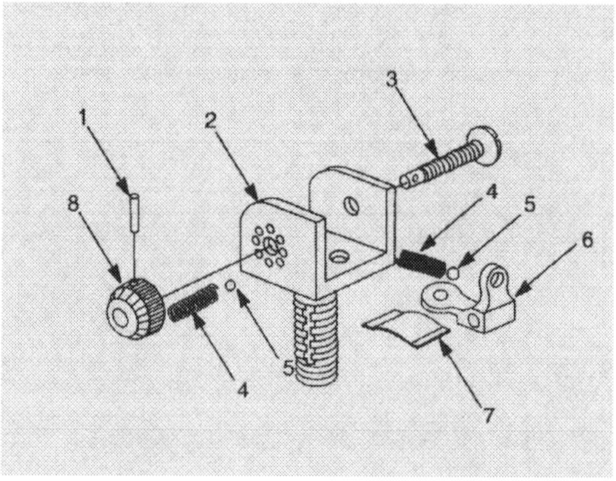

Figure C-10. (M16A2) Rear Sight Assembly 9349072.

(1) ITEM NO	(2) SMR CODE	(3) NSN	(4) CAGEC	(5) PART NUMBER	(6) DESCRIPTION AND USABLE ON CODE (UOC)	(7) QTY
					GROUP 030202 (M16A2) REAR SIGHT ASSEMBLY 9349072 FIG. C-10. REAR SIGHT ASSEMBLY	
1	PAFZZ	5315000586678	96906	MS16562-103	PIN,SPRING ,WINDAGE	1
2	PAFZZ	1005011343631	19200	9349074	BASE,REAR SIGHT UOC:ARB	1
3	PAFZZ	5305011441490	19200	9349076	SCREW,EXTERNALLY RE	1
4	PAFZZ	5360011481751	19200	9349069	SPRING,HELICAL,COMP ,REAR SIGHT UOC:AR8	2
5	PAFZZ	3110001839175	96906	M5S19060-4808	BALL,BEARING UOC:AR8	2
6	PAFZZ	1005011353697	19200	9349075	APERTURE,SIGHT	1
7	PAFZZ	5360013816183	19200	12011987	SPRING,FLAT REAR SIGHT	1
8	PAFZZ	5355011343627	19200	9349077	KNOB,WINDAGE	1

END OF FIGURE

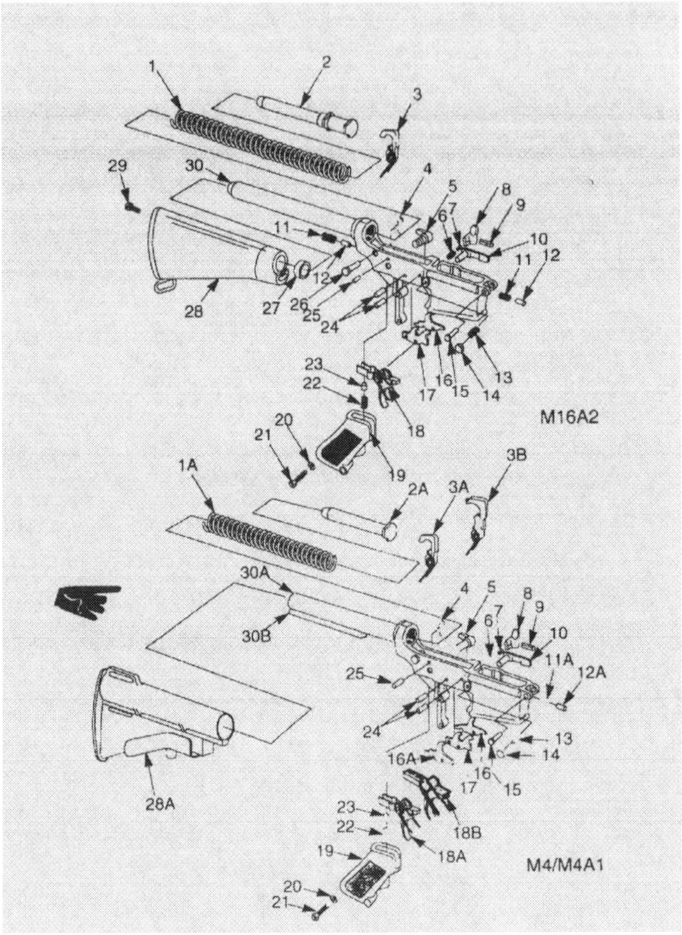

Figure C-11. Lower Receiver and Buttstock Assembly (MI 6A2) 9349100, (M4) 9390011 and (M4A1) 12972690.

(1) ITEM NO	(2) SMR CODE	(3) NSN	(4) CAGEC	(5) PART NUMBER	(6) DESCRIPTION AND USABLE ON CODE (UOC)	(7) QTY
					GROUP 04 LOWER RECEIVER AND BUTTSTOCK ASSEMBLY (M16A2) 9349100, (M4) 9390011, AND (M4A1) 12972690 FIG C-11 LOWER RECEIVER AND BUTTSTOCK ASSEMBLY	
1	PAOZZ	5360009926665	19204	8448629	SPRING, HELICAL, COMP ACTION (M16A2) UOC:AR8	1
1A	PAOZZ	5360012338617	19200	9390022	SPRING, HELICAL, COMP COMPRESSION (M4 AND M4A1) UOC:AS1, AY6	1
2	PAOZZ	1005009373078	19200	8448615	BUFFER ASSEMBLY (MI6A2) UOC:AR8	1
2A	PAOZZ	1005012313138	19200	9390023	BUFFER ASSEMBLY (M4 AND M4AI) UOC:AS1, AY6	1
3	AFFFF		19200	9349106	HAMMER ASSEMBLY (M16A2) UOC:AR8	1
3A	AFFFD		19200	9390032	HAMMER ASSEMBLY (M4) UOC:AS1	1
3B	AFFFF		19204	8448610	HAMMER ASSY (M4A1) UOC:AY6	1
4	PAFZZ	1005009926649	19200	8448595	SEAR	1
5	PAFZZ	5340012258339	19200	9381367	SELECTOR, FIRE CONTROL	1
6	PAFZZ	5360000562246	19204	8448633	SPRING, HELICAL, COMP COMPRESSION, BOLT CATCH	1
7	PAFZZ	1005000562247	19204	8448634	PLUNGER, BOLT CATCH	1
8	PAFZZ	1005000179548	19200	8448628	CATCH, BOLT	1
9	PAFZZ	5315008123312	96906119	MS16562-119	PIN, SPRING , BOLT CATCH	1
10	PAFZZ	1005000562201	19204	8448638	CATCH, MAGAZINE	1
11	PAOZZ	5360009926655	19204	8448586	SPRING, HELICAL, COMP COMPRESSION, TAKE DOWN/PIVOT PIN (M16A2) UOC:AR8	2
11A	PAOZZ	5360009926655	19204	8448586	SPRING, HELICAL, COMP COMPRESSION (M4 AND M4AI) UOC:AS1	1
12	PAOZZ	5315009926654	19204	8448585	PIN, STRAIGHT, HEADLE DETENT, TAKEDOWN PIN (M16A2) UOC:ARS	2
12A	PAOZZ	5315009926654	19204	8448585	PIN, STRAIGHT, HEADLE HEADLESS (M4 AND M4A1) UOC:AS1, AY6	1
13	PAFZZ	5360009927301	19204	8448637	SPRING, HELICAL, COMP COMPRESSION, MAGAZINE CATCH	1
14	PAFZZ	1005009927302	19204	8448636	BUTTON, MAGAZINE CAT CATCH	1
15	PAOZZ	5315000179537	19204	8448621	PIN, GROOVED, HEADED (PIVOT PIN)	1
16	PAFZZ	5340011457910	19200	9349114	LEVER, LOCK-RELEASE SEMI (M16A2 AND M4) UOC:AR8, AS1	1
16A	PAFZZ	1005009990406	19200	8448635	DISCONNECTOR (M4A1) UOC:AY6	1
17	PAFZZ	5340011441499	19200	9349113	LEVER, LOCK-RELEASE , BURST (MI6A2 AND M4)	1

(1) ITEM NO	(2) SMR CODE	(3) NSN	(4) CAGEC	(5) PART NUMBER	(6) DESCRIPTION AND USABLE ON CODE (UOC)	(7) QTY
					UOC:AR8, AS1	
18	AFFFF		19200	9349115	TRIGGER ASSEMBLY (M16A2)..... UOC:AR8.....	1
18A	AFFFF		19200	12972697	TRIGGER ASSEMBLY (M4)..... UOC:AS1.....	1
18B	AFFFF		19200	12972698	TRIGGER ASSEMBLY (M4A1)..... UOC:AY6.....	1
19	PAOZZ	1005011484805	19200	9349127	GRIP, RIFLE PLASTIC, BLACK	1
20	PAOZZ	5310005273634	96906	MS35335-61	WASHER, LOCK , RIFLE GRIP	1
21	PAOZZ	5305012681191	88044	AN501D416-18	SCREW, MACHINE , RIFLE GRIP	1
22	PAOZZ	5360009927292	19204	8448516	SPRING, HELICAL, COMP SAFETY.....	1
23	PAOZZ	1005009926667	19204	8448631	DETENT, SAFETY	1
24	PAFZZ	5315009927309	19204	8448609	PIN, GROOVED, HEADLES TRIGGER AND ..... HAMMER.....	2
25	PAFZZ	5315009926650	19204	8448599	PIN, GROOVED, HEADLES , AUTOMATIC SEAR.....	1
26	PAFZZ	5315009926653	19204	8448584	PIN, GROOVED, HEADED (M16A2)	1
27	PAOZZ	5365012672169	19200	12597640	SPACER, STEPPED (M16A2)..... UOC:AR8.....	1
28	PAOOO	1005011354973	19200	9349119	BUTTSTOCK ASSEMBLY, (M16A2) ..... UOC:AR8.....	1
28A	AOOOO		19200	9390012	BUTTSTOCK ASSY (M4 AND M4A1)..... UOC:AS1, AY6	1
29	PAOZZ	5305011478585	19200	9349128	SCREW, MACHINE , BUTTCAP (M16A2)..... UOC:AR8.....	1
30	XAFFA		19200	9349101	LOWER RECEIVER AND RECEIVER..... EXTENSION ASSEMBLY (M16A2) ..... UOC:AR8.....	1
30A	XAFFA		99999	NPN	LOWER RECEIVER ASSY ASSEMBLY (M4) ..... UOC:AS1.....	1
30B	XAFFA		99999	NPN	LOWER RECEIVER ASSY (M4AI) ..... UOC:AY6.....	1
					END OF FIGURE.....	

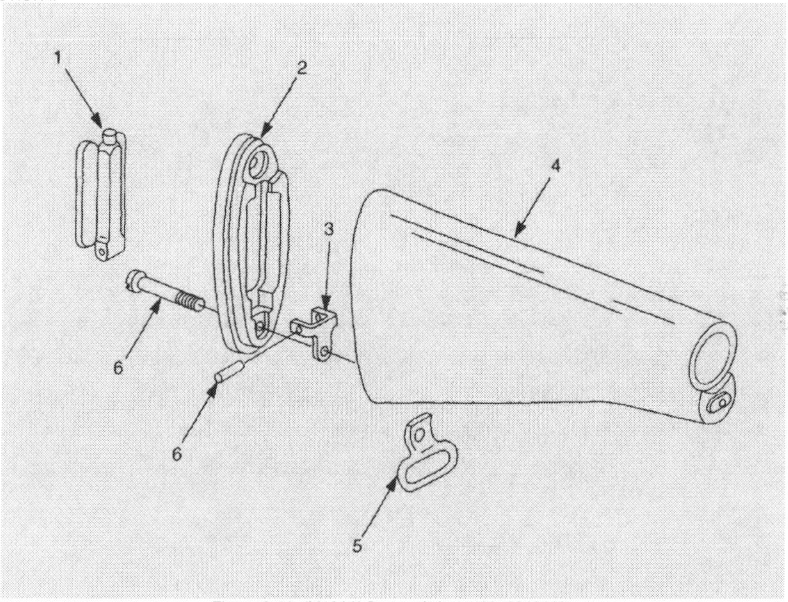

Figure C-12. (M16A2) Buttstock Assembly 9349119.

(1) ITEM NO	(2) SMR CODE	(3) NSN	(4) CAGEC	(5) PART NUMBER	(6) DESCRIPTION AND USABLE ON CODE (UOC)	(7) QTY
					GROUP 0401 (M16A2) BUTTSTOCK ASSEMBLY 9349119 FIG C-12 (M16A2) BUTTSTOCK ASSEMBLY	
1	PAOZZ	1005012288504	19200	9381380	DOOR ASSEMBLY, THUMB UOC:AR8	1
2	PAOZZ	1005011467685	19200	9349130	PLATE, BUTT, SHOULDER GUN STOCK UOC:AR8	1
3	PAOZZ	5340004633892	19200	8448653	HINGE, ACCESS DOOR BUTT PLATE................................. UOC:AR8	1
4	XAOZZ		19200	9349121	BUTTSTOCK.. UOC:AR8	1
5	PAOZZ	1005004030964	19204	8448652	SWIVEL, SLING, SMALL... UOC:AR8	1
6	PAOZZ	5315004633894	19204	8448655	PIN, STRAIGHT, HEADLE HEADLESS, ACCESS DOOR UOC:AR8	1
7	PAOZZ	5305011441494	19200 9349120		SCREW, MACHINE BUTT PLATE UOC:AR8	1
					END OF FIGURE	

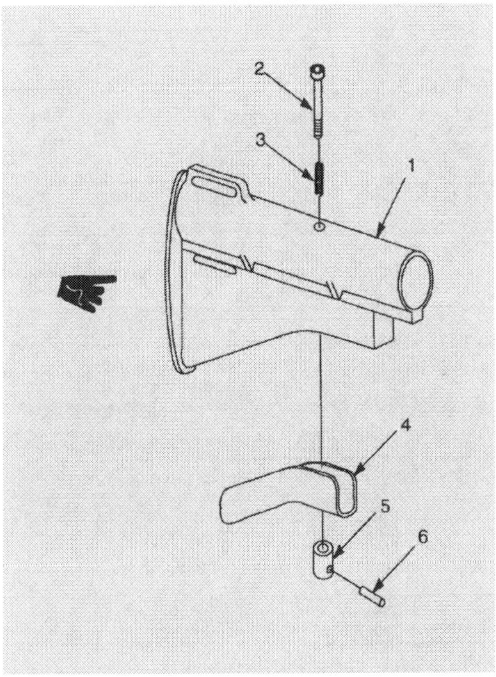

Figure C12A. (M4 and M4A1) Buttstock Assembly 9390012.

(1) ITEM NO	(2) SMR CODE	(3) NSN	(4) CAGEC	(5) PART NUMBER	(6) DESCRIPTION AND USABLE ON CODE (UOC)	(7) QTY
					GROUP 0401A (M4 AND M4A1) 9390012 BUTTSTOCK ASSEMBLY FIG C12A (M4 AND M4A1) BUTTSTOCK ASSEMBLY	
1	PAOZZ	1005012338636	19200	9390013	STOCK, GUN, SHOULDER UOC:AS1, AY6	1
2	PAOZZ	5315012338608	19200	9390025	PIN, SHOULDER, HEADLE UOC:AS1, AY6	1
3	PAOZZ	5360012338616	19200	9390027	SPRING, HELICAL, COMP COMPRESSION UOC:AS1, AY6	1
4	PAOZZ	5340012338638	19200	9390014	LEVER, LOCK-RELEASE UOC:AS1, AY6	1
5	PAOZZ	5310012338626	19200	9390026	NUT, SELF-LOCKING, RO UOC:AS1, AY6	1
6	PAOZZ	5315008439487	96906	MS16562-202	PIN, SPRING UOC:AS1, AY6	1

END OF FIGURE

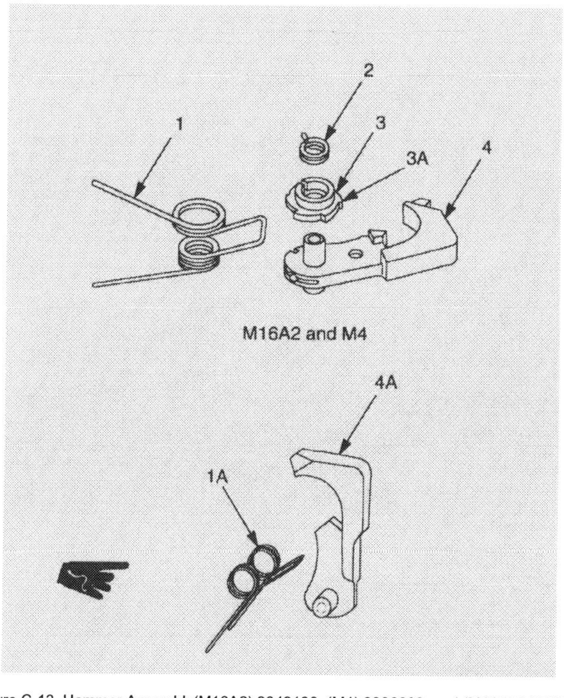

Figure C-13. Hammer Assembly(M16A2) 9349106, (M4) 9390032, and (M4A1) 8448610.

(1) ITEM NO	(2) SMR CODE	(3) NSN	(4) CAGEC	(5) PART NUMBER	(6) DESCRIPTION AND USABLE ON CODE (UOC)	(7) QTY
					GROUP 0402 HAMMER ASSEMBLY (M16A2) 9349106, (M4) 9390032, AND (M4A1) 8448610 FIG C-13 HAMMER ASSEMBLY	
1	PAFZZ	5360011441492	19200	9349107	SPRING, HELICAL, TORS HAMMER (MI6A2 AND M4) UOC:AR8, AS1	1
1A	PAFZZ	5360009926648	19204	8448611	SPRING, HELICAL, TORS (M4A1) UOC:AY6	1
2	PAFZZ	5360011365471	19200	9349109	SPRING, HELICAL, TORS BURST CAM (M16A2 AND M4) UOC:AR8, AS1	1
3	PAFZZ	1005011480172	19200	9349108	CAM, BURST (M16A2) (BLACK) UOC:AR8	1
3A	PAFZZ	3040012477969	19200	9390031	CAM, CONTROL (M4) (NICKEL/SHINY) UOC:AS1	1
4	PAFZZ	1005011343630	19200	9349110	HAMMER AND HAMMER P PIN RETAINER ASSEMBLY (MI6A2 AND M4) UOC:AR8, AS1	1
4A	PAFZZ	1005000179551	19200	8448612	HAMMER, FIRING, SMALL (M4A1) UOC:AY6	1
					END OF FIGURE	

SECTION II

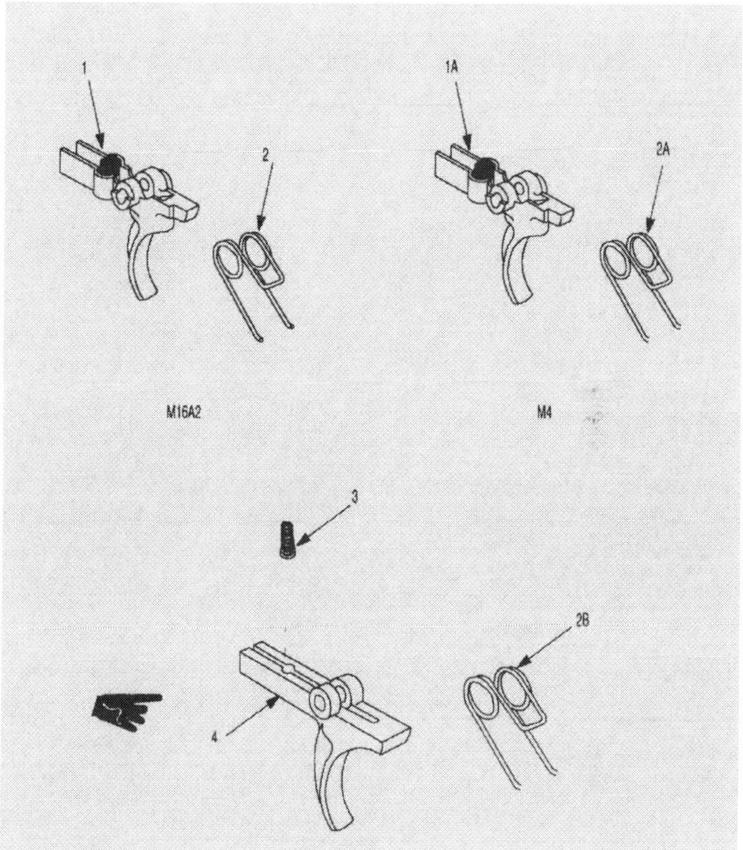

Figure C-14. Trigger Assembly(M16A2) 9349115, (M4) 12972697, and (M4A1) 12972676

(1) ITEM NO	(2) SMR CODE	(3) NSN	(4) CAGEC	(5) PART NUMBER	(6) DESCRIPTION AND USABLE ON CODE (UOC)	(7) QTY
					GROUP 0403 TRIGGER ASSEMBLY (M16A2) 9349115, (M4) 12972697, AND (M4A1) 12972676 FIG C-14 TRIGGER ASSEMBLY	
1	PAFFF	1005012192402	19200	9392518	TRIGGER SUB ASSEMBL (M16A2)........ UOC:AR8	1
1A	PAFFF	1005013954257	19200	12972696	TRIGGER SUBASSEMBLY (M4)........ UOC:AS1	1
2	PAFZZ	5360009927308	19204	8448593	SPRING, HELICAL, TORS TORSION, ... TRIGGER (MI6A2)........ UOC:AR8	1
2A	PAFZZ	5360009927308	19204	8448593	SPRING, HELICAL, TORS (M4)........ UOC:ASI	1
2B	PAFZZ	5360009927308	19204	8448593	SPRING, HELICAL, TORS (M4A1)........ UOC:AY6	1
3	PAFZZ	5360013960256	19200	12972695	SPRING, HELICAL, COMP DISCONNECT .... (M4A1)........ UOC:AY6	1
4	PAFZZ	1005009927307	19204	8448592	TRIGGER (M4A1)........ UOC:AY6	1

END OF FIGURE

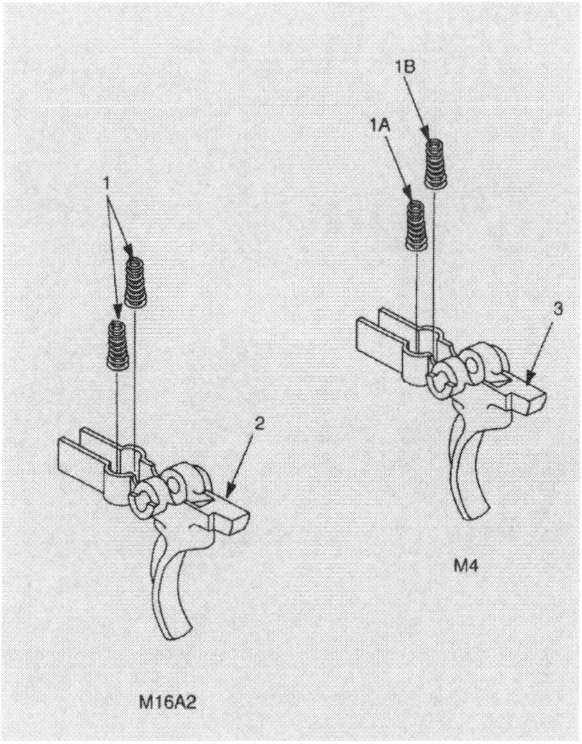

Figure C-15. Trigger Subassembly(M16A2) 9392518, (M4) 12972696.

(1) ITEM NO	(2) SMR CODE	(3) NSN	(4) CAGEC	(5) PART NUMBER	(6) DESCRIPTION AND USABLE ON CODE (UOC)	(7) QTY
					GROUP 040301 TRIGGER SUBASSEMBLY (M6A2) 9392518 AND (M4) 12972696 FIG C-15 TRIGGER SUBASSEMBLY	
1	PAFZZ	5360011350353	19200	9349116	SPRING, HELICAL, COMP DISCONNECT .... (M16A2).. UOC:AR8	2
1A	PAFZZ	5360011350353	19200	9349116	SPRING, HELICAL, COMP DISCONNECT .... (M4) (NICKEL/SHINY).. UOC:AS1	1
1B	PAFZZ	5360013960256	19200	12972695	SPRING, HELICAL, COMP DISCONNECT .... (M4) (BLACK)... UOC:AS1	1
2	XAFZZ		19200	9390736	TRIGGER (M16A2).. UOC:AR8	1
3	XAFZZ		19200	9390736	TRIGGER (M4).. UOC:AS1	1
					END OF FIGURE	

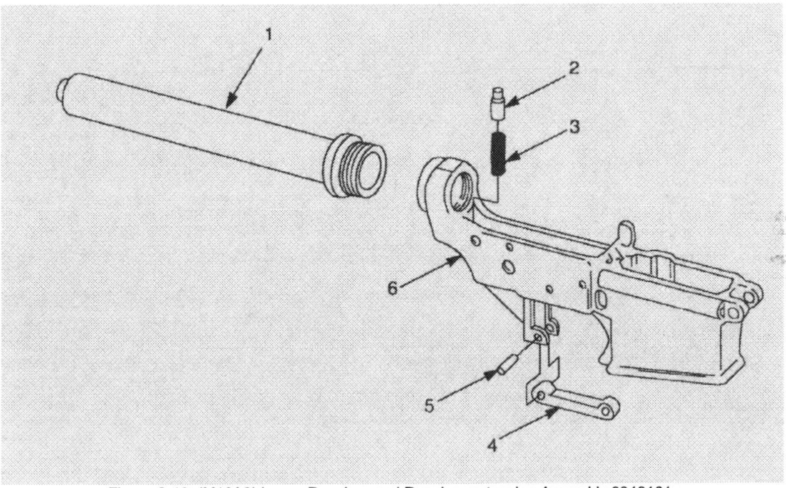

Figure C-16. (M16A2) Lower Receiver and Receiver extension Assembly 9349101.

(1) ITEM NO	(2) SMR CODE	(3) NSN	(4) CAGEC	(5) PART NUMBER	(6) DESCRIPTION AND USABLE ON CODE (UOC)	(7) QTY
					GROUP 0404 (M16A2) LOWER RECEIVER AND RECEIVER EXTENSION ASSEMBLY 9349101 FIG C-16 (M16A2) LOWER RECEIVER AND RECEIVER EXTENSION ASSEMBLY	
1	PAFZZ	5340009927297	19200	8448581	EXTENSION, LOWER REC RECEIVER .... (LOWER RECEIVER EXTENSION) (M1SA2)........................ UOC:AR8	1
2	PAFZZ	5315009926651	19204	8448582	PIN, SHOULDER, HEADLE BUFFER.................................... RETAINER (M16A2)... UOC:AR8	1
3	PAFZZ	5360009926652	19200	8448583	SPRING, HELICAL, COMP COMPRESSION, ... BUFFER RETAINER (NI6A2) UOC:AR8	1
4	PAFZZ	1005009927299	19204	8448587	GUARD, TRIGGER (M16A2)...	1
5	PAFZZ	5315000586081	96906	MS16562-129	PIN, SPRING TRIGGER GUARD (M16A2)...........................	1
6	XAFDA		19200	9349102	RECEIVER, (M16A2).. UOC:AR8	1

END OF FIGURE

SECTION II

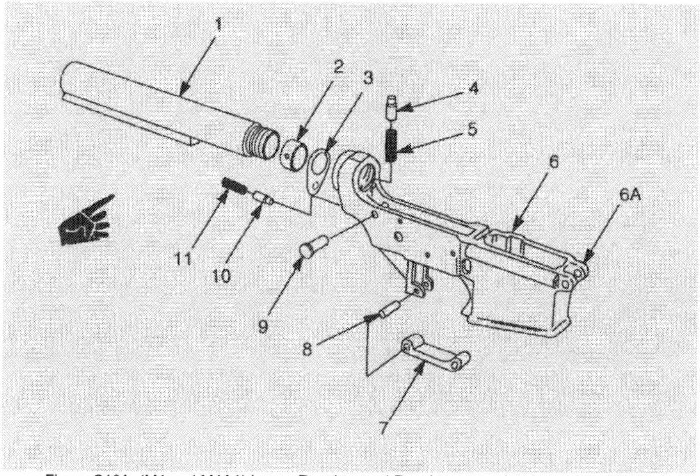

Figure C16A. (M4 and M4A1) Lower Receiver and Receiver extension Assembly, (NPN)

(1) ITEM NO	(2) SMR CODE	(3) NSN	(4) CAGEC	(5) PART NUMBER	(6) DESCRIPTION AND USABLE ON CODE (UOC)	(7) QTY
					GROUP 0404A (M4 AND M4A1) LOWER RECEIVER AND RECEIVER EXTENSION ASSEMBLY (NPN) FIG C16A (M4 AND M4A1) LOWER RECEIVER AND RECEIVER EXTENSION ASSEMBLY	
1	PAFZZ	1005012338531	19200	9390019	EXTENSION, LOWER RECEIVER (M4 AND M4A1) UOC:AS1, AY6	1
2	PAFZZ	5310012338625	19200	9390020	NUT, PLAIN, ROUND (M4 AND M4A1) UOC:AS1, AY6	1
3	PAFZZ	105012338530	19200	9390021	PLATE, RECEIVER END (M4 AND M4A1) UOC:AS1, AY6	1
4	PAFZZ	5315009926651	19204	8448582	PIN, SHOULDER, HEADLE (M4 AND M4A1) UOC:AS1, AY6	1
5	PAFZZ	5360009926652	19200	8448583	SPRING, HELICAL, COMP (M4 AND M4A1) UOC:AS1, AY6	1
6	XAFDA		19200	9390015	RECEIVER, (M4) UOC:AS1	1
6A	XAFDA		19200	12972652	LOWER RECEIVER (M4A1) UOC:AY6	1
7	PAFZZ	1005009927299	19204	8448587	GUARD, TRIGGER (M4 AND M4A1)	1
8	PAFZZ	5315000586081	96906	MS16562-129	PIN, SPRING (TRIGGER GUARD PIN) (M4 AND M4A1)	1
9	PAFZZ	5315009926653	19204	8448584	PIN, GROOVED, HEADED (TAKE DOWN) (M4 AND M4A1)	1
10	PAFZZ	5315009926654	19204	8448585	PIN, STRAIGHT, HEADLE HEADLESS (M4 AND M4A1) UOC:AS1, AY6	1
11	PAFZZ	5360009926655	19204	8448586	SPRING, HELICAL, COMP COMPRESSION (M4 AND M4A1) UOC:AS1, AY6	1

END OF FIGURE

SECTION II

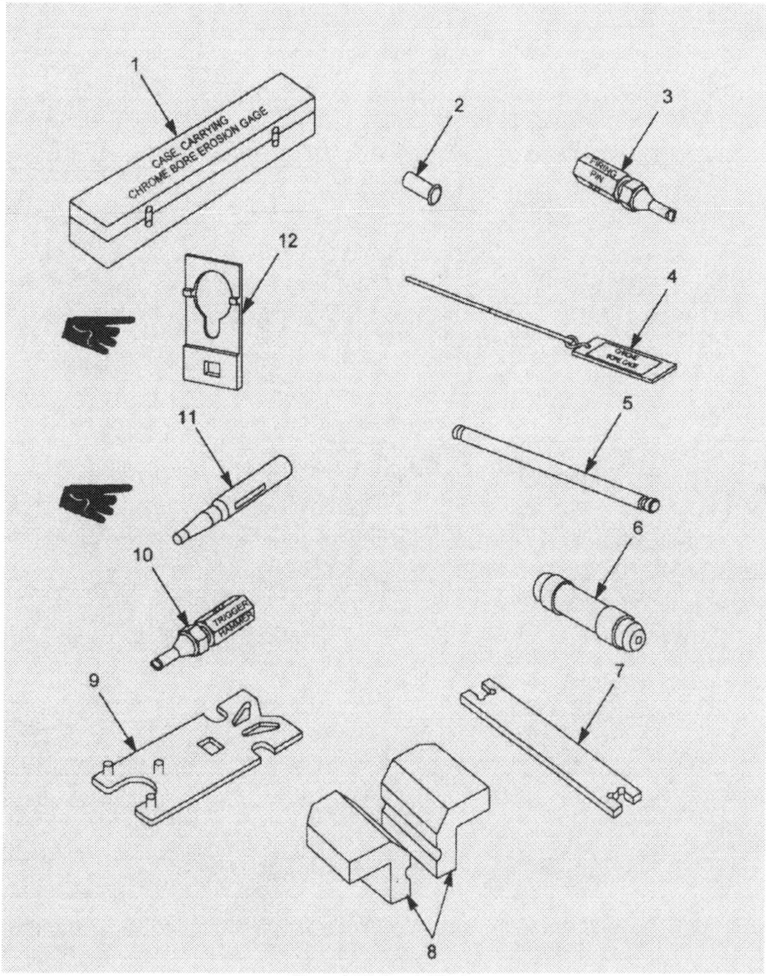

Figure C-17. Special Tools

(1) ITEM NO	(2) SMR CODE	(3) NSN	(4) CAGEC	(5) PART NUMBER	(6) DESCRIPTION AND USABLE ON CODE (UOC)	(7) QTY
					GROUP 9500 SPECIAL TOOLS FIG C-17 SPECIAL TOOLS	
	PEFZZ	1005000567106	19204	8426685	MAINTENANCE KIT, GUN DS/GS SUPPORT MAINTENANCE FOR 5.S6MM RIFLE, M16 RIFLE SERIES BOI: 2 PER SUPPORTING DSU/GSU	
1	PAFZZ	4933010355607	19204	12006359	CASE, BORE GAGE PART OF KIT P/N 8426685........	
2	PAFZZ	4933008007508	19204	8448201	..REFLECTOR TOOL, CHAMBER, PART OF KIT P/N 8426685	
3	PAFZZ	5220010755004	19200	12620101	..GAGE, PLUG, PLAIN PART OF KIT P/N 8426685........	
4	PAFZZ	5220010148183	19204	8448496	..GAGE, BARREL, EROSION BARREL EROSION PART OF KIT P/N 8426685(CHROME BARREL)........	
5	PAFZZ	5220002219391	19204	8448202	..GAGE, STRAIGHTNESS PART OF KIT P/ N 8426685	
6	PAFZZ	5220000707814	19204	7799734	..GAGE, HEADSPACE PART OF KIT P/N 8426685........	
7	PAFZZ	6695000707815	19204	7799735	..GAGE, FIRING PIN PROTRUSION PART OF KIT P/N 8426685	
8	PAFZZ	4933000709151	19204	11010032	..FIXTURE, BARREL REMOVAL PART OF KIT P/N 8426685	
9	PAFZZ	5120000709152	19204	11010033	..WRENCH, COMBINATION PART OF KIT P/ N 8426685	
10	PAFZZ	5220010439473	19204	12006472	GAGE, PLUG, TAPER CYLINDER PART OF KIT P/N 8426685	
11	PAOZZ	5315013100370	19200	12926769	KEY, MACHINE KEY, 2 PER ORGANIZATIONAL OR DIRECT SUPPORT SHOP	
12	PAFZZ	5120013246631	19200	9390035	WRENCH, SPANNER 2 PER DS/GS SUPPORT SHOP A/R FOR M4/M4A1 CARBINE UOC:AS1, AY6	
					END OF FIGURE	

CROSS-REFERENCE INDEXES

NATIONAL STOCK NUMBER INDEX

STOCK NUMBER	FIG	ITEM	STOCK NUMBER	FIG	ITEM
5315-00-017-9537	C-11	15	5315-00-979-3930	C-7	2
3040-00-017-9539	C-9	5	5360-00-979-3931	C-7	3
1005-00-017-9540	C-9	4	5360-00-992-6648	C-13	1A
5360-00-017-9541	C-8	13	1005-00-992-6649	C-11	4
1005-00-017-9543	C-7	4	5315-00-992-6650	C-11	25
1005-00-017-9546	C-1	1	5315-00-992-6651	C-16	2
1005-00-017-9547	C-2	1		C16A	4
1005-00-017-9548	C-11	8	5360-00-992-6652	C-16	3
1005-00-017-9551	C-13	4A		C16A	5
1005-00-056-2201	C-11	10	5315-00-992-6653	C-11	26
5360-00-056-2246	C-11	6		C16A	9
1005-00-056-2247	C-11	7	5315-00-992-6654	C-11	12
1005-00-056-7106	C-17			C-11	12A
5315-00-058-6044	C-6	7		C16A	10
5315-00-058-6081	C-16	5	5360-00-992-6655	C-11	11
	C16A	8		C-11	11A
5315-00-058-6678	C-1B	1		C16A	11
	C-10	1	5360-00-992-6665	C-11	1
5220-00-070-7814	C-17	6	1005-00-992-6667	C-11	23
6695-00-070-7815	C-17	7	1005-00-992-7283	C-4	2
4933-00-070-9151	C-17	8	5305-00-992-7284	C-4	1
5120-00-070-9152	C-17	9	1005-00-992-7285	C-2	4
1005-00-087-8998	C-6	8	1005-00-992-7287	C-3	1
3110-00-183-9175	C-8	4	1005-00-992-7288	C-3	7
	C-1A	6	1005-00-992-7290	C-3	3
	C-1B	5	1005-00-992-7291	C-3	6
	C-10	5	5360-00-992-7292	C-3	5
5220-00-221-9391	C-17	5		C-11	22
1005-00-403-0964	C-12	5	5315-00-992-7294	C-2	3
5340-00-463-3892	C-12	3	5340-00-992-7297	C-16	1
5315-00-463-3894	C-12	6	1005-00-992-7299	C-16	4
5360-00-523-8084	C-9	3		C16A	7
5310-00-527-3634	C-11	20	5360-00-992-7301	C-11	13
5315-00-597-5086	C-3	4	1005-00-992-7302	C-11	14
5315-00-690-0544	C-7	8	1005-00-992-7307	C-14	4
1005-00-738-6213	C-4	3	5360-00-992-7308	C-14	2
1005-00-760-3768	C-3	8		C-14	2A
4933-00-800-7508	C-17	2		C-14	2B
5315-00-812-3312	C-11	9	5315-00-992-7309	C-11	24
5315-00-840-3812	C-8	11	5360-00-999-0404	C-5	3
	C-8	12	5340-00-999-0405	C-5	2
	C-1A	8	1005-00-999-0406	C-11	16A
5315-00-843-9487	C12A	6	5325-00-999-0863	C-6	10
1005-00-921-5004	C-1	4	5325-00-999-0864	C-8	7
1005-00-937-3078	C-11	2	1005-00-999-1509	'C-2	2
1005-00-978-1022	C-8	9	5220-01-014-8183	C-17	4
5315-00-978-1023	C-8	8	4933-01-035-5607	C-17	1
5360-00-978-1025	C-8	10	5220-01-043-9473	C-17	10
5360-00-978-1036	C-6	9	5315-01-048-9372	C-5	1
4710-00-978-1038	C-6	3		C-9	2

CROSS-REFERENCE INDEXES

NATIONAL STOCK NUMBER INDEX

STOCK NUMBER	FIG	ITEM
5320-01-063-7635	C-7	5
5220-01-075-5004	C-17	3
1005-01-134-3621	C-8	1
5305-01-134-3622	C-8	2
	C-1A	2
1005-01-134-3625	C-7	1
5355-01-134-3627	C-1B	8
	C-10	8
1005-01-134-3629	C-6	2
1005-01-134-3630	C-13	4
1005-01-134-3631	C-10	2
1005-01-134-3633	C-6	6
1005-01-134-3701	C-8	6
5360-01-134-3710	C-8	15
	C-1A	5
5360-01-135-0353	C-15	1
	C-15	1A
1005-01-135-3697	C-1B	6
	C-10	6
5355-01-135-4972	C-8	16
1005-01-135-4973	C-11	28
5360-01-136-5471	C-13	2
1005-01-144-1468	C-9	1
5305-01-144-1490	C-1B	3
	C-10	3
5360-01-144-1492	C-13	1
5305-01-144-1494	C-12	7
5365-01-144-1496	C-6	5
5340-01-144-1499	C-11	17
5340-01-145-7910	C-11	16
1005-01-146-7684	C-6	4
1005-01-146-7685	C-12	2
5305-01-147-8585	C-11	29
1005-01-148-0172	C-13	3
5360-01-148-1751	C-8	5
	C-1B	4
	C-10	4
1005-01-148-4805	C-11	19
1005-01-216-4510	C-1	5
1005-01-219-2402	C-14	1
5340-01-225-8339	C-11	5
1005-01-228-8504	C-12	1
1005-01-231-3138	C-11	2A
1005-01-233-8529	C-6	4A
1005-01-233-8530	C16A	3
1005-01-233-8531	C16A	1
5315-01-233-8608	C12A	2
5360-01-233-8616	C12A	3
5360-01-233-8617	C-11	1A
5310-01-233-8625	C16A	2
5310-01-233-8626	C12A	5
1005-01-233-8636	C12A	1
4710-01-233-8637	C-6	3A
5340-01-233-8638	C12A	4
1005-01-234-2297	C-6	2A
3040-01-247-7969	C-13	3A
1010-01-264-6517	C-7	9
5340-01-264-6530	C-7	7
5365-01-267-2169	C-11	27
5305-01-268-1191	C-11	21
5315-01-310-0370	C-17	11
5120-01-324-6631	C-17	12
1005-01-368-9852	C-1	5A
5360-01-381-6183	C-1B	7
	C-10	7
5340-01-382-3201	C-1A	10
5310-01-382-6793	C-1A	9
1005-01-382-6795	C-8	6A
5355-01-382-6801	C-1A	4
5360-01-382-6802	C-1A	7
1005-01-382-7083	C-1A	11
1005-01-382-7086	C-1B	2
1005-01-382-7089	C-1A	3
1005-01-395-4257	C-14	1A
5360-01-396-0256	C-14	3
	C-15	1B
1005-01-422-3770	C-2	4A
1005-01-424-5899	C-3	9

CROSS-REFERENCE INDEXES

NATIONAL STOCK NUMBER INDEX

STOCK NUMBER	FIG	ITEM	STOCK NUMBER	FIG	ITEM
88044	AN5010416-18		5305-01-268-1191	C-11	21
96906	MS16562-103		5315-00-058-6678	C-1B	1
				C-10	1
96906	MS16562-106		5315-00-058-6044	C-6	7
96906	MS16562-119		5315-00-812-3312	C-11	9
96906	MS16562-121		5315-00-840-3812	C-8	11
				C-8	12
				C-1A	8
96906	MS16562-129		5315-00-058-6081	C-16	5
				C16A	8
96906	MS16562-202		5315-00-843-9487	C12A	6
96906	MS16562-98		5315-00-597-5086	C-3	4
96906	MS16626-3137		5325-00-999-0863	C-6	10
96906	MS16632-3012		5325-00-999-0864	C-8	7
96906	MS19060-4808		3110-00-183-9175	C-8	4
				C-1A	6
				C-1B	5
				C-10	5
96906	MS35335-61		5310-00-527-3634	C-11	20
96906	MS39086-93		5315-00-690-0544	C-7	8
99999	NPN			C-11	30A
				C-11	30B
19204	11010032		4933-00-070-9151	C-17	8
19204	11010033		5120-00-070-9152	C-17	9
19204	12006359		4933-01-035-5607	C-17	1
19204	12006472		5220-01-043-9473	C-17	10
19200	12011849			C-1	2A
19200	12011987		5360-01-381-6183	C-1B	7
				C-10	7
19200	12011996		1005-01-368-9852	C-1	5A
19200	12597640		5365-01-267-2169	C-11	27
19200	12598617		1010-01-264-6517	C-7	9
19200	12598618		5340-01-264-6530	C-7	7
19200	12620101		5220-01-075-5004	C-17	3
19204	12624561		1005-01-216-4510	C-1	5
19200	12926769		5315-01-310-0370	C-17	11
19200	12951011			C-1	7
19200	12951017		5340-01-382-3201	C-1A	10
19200	12951018		1005-01-382-7089	C-1A	3
19200	12951019		5355-01-382-6801	C-1A	4
19200	12951020		5360-01-382-6802	C-1A	7
19200	12951021		1005-01-382-7083	C-1A	11
19200	12951023		5310-01-382-6793	C-1A	9
19200	12951026			C-1A	1
19200	12951028		1005-01-382-7086	C-1B	2
19200	12972652			C16A	6A
19200	12972670		1005-01-382-6795	C-8	6A
19200	12972675			C-6	1A
19200	12972680			C-1	3A
19200	12972690			C-1	6B
19200	12972691		1005-01-422-3770	C-2	4A

CROSS-REFERENCE INDEXES

PART NUMBER INDEX

CAGEC	PART NUMBER	STOCK NUMBER	FIG	ITEM
19200	12972692	1005-01-424-5899	C-3	9
19200	12972695	5360-01-396-0256	C-14	3
			C-15	1B
19200	12972696	1005-01-395-4257	C-14	1A
19200	12972697		C-11	18A
19200	12972698		C-11	18B
19204	7799734	5220-00-070-7814	C-17	6
19204	7799735	6695-00-070-7815	C-17	7
19204	8426685	1005-00-056-7106	C-17	
19204	8448201	4933-00-800-7508	C-17	2
19204	8448202	5220-00-221-9391	C-17	5
19204	8448496	5220-01-014-8183	C-17	4
19204	8448501		C-1	2
19204	8448502	5315-00-992-7294	C-2	3
19204	8448503	1005-00-017-9547	C-2	1
19204	8448504	1005-00-999-1509	C-2	2
19204	8448505		C-2	5
19200	8448506	1005-00-992-7283	C-4	2
19200	8448507	1005-00-738-6213	C-4	3
19204	8448508	5305-00-992-7284	C-4	1
19200	8448509	1005-00-992-7285	C-2	4
19204	8448510		C-3	2
19204	8448511	1005-00-992-7287	C-3	1
19204	8448512	1005-00-992-7288	C-3	7
19204	8448513	1005-00-992-7290	C-3	3
19204	8448515	1005-00-992-7291	C-3	6
19204	8448516	5360-00-992-7292	C-3	5
			C-11	22
19204	8448517	1005-00-017-9546	C-1	1
19204	8448518		C-5	4
19200	8448519	5340-00-999-0405	C-5	2
19204	8448520	5360-00-999-0404	C-5	3
19204	8448521-2	5315-01-048-9372	C-5	1
			C-9	2
19204	8448525	1005-00-978-1022	C-8	9
19204	8448532	5360-00-978-1025	C-8	10
19204	8448533	5315-00-978-1023	C-8	8
19200	8448540	5360-00-017-9541	C-8	13
19200	8448542	5360-00-523-8084	C-9	3
19204	8448543	3040-00-017-9539	C-9	5
19204	8448544	1005-00-017-9540	C-9	4
19204	8448555	5360-00-978-1036	C-6	9
19200	8448567	4710-00-978-1038	C-6	3
19204	8448571	1005-00-017-9543	C-7	4
19204	8448573	5315-00-979-3930	C-7	2
19204	8448574	5360-00-979-3931	C-7	3
19200	8448581	5340-00-992-7297	C-16	1
19204	8448582	5315-00-992-6651	C-16	2
			C16A	4
19200	8448583	5360-00-992-6652	C-16	3
			C16A	5

CROSS-REFERENCE INDEXES

PART NUMBER INDEX.

CAGEC	PART NUMBER	STOCK NUMBER	FIG	ITEM
19204	8448584	5315-00-992-6653	C-11	26
			C16A	9
19204	8448585	5315-00-992-6654	C-11	12
			C-11	12A
			C16A	10
19204	8448586	5360-00-992-6655	C-11	11
			C-11	11A
			C16A	11
19204	8448587	1005-00-992-7299	C-16	4
			C16A	7
19204	8448592	1005-00-992-7307	C-14	4
19204	8448593	5360-00-992-7308	C-14	2
			C-14	2A
			C-14	2B
19200	8448595	1005-00-992-6649	C-11	4
19204	8448599	5315-00-992-6650	C-11	25
19204	8448609	5315-00-992-7309	C-11	24
19204	8448610		C-11	3B
19204	8448611	5360-00-992-6648	C-13	1A
19200	8448612	1005-00-017-9551	C-13	4A
19200	8448615	1005-00-937-3078	C-11	2
19204	8448621	5315-00-017-9537	C-11	15
19200	8448628	1005-00-017-9548	C-11	8
19204	8448629	5360-00-992-6665	C-11	1
19204	8448631	1005-00-992-6667	C-11	23
19204	8448633	5360-00-056-2246	C-11	6
19204	8448634	1005-00-056-2247	C-11	7
19200	8448635	1005-00-999-0406	C-1l	11A
19204	8448636	1005-00-992-7302	C-11	14
19204	8448637	5360-00-992-7301	C-11	13
19204	8448638	1005-00-056-2201	C-11	10
19204	8448652	1005-00-403-0964	C-12	5
19200	8448653	5340-00-463-3892	C-12	3
19204	8448655	5315-00-463-3894	C-12	6
19200	8448670	1005-00-921-5004	C-1	4
19204	8448697	5320-01-063-7635	C-7	5
19204	8448712	1005-00-087-8998	C-6	8
19200	8448755	1005-00-760-3768	C-3	8
19200	9349050		C-1	3
19200	9349051	1005-01-134-3633	C-6	6
19200	9349052	5365-01-144-1496	C-6	5
19200	9349054		C-7	6
19200	9349056	1005-01-134-3625	C-7	1
19200	9349059	1005-01-134-3629	C-6	2
19200	9349062		C-6	1
19200	9349063	1005-01-134-3701	C-8	6
19200	9349065	5305-01-134-3622	C-8	2
			C-1A	2
19200	9349066	1005-01-134-3621	C-8	1
19200	9349067	5355-01-135-4972	C-8	16
19200	9349069	5360-01-148-1751	C-8	5

CROSS-REFERENCE INDEXES

PART NUMBER INDEX

CAGEC	PART NUMBER	STOCK NUMBER	FIG	ITEM
19200	9349069	5360-01-148-1751	C-1B	4
			C-10	4
19200	9349070	5360-01-134-3710	C-8	15
			C-1A	5
19200	9349072		C-8	3
19200	9349074	1005-01-134-3631	C-10	2
19200	9349075	1005-01-135-3697	C-1B	6
			C-10	6
19200	9349076	5305-01-144-1490	C-1B	3
			C-10	3
19200	9349077	5355-01-134-3627	C-1B	8
			C-10	8
19200	9349085	1005-01-144-1468	C-9	1
19204	9349086		C-8	14
19200	9349100		C-1	6
19200	9349101		C-11	30
19200	9349102		C-16	6
19200	9349106		C-11	3
19200	9349107	5360-01-144-1492	C-13	1
19200	9349108	1005-01-148-0172	C-13	3
19200	9349109	5360-01-136-5471	C-13	2
19200	9349110	1005-01-134-3630	C-13	4
19200	9349113	5340-01-144-1499	C-11	17
19200	9349114	5340-01-145-7910	C-11	16
19200	9349115		C-11	18
19200	9349116	5360-01-135-0353	C-15	1
			C-15	1A
19200	9349119	1005-01-135-4973	C-11	28
19200	9349120	5305-01-144-1494	C-12	7
19200	9349121		C-12	4
19200	9349124	1005-01-146-7684	C-6	4
19200	9349127	1005-01-148-4805	C-11	19
19200	9349128	5305-01-147-8585	C-11	29
19200	9349130	1005-01-146-7685	C-12	2
19200	9381367	5340-01-225-8339	C-11	5
19200	9381380	1005-01-228-8504	C-12	1
19200	9390003	1005-01-234-2297	C-6	2A
19200	9390007	1005-01-233-8529	C-6	4A
19200	9390009		C-7	6A
19200	9390011		C-1	6A
19200	9390012		C-11	28A
19200	9390013	1005-01-233-8636	C12A	1
19200	9390014	5340-01-233-8638	C12A	4
19200	9390015		C16A	6
19200	9390016	4710-01-233-8637	C-6	3A
19200	9390019	1005-01-233-8531	C16A	1
19200	9390020	5310-01-233-8625	C16A	2
19200	9390021	1005-01-233-8530	C16A	3
19200	9390022	5360-01-233-8617	C-11	1A
19200	9390023	1005-01-231-3138	C-11	2A
19200	9390025	5315-01-233-8608	C12A	2

CROSS-REFERENCE INDEXES

PART NUMBER INDEX

CAGEC	PART NUMBER	STOCK NUMBER	FIG	ITEM
19200	9390026	5310-01-233-8626	C12A	5
19200	9390027	5360-01-233-8616	C12A	3
19200	9390031	3040-01-247-7969	C-13	3A
19200	9390032		C-11	3A
19200	9390035	5120-01-324-6631	C-17	12
19200	9390736		C-15	2
			C-15	3
19200	9392518	1005-01-219-2402	C-14	1

APPENDIX D
EXPENDABLE/DURABLE SUPPLIES AND MATERIALS LIST

Section I. INTRODUCTION

D-1. SCOPE. This appendix lists expendable durable supplies and materials you will need to operate and maintain the M16A2 Rifle This listing is for informational purposes only and Is not authority to requisition the listed Items. These items are authorized to you by CTA 50-970, Expendable, Durable Items (Except Medical, Class V, Repair Parts, and Heraldic Items), or CTA 8-100, Army Medical Department Expendable/Durable Items.

D-2. EXPLANATION OF COLUMNS.

a. Column (1)-Item Number. This number is assigned to the entry in the listing and is referenced In the narrative instructions to identify the material (e.g., "Use cloth, abrasive, crocus, Item 12, app D").

b. Column (2)-Level. This column identifies the lowest level of maintenance that requires the listed Item.

C - Operator/Crew Maintenance
O - Unit Maintenance
F - Direct Support Maintenance

c. Column (3)-National Stock Number. This is the National stock number assigned to the Item, use It to request or requisition the Item.

d. Column (4)-Description. Indicates the Federal Item name and, if required, a description to identify the item. The last line for each Item indicates the Commercial and Government Entity Code (CAGEC) in parentheses followed by the part number.

e. Column (5)-Unit of Measure (U/M). Indicates the measure used in performing the actual maintenance function. This measure is expressed by a two-character alphabetical abbreviation (e g., ea, In , pr). If the unit of measure differs from the unit of issue, requisition the lowest unit of issue that will satisfy your requirements.

Section II. EXPENDABLE/DURABLE SUPPLIES AND MATERIALS LIST

(1) ITEM NUMBER	(2) LEVEL	(3) NATIONAL STOCK NUMBER	(4) DESCRIPTION PART NUMBER AND CAGE	(5) UNIT OF MEAS
1	F	8040 00-944 7292	ADHESIVE, I<IT, (81348) MMM A 1754	KT
2	0	8020 00 244 0153	BRUSH, ARTIST'S metal ferrule, flat chisel ed(i(e 7 16 w, 1 1/8, ex-posed bristle (813481 H B 241	EA
3	F	1005-00-716 2702	BRUSH, CLEANING, SMALL ARMS (192051 7162702	EA
4	C	1005-00-903 1296	BRUSH, CLEANING, SMALL ARMS bore (19204) 11686340	EA
5	C	1005-00 999-1435	BRUSH, CLEANING, SMALL ARMS chamber (19204) 8432358	EA
6	C	1005 00-444-6602	BRUSH, CLEANING, SMALL ARMS tooth 1192041 8448462	EA
7	O	7920-00 205-2401	BRUSHCLEANING, TOOLS AND PARTS (813491 MILS43871	EA
8	O	6850-00-965-2332	CARBON REMOVING COMPOUND (813481 P-C-111	GL
9			CLEANER, LUBRICANT AND PRESERV-ATIVE (27412)	
	O	9150-01-079-6124	CLP- 4 oz (118 30 ml) bottle	EA
	O	9150-01-054 6453	CLP- 1 pt (0 47 1l bottle	EA
	O	9150-01 053-6688	CLP 7 gal (26 50 11 bottle	EA
10	C	9150 01-102 1473	CLEANER, LUBRICANT AND PRESERV-ATIVE (81349i MIL L 63460 1/2 oz (14 79 ml) bottle	EA
11	C	9920-00-292 9946	CLEANER, TOBACCO PIPE cotton turf, wire core (89855) DILLSPIPE cleaner (6 per pkg)	
			CLEANING COMPOUND, RIFLE BORE small arms bore cleaner solution (RBC) 1813491 MIL C 372	
	C	6850-00-224-6656	2 oz 159.15 mil) bottle	OZ
	O	6850-00-224-6657	8 oz (236 59 ml) can	CN
	O	6850-00-224-6663	1 gal (3.79 l) can	CN

EXPENDABLE/DURABLE SUPPLIES AND MATERIALS LIST (CONT)

(1) ITEM NUMBER	(2) LEVEL	(3) NATIONAL STOCK NUMBER	(4) DESCRIPTION PART NUMBER AND CAGE	(5) UNIT OF MEAS
13	0	5350-00-221-0872	CLOTH, ABRASIVE: (58536) A-A-1206	SH
14			Deleted See Item 23.2	
15	F		DICHLOROMETHANE, TECHNICAL: (81346) ASTM D 4701	
		6810-00-244-0290	5 gal (18.93 1) pail	CN
		6810-00-616-9188	600 lb (272.16 kg) drum	DR
16	0	6850-00-281-1985	DRY CLEANING SOLVENT: (58536) A-A-711 1 gal(3.79 l1) can	GL
17	0	8010-00-297-0560	ENAMEL: olive drab no 3407 (81348) TT-E-527 1 gal (3.791) can	GL
18	0		GLOVES, CHEMICAL AND OIL PROTECTIVE: (81348) ZZ-G-381	
		8415-00-823-7455	Size 9	PR
		8415-00-823-7456	Size10	PR
		8415-00-823-7457	Size 11	PR
19	F	9150-00-754-2595	GREASE, MOLYBDENUM DISULFIDE: (81349) MIL-G-21164	LB
20	C	1005-01-113-0321	HANDLE SECTION, CLEANING ROD, SMALL ARMS: (19204)8436776	EA
21	0	9150-01-260-2534	LUBRICANT, SOLID FILM: (81349) MIL-L-23398 16 OZ (473.18 ml) spray can	OZ
22	C	9150-00-292-9689	LUBRICATING OIL, WEAPONS: (LAW) (81349) MIL-L-14107 1 qt (0.95 1) can	QT

EXPENDABLE/DURABLE SUPPLIES AND MATERIALS LIST (CONT)

(1) ITEM NUMBER	(2) LEVEL	(3) NATIONAL STOCK NUMBER	(4) DESCRIPTION PART NUMBER AND CAGE	(5) UNIT OF MEAS
23			LUBRICATING OIL, WEAPONS: (LSA), semifluid (81349) MIL-L-46000	
	C	9150-00-935-6597	2 oz (59.15 ml) plastic btl	OZ
	C	9150-00-889-3522	4 oz (118.30 ml) plastic btl	OZ
	O	9150-00-687-4241	1 qt (0.95 l1) can	CN
	O	9150-00-753-4686	1 gal(3.79 l1) can	CN
23.1	0	5340-01-230-3181	MOUNTING BRACKET (M4/M4A1 ONLY) (19200)12556036	EA
23.2	0	8010-00-087-0102	PAINT, ENAMEL, SEMIGLOSS: paint for blank firing attachment (M1 5A2) (81348) TT-E-529 1 qt can (RED - M16A2)	EA
23.3	0	8010-01-031-1274	PAINT, ENAMEL, SEMIGLOSS: paint for blank firing attachment (M23) (81348) TT-E-529 1 pt can (YELLOW - M4/M4A1)	EA
24	0	4940-00-795-3595	PAN, WASH: (94453) 1211	EA
25	F	6850-00-826-0981	PENETRANT KIT: (81349) MIL-I-25135	KT
26	F	8135-01-019-1691	POLYETHYLENE SHEET: (84744) PE88-80-2	EA
26.1	0	1005-01-394-7677	PROTECTOR, RAIL (M4/M4A1 ONLY) (19200)12972676	EA
27	C	7920-00-205-1711	RAG, WIPING: (58536) A-A-531 50 lb (22.68 kg) bdl	LB
28	C	1005-00-050-6357	ROD SECTION, CLEANING SMALL ARMS: (19204) 8436775 (3 REQUIRED)	EA
29	F	8030-00-670-8553	SEALING COMPOUND (16059) DEVCONF	KT

EXPENDABLE/DURABLE SUPPLIES AND MATERIALS LIST (CONT)

(1) ITEM NUMBER	(2) LEVEL	(3) NATIONAL STOCK NUMBER	(4) DESCRIPTION PART NUMBER AND CAGE	(5) UNIT OF MEAS
30	C	1005-00-937-2250	SWAB HOLDER SECTION, CLEANING ROD, SMALL ARMS: (19204) 11686327	EA
31	C	1005-00-912-4248	SWAB, SMALL ARMS: (19204)11686408	SH
32	C	6920-01-395-2949	TARGET, 25 METER ZEROING, M16A2: (19200)12012024	BX

APPENDIX E

ILLUSTRATED LIST OF MANUFACTURED ITEMS

E-1. INTRODUCTION.

a. This appendix includes complete instructions for making items authorized to be manufactured or fabricated at unit or direct support maintenance.

b. A part number in alphanumeric order is provided for cross-referencing the part number of the item to be manufactured to the figure which covers fabrication criteria.

c. All bulk matenals needed for manufacture of an item are listed by part number or specification number In a tabular list on the illustration.

E-2. MANUFACTURED ITEMS PART NUMBER INDEX.

INDEX

Item	Figure Number
Front sight detent depressor	E-1
Front sight post removal and installation tool	E-2
Pivot pin removal tool	E-3
Pivot pin installation tool	E-4
Slave pin	E-5
Adapter bar for M12 arms rack	E-6
Modified needle nose pliers	E-7

E-3. MANUFACTURED ITEMS ILLUSTRATIONS - FORMAT AND CONTENT.

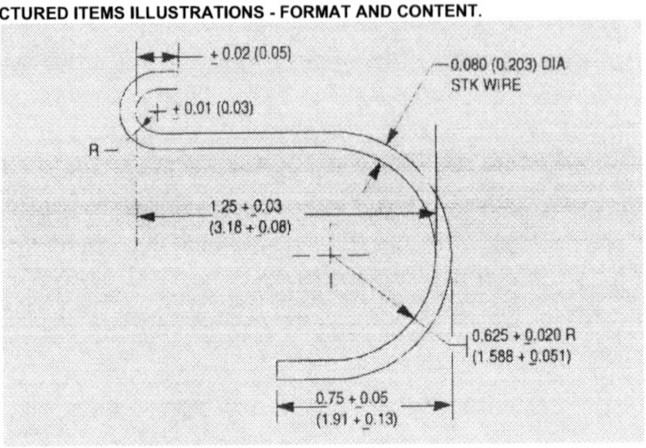

FABRICATE FROM 0.08 IN. MUSIC WIRE OR EQUIVALENT.
FINISH: NO. 5.3.1.2 OR 5.3.2.2 OF MIL-STD 17.

NOTE: ALL DIMENSIONS SHOWN ARE IN INCHES WITH THE METRIC CONVERSION TO CENTIMETERS IN PARENTHESIS.

Figure E-1. Front sight detent depressor.

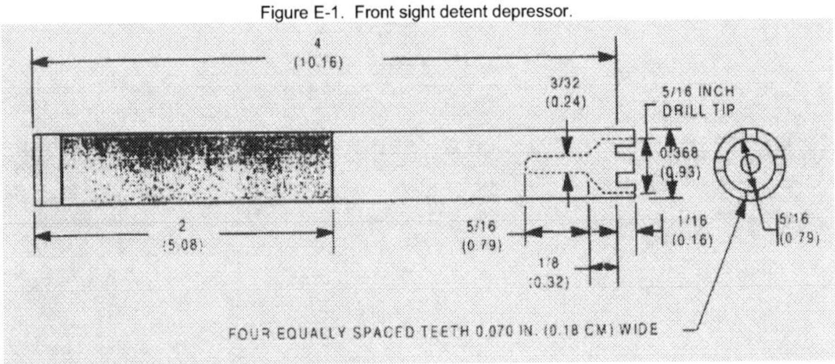

FABRICATE FROM 0.375 INCH ROUND METAL BAR,
ASTM A686. FSCM 81346, GRADE C. CLASS W2-09.
NSN 9510-00-640-4407 OR EQUIVALENT.

NOTES:
1. ALL DIMENSIONS SHOWN ARE IN INCHES WITH THE METRIC CONVERSION TO CENTIMETERS IN PARENTHESIS.
2. TEETH MUST BE HAND FILED TO FIT FRONT SIGHT POST.

Figure E-2. Front sight post removal and Installation tool.

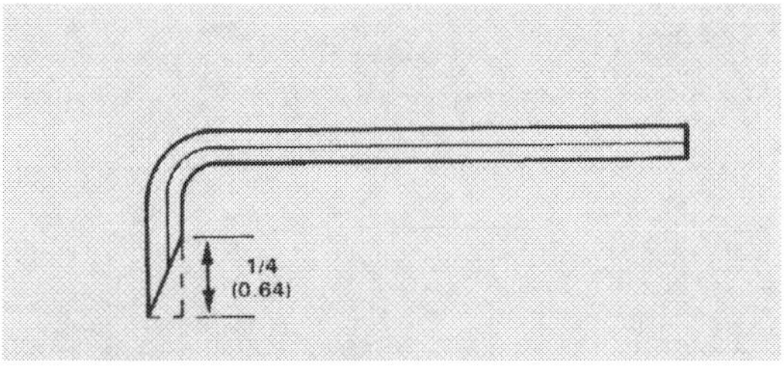

FABRICATE FROM 1'16 IN. SOCKET HEAD SCREW KEY
NSN 5120-00-198-5398 OR EQUIVALENT.

NOTE: ALL DIMENSIONS SHOWN ARE IN INCHES WITH THE METRIC CONVERSION TO CENTIMETERS IN PARENTHESES

Figure E-3. Pivot pin removal tool.

E-3. MANUFACTURED ITEMS ILLUSTRATIONS-FORMAT AND CONTENT (CONT).

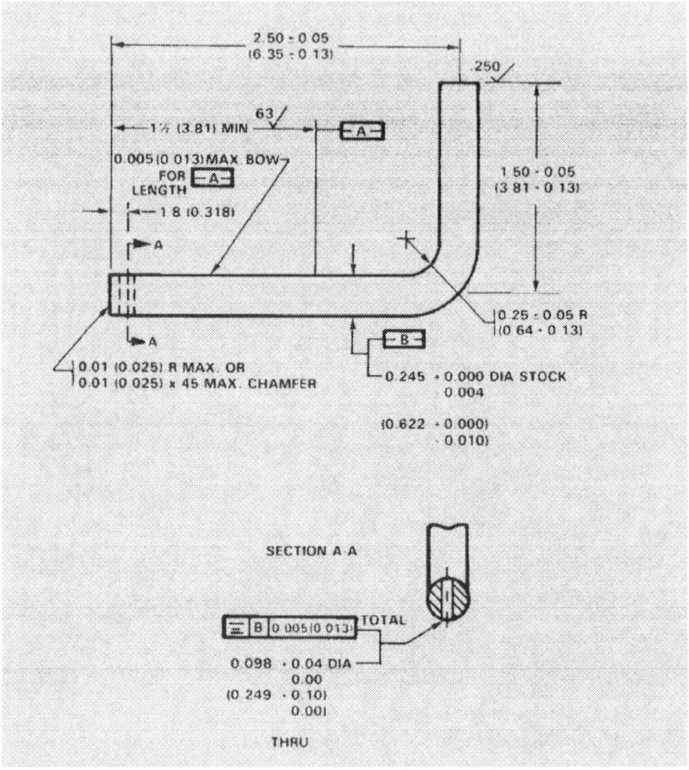

FABRICATE FROM 0 245 IN STEEL AISI 1095 OR EQUIVALENT
HARDEN AND TEMPER TO RC-57-61 FOR LENGTH -A-
FINISH 5 3 1.2 OR 5 3.2.2 OF MIL-STD-171

NOTE: ALL DIMENSIONS SHOWN ARE IN INCHES WITH THE METRIC CONVERSION TO CENTIMETERS IN PARENTHESES

Figure E-4. Pivot pin installation tool.

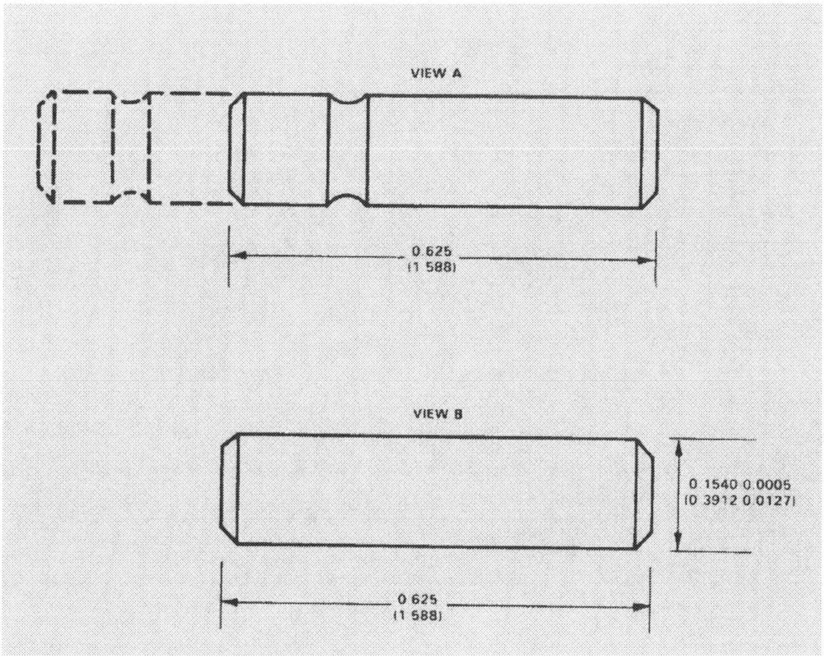

FABRICATE FROM OLD TRIGGER PIN (VIEW AI P,/N 8448609 OR FABRICATE SLAVE PIN (VIEW B) FROM MATERIAL BLOCK, WIRE, STEEL ALLOY, GRADE 4140, ASTM-A547 OR EQUIVALENT

NOTE: ALL DIMENSIONS SHOWN ARE IN INCHES WITH THE METRIC CONVERSION TO CENTIMETERS IN PARENTHESES

Figure E-5. Slave pin.

E-3. MANUFACTURED ITEMS ILLUSTRATIONS -- FORMAT AND CONTENT (CONT).

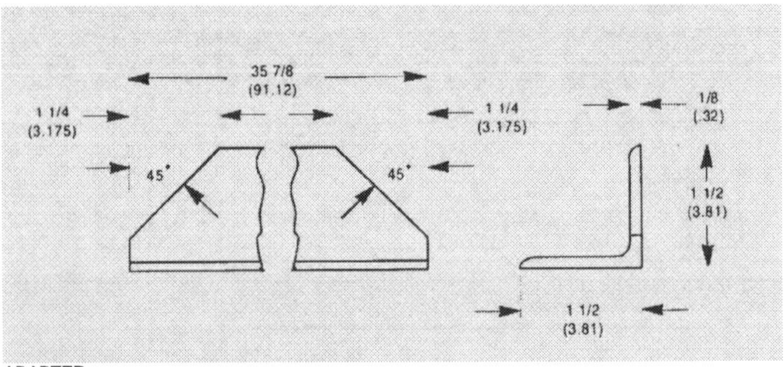

ADAPTER
FABRICATE FROM 1 1/2 INCH BY 1 1/2 INCH
BY 1/8 INCH ANGLE IRON NSN 9520-00-277-4902
OR EQUIVALENT. PAINT WITH OLIVE DRAB
ENAMEL PAINT, NSN 8010-01-350-5249, OR
EQUIVALENT.

NOTE: ALL DIMENSIONS SHOWN ARE IN INCHES WITH THE METRIC CONVERSION TO CENTIMETERS IN PARENTHESES.

Figure E-6. Adapter bar for M12 arms rack.

PIN: 060514-004

***U S Government Printing Office 1995- 648-378**

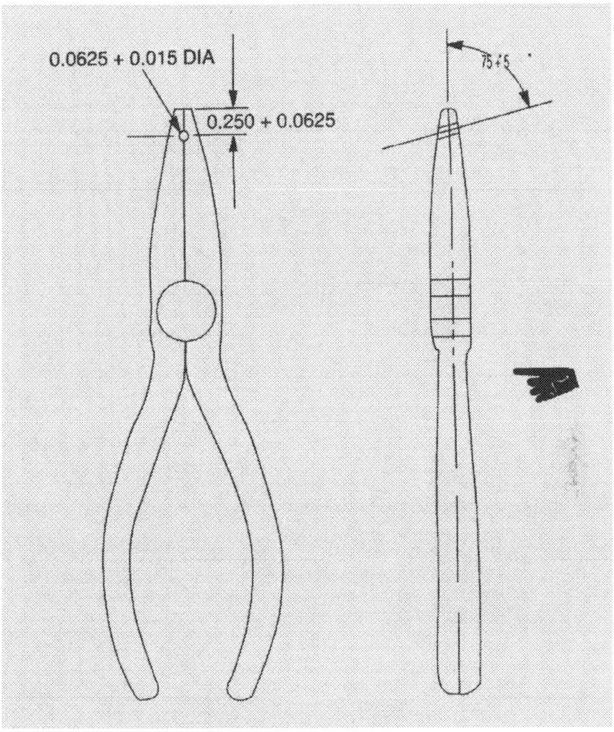

FABRICATE FROM NEEDLE NOSE PLIERS,
NSN 5120-00-268-3579 OR EQUIVALENT.

NOTE: ALL DIMENSIONS ARE IN INCHES.

Figure E-7. Modified needle nose pliers.

***U. S. GOVERNMENT PRINTING OFFICE 1997 - 545-039/4 077**

ALPHABETICAL INDEX

Subject	Page
A	
Auxiliary Equipment	4-1
B	
Barrel Assembly (See Upper Receiver and Barrel Assembly, Barrel Assembly, and Upper Receiver Assembly)	
Blank Firing Attachment M 15A12	
Cleaning	4-11
Inspection	4-11
Installation	4-9
Removal	4-10
Repainting	4-11
Replacement	4-11
Bolt Assembly	
Cleaning	2-40
Disassembly	2-39, 3-21
Inspection Repair	2-40, 3-21
Lubrication	2-41
Reassembly	2-41 3 24
Test	3-24
Bolt Carrier Assembly	
Cleaning	2-36, 3-16
Disassembly	2-35, 3-16
Inspection	3-17
Inspection Repair	2-36
Lubrication	2-37
Reassembly	2-37, 3-20
Repair	3-20
Test	3-20
Buttstock Assembly	
Cleaning	2-65
Disassembly	2-65
Inspection	2-66, 3-66
Lubrication	2-67
Reassembly	2-67
Repair	2-67, 3-67
C	
Charging Handle Assembly	
Cleaning	2-44
Disassembly	2-43
Inspection Repair	2-44
Lubrication	2-44
Reassembly	2-44
Common Tools and Equipment	2-1, 3-1
Corrosion Prevention and Control ,CPCI	1-2

ALPHABETICAL INDEX (CONT)

Subject	Page
D	
Destruction of Army Materiel to Prevent Enemy Use	1-1
E	
Equipment Characteristics, Capabilities, and Features	1-3
Equipment Data	1-5
Equipment Description and Data	1-3
Expendable Durable Supplies and Materials List	D-1
F	
Forward Assist Assembly	
Disassembly	3-59
Inspection	3-60
Lubrication	3-61
Reassembly	3-61
Repair	3-60
Functional Theory of Three-Round Burst Control (See M16A2 Rifle Final Inspection)	
G	
Gauging Requirements (See M16A2 Rifle Annual Gauging Requirements)	
General Information	1-1
General Inspection Criteria	3-97
H	
Hammer Assembly	
Disassembly	3-73
Inspection Repair	3-74
Reassembly	3-74
How to Use This Manual	iii
I	
Illustrated List of Manufactured Items	E 1
Initial Setup	2 33
K	
Key and Bolt Carrier Assembly	
Disassembly	3-26
Reassembly	3-27
Repair	3-26

ALPHABETICAL INDEX (CONT)

Subject	Page
L	
Location and Description of Major Components	1-4
Lock Plate	
Inspection	4-5
Installation	4-2
Removal	4-4
Lower Receiver and Buttstock Assembly	
Cleaning	2-59
Disassembly	2-57, 3-62
Inspection	6-0, 3-66
Lubrication	2-61
Reassembly	2-62, 3-70
Repair	2-61, 3-67
Test	3-68
Lower Receiver and Receiver Extension Assembly	
Disassembly	3-77
Inspection	3-79
Reassembly	3-80
Repair/Modify	3-79
Lubrication General	2-33
M	
Maintenance Allocation Chart	B-1
Maintenance Forms, Records, and Reports	1-1
Maintenance Procedures	2-33, 3-15
Major Components of M16A2 Rifle (See M16A2 Rifle)	
Mechanical Zero Procedures	2-54, 3-58
M15A12 Blank Firing Attachment (See Blank Firing Attachment M15A12)	
M16A2 Rifle	
Disassembly	2-34, 3-15
Inspection	2-69
Reassembly	2-68, 3-83
Stowage	2-70
Test	3-83
M16A2 Rifle Annual Gauging Requirements	3-94
M16A2 Rifle Final Inspection	
Final Inspection	3-85
Functional Theory of Three-Round Burst Control	3-88
Test	3-88
O	
Official Nomenclature, Names, and Designations	1-1
P	
Preembarkation Inspection of Materiel in Units Alerted for Overseas Movement	3-95
Preparation for Storage or Shipment	1-1, 2-70
Preventive Maintenance Checks and Services	2-3
Principles of Operation-	1-6

ALPHABETICAL INDEX (CONT)

Subject **Page**

R

Rear Sight Assembly (See Upper Receiver Assembly and Rear Sight Assembly)
References A-1
Repair Parts 2-1, 3-1
Repair Parts and Special Tools List C-1
Repair Parts List C-1 1
Repair Parts, Special Tools, TMDE, and Support Equipment 2-1, 3-1
Reporting Equipment Improvement Recommendations (EIR) 1-2
Reporting Errors and Recommending Improvements

S

Service Upon Receipt 2-2
Special Tools List C-17-1
Special Tools, TMDE, and Support Equipment 2-1
Symptom Index 2-21, 3-2

T

Top Sling Adapter
  Inspection 4-8
  Installation 4-6
  Removal 4-7
Trigger Assembly and Trigger Subassembly
  Disassembly 3-75
  Inspection Repair 3-76
  Reassembly 3-76
Trigger Subassembly (See Trigger Assembly and Trigger Subassembly)
Troubleshooting 2-21, 3-2

U

Unit Level Auxiliary Equipment Repair 4-1
Upper Receiver and Barrel Assembly
  Disassembly 3-30
  Inspection Cleaning 3-33
  Reassembly 3-39
  Repair 3-35
  Test 3-45
Upper Receiver and Barrel Assembly, Rifle Barrel Assembly, and Upper Receiver Assembly
  Cleaning 2-48
  Disassembly 2-45
  Inspection 2-49
  Lubrication 2-52
  Reassembly 2-52
  Repair 2-52

Subject	Page

Upper Receiver Assembly (See Upper Receiver and Barrel Assembly, Barrel Assembly, and Upper Receiver Assembly or Upper Receiver Assembly and Rear Sight Assembly)
Upper Receiver Assembly and Rear Sight Assembly
Disassembly 3-48
Inspection 3-51
Lubrication 3-53
Reassembly 3-54
Repair 3-53

W

Warnings .... a

Made in United States
Troutdale, OR
11/03/2026

56844374R00235